Systems Biology Modelling and Analysis

Systems Biology Modelling and Analysis

Formal Bioinformatics Methods and Tools

Edited by

Elisabetta De Maria
Université Côte d'Azur
France

This edition first published 2023
© 2023 John Wiley & Sons, Inc.

All rights reserved. No part of this publication may be reproduced, stored in a retrieval system, or transmitted, in any form or by any means, electronic, mechanical, photocopying, recording or otherwise, except as permitted by law. Advice on how to obtain permission to reuse material from this title is available at http://www.wiley.com/go/permissions.

The right of Elisabetta De Maria to be identified as the author of the editorial material in this work has been asserted in accordance with law.

Registered Office
John Wiley & Sons, Inc., 111 River Street, Hoboken, NJ 07030, USA

For details of our global editorial offices, customer services, and more information about Wiley products visit us at www.wiley.com.

Wiley also publishes its books in a variety of electronic formats and by print-on-demand. Some content that appears in standard print versions of this book may not be available in other formats.

Trademarks: Wiley and the Wiley logo are trademarks or registered trademarks of John Wiley & Sons, Inc. and/or its affiliates in the United States and other countries and may not be used without written permission. All other trademarks are the property of their respective owners. John Wiley & Sons, Inc. is not associated with any product or vendor mentioned in this book.

Limit of Liability/Disclaimer of Warranty
In view of ongoing research, equipment modifications, changes in governmental regulations, and the constant flow of information relating to the use of experimental reagents, equipment, and devices, the reader is urged to review and evaluate the information provided in the package insert or instructions for each chemical, piece of equipment, reagent, or device for, among other things, any changes in the instructions or indication of usage and for added warnings and precautions. While the publisher and authors have used their best efforts in preparing this work, they make no representations or warranties with respect to the accuracy or completeness of the contents of this work and specifically disclaim all warranties, including without limitation any implied warranties of merchantability or fitness for a particular purpose. No warranty may be created or extended by sales representatives, written sales materials or promotional statements for this work. The fact that an organization, website, or product is referred to in this work as a citation and/or potential source of further information does not mean that the publisher and authors endorse the information or services the organization, website, or product may provide or recommendations it may make. This work is sold with the understanding that the publisher is not engaged in rendering professional services. The advice and strategies contained herein may not be suitable for your situation. You should consult with a specialist where appropriate. Further, readers should be aware that websites listed in this work may have changed or disappeared between when this work was written and when it is read. Neither the publisher nor authors shall be liable for any loss of profit or any other commercial damages, including but not limited to special, incidental, consequential, or other damages.

Library of Congress Cataloging-in-Publication Data Applied for:

ISBN 9781119716532 (hardback)

Cover Design: Wiley
Cover Image: © Ekaterina Goncharova/Getty Images

Set in 9.5/12.5pt STIXTwoText by Straive, Chennai, India

To my children Arthur and Juliette, who are eager to grow up to read this book, which seems magical to them.

Contents

List of Contributors *xv*
Preface *xix*
Acknowledgments *xxv*

1   **Introduction** *1*
           Elisabetta De Maria
1.1   Why Writing Models *2*
1.2   Modelling and Validating Biological Systems: Three Steps *4*
1.2.1   Modelling Biological Systems *4*
1.2.2   Specifying Biological Systems *7*
1.2.3   Validating Biological Systems *8*
           References *9*

2   **Petri Nets for Systems Biology Modelling and Analysis** *15*
           Fei Liu, Hiroshi Matsuno, and Monika Heiner
2.1   Introduction *15*
2.2   A Running Example *16*
2.3   Petri Nets *16*
2.3.1   Modelling *17*
2.3.2   Analysis *18*
2.3.3   Applications *20*
2.4   Extended Petri Nets *20*
2.5   Stochastic Petri Nets *20*
2.5.1   Modelling *21*
2.5.2   Stochastic Simulation *21*
2.5.3   CSL Model Checking *22*
2.5.4   Applications *23*
2.6   Continuous Petri Nets *24*

2.6.1	Modelling	*24*
2.6.2	Deterministic Simulation	*24*
2.6.3	Simulative Model Checking	*25*
2.6.4	Applications	*27*
2.7	Fuzzy Stochastic Petri Nets	*27*
2.7.1	Modelling	*27*
2.7.2	Fuzzy Stochastic Simulation	*27*
2.7.3	Applications	*29*
2.8	Fuzzy Continuous Petri Nets	*29*
2.8.1	Modelling	*29*
2.8.2	Fuzzy Deterministic Simulation	*29*
2.8.3	Applications	*30*
2.9	Conclusions	*30*
	Acknowledgment	*31*
	References	*31*
3	**Process Algebras in Systems Biology**	**35**
	Paolo Milazzo	
3.1	Introduction	*35*
3.2	Process Algebras in Concurrency Theory	*36*
3.2.1	π-Calculus	*38*
3.3	Analogies between Biology and Concurrent Systems	*42*
3.3.1	Elements of Cell Biology	*43*
3.3.2	Cell Pathways	*44*
3.3.3	"Molecules as Processes" Abstraction	*48*
3.4	Process Algebras for Qualitative Modelling	*51*
3.4.1	Formal Analysis Techniques	*51*
3.5	Process Algebras for Quantitative Modelling	*53*
3.5.1	Chemical Kinetics	*54*
3.5.2	Stochastic Process Algebras	*59*
3.6	Conclusions	*61*
	Acknowledgments	*61*
	References	*62*
4	**The Rule-Based Model Approach:** *A Kappa Model for Hepatic Stellate Cells Activation by TGFB1*	**69**
	Matthieu Bouguéon, Pierre Boutillier, Jérôme Feret, Octave Hazard, and Nathalie Théret	
4.1	Introduction	*69*
4.1.1	Modelling Systems of Biochemical Interactions	*69*
4.1.2	Modelling Languages	*70*

4.1.3	Kappa 71
4.1.3.1	Overview 71
4.1.3.2	Semantics of Kappa 72
4.1.3.3	Kappa Ecosystem 73
4.1.3.4	Main Limitations 75
4.1.4	Modelling a Population of Hepatic Stellate Cells 76
4.1.5	Outline 78
4.2	Kappa 78
4.2.1	Site Graphs 78
4.2.1.1	Signature 79
4.2.1.2	Complexes 81
4.2.1.3	Patterns 82
4.2.1.4	Embeddings Between Patterns 84
4.2.2	Site Graph Rewriting 86
4.2.2.1	Interaction Rules 86
4.2.2.2	Reactions Induced by an Interaction Rule 87
4.2.2.3	Underlying Reaction Network 88
4.3	Model of Activation of Stellate Cells 91
4.3.1	Overview of Model 91
4.3.2	Some Elements of Biochemistry 91
4.3.2.1	Reaction Half-Time 92
4.3.2.2	Conversion 93
4.3.2.3	Production Equilibrium 93
4.3.2.4	Erlang Distributions 94
4.3.3	Interaction Rules 94
4.3.3.1	Behavior of TGFB1 Proteins 95
4.3.3.2	Renewal of Quiescent HSCs 96
4.3.3.3	Activation and Differentiation 97
4.3.3.4	Proliferation of Activated Hepatic Stellate Cells 99
4.3.3.5	Proliferation of Myofibroblasts 100
4.3.3.6	Apoptosis and Senescence of Myofibroblasts 101
4.3.3.7	Inactivation of Myofibroblasts 102
4.3.3.8	Behavior of Inactivated Hepatic Stellate Cells 102
4.3.3.9	Proliferation of Reactivated Cells 105
4.3.3.10	Degradation of Reactivated *MFB* 106
4.3.3.11	Behavior of Receptors 106
4.3.4	Parameters 108
4.4	Results 109
4.4.1	Static Analysis 109
4.4.2	Underlying Reaction Network 111
4.4.3	Simulations 111

4.5	Conclusion *113*
	References *116*

5	**Pathway Logic: Curation and Analysis of Experiment-Based Signaling Response Networks** *127*
	Merrill Knapp, Keith Laderoute, and Carolyn Talcott
5.1	Introduction *127*
5.2	Pathway Logic Overview *130*
5.3	PL Representation System *133*
5.3.1	Rewriting Logic and Maude *133*
5.3.2	Pathway Logic Language *134*
5.3.3	Petri Net Representation *140*
5.3.4	Computing with Petri Nets *142*
5.4	Pathway Logic Assistant *144*
5.5	Datum Curation and Model Development *150*
5.5.1	Datum Curation *150*
5.5.2	Model Development – Inferring Rules *153*
5.6	STM8 *155*
5.6.1	LPS Response Network *156*
5.6.2	Combining Network Analyses *158*
5.6.3	Death Map: A Review Model *159*
5.6.3.1	Review Map as a Summary of the State of the Art *163*
5.7	Conclusion *163*
	Acknowledgments *164*
	Appendix 5.A: Summary of STM8 Networks *164*
	References *168*

6	**Boolean Networks and Their Dynamics: The Impact of Updates** *173*
	Loïc Paulevé and Sylvain Sené
6.1	Introduction *173*
6.1.1	General Notations and Definitions *178*
6.2	Boolean Network Framework *179*
6.2.1	On the Simplicity of Boolean Networks *179*
6.2.2	Boolean Network Specification *181*
6.2.3	Boolean Network Dynamics *183*
6.2.3.1	Updates *183*
6.2.3.2	Transitions and Trajectories *185*
6.2.3.3	Updating Mode and Transition Graph *186*
6.2.3.4	Deterministic Updating Modes *187*
6.2.3.5	Non-deterministic Updating Modes *199*

6.3	Biological Case Studies	*208*
6.3.1	Floral Morphogenesis of *A. thaliana*	*209*
6.3.2	Cell Cycle	*211*
6.3.3	Vegetal and Animal Zeitgebers	*212*
6.3.4	Abstraction of Quantitative Models	*214*
6.4	Fundamental Knowledge	*216*
6.4.1	Structural Properties and Attractors	*216*
6.4.1.1	Fixed Points Stability	*216*
6.4.1.2	Feedback Cycles as Engines of Dynamical Complexity	*217*
6.4.1.3	About Signed Feedback Cycles	*219*
6.4.2	Computational Complexity	*224*
6.4.2.1	Existence of a Fixed Point	*225*
6.4.2.2	Reachability Between Configurations	*227*
6.4.2.3	Limit Configurations	*229*
6.5	Conclusion	*232*
6.5.1	Updating Modes and Time	*232*
6.5.1.1	Modelling Durations	*233*
6.5.1.2	Modelling Precedence	*234*
6.5.1.3	Modelling Causality	*234*
6.5.2	Toward an Updating Mode Hierarchy	*235*
6.5.2.1	Software Tools	*235*
6.5.3	Opening on Intrinsic Simulations	*236*
	Acknowledgments	*238*
	References	*238*
7	**Analyzing Long-Term Dynamics of Biological Networks With Answer Set Programming**	***251***
	Emna Ben Abdallah, Maxime Folschette, and Morgan Magnin	
7.1	Introduction	*251*
7.2	State of the Art	*253*
7.2.1	Qualitative Modelling of Biological Systems	*253*
7.2.2	Identifying Attractors: A Major Challenge	*255*
7.2.3	Answer Set Programming for Systems Biology	*257*
7.2.4	Enumerating Attractors of a Biological Model Using Answer Set Programming	*258*
7.3	Basic Notions of Answer Set Programming	*259*
7.3.1	Syntax and Rules	*259*
7.3.2	Predicates	*261*
7.3.3	Scripting	*263*
7.4	Dynamic Modelling Using Asynchronous Automata Networks	*264*

7.4.1	Motivation: Using ASP to Analyze the Dynamics	*264*
7.4.2	Definition of Asynchronous Automata Networks	*264*
7.4.3	Semantics and Dynamics of Asynchronous Automata Networks	*267*
7.4.4	Stable States and Attractors in Asynchronous Automata Networks	*271*
7.5	Encoding into Answer Set Programming	*275*
7.5.1	Translating Asynchronous Automata Networks into Answer Set Programs	*276*
7.5.2	Stable-State Enumeration	*278*
7.5.3	Attractors	*280*
7.5.3.1	Cycle Enumeration	*281*
7.5.3.2	Attractor Enumeration	*285*
7.5.3.3	Python Scripting	*288*
7.6	Case Studies	*290*
7.6.1	Toy Example	*290*
7.6.2	Bacteriophage Lambda	*292*
7.6.3	Benchmarks on Models Coming from the Literature	*293*
7.7	Conclusion	*297*
	Acknowledgments	*299*
	References	*299*

8 Hybrid Automata in Systems Biology *305*
Alberto Casagrande, Raffaella Gentilini, Carla Piazza, and Alberto Policriti

8.1	Introduction	*305*
8.2	Basics	*307*
8.2.1	Languages and Theories	*308*
8.3	Events	*313*
8.3.1	Temporal Logics	*316*
8.3.2	Model Checking	*318*
8.4	Events and Time	*318*
8.4.1	Hybrid Automata and Gene Regulatory Networks	*319*
8.4.2	Expressibility and Decidability Issues	*323*
8.5	Events, Time, and Uncertainty	*327*
8.6	Conclusions	*331*
	Acknowledgement	*332*
	References	*332*

9	**Kalle Parvinen: Ordinary Differential Equations** *339*
	Kalle Parvinen
9.1	Introduction *339*
9.2	Analyzing and Solving Ordinary Differential Equations *340*
9.2.1	Solving Ordinary Differential Equations Analytically *340*
9.2.2	Equilibria and Their Stability *341*
9.2.3	Solving Differential Equations Numerically *344*
9.3	Mechanistic Derivation of Ordinary Differential Equations *345*
9.3.1	Elementary Unimolecular Reaction (EUR) *346*
9.3.2	Elementary Bimolecular Reaction (EBR) *347*
9.3.3	Elementary Bimolecular Reaction of Two Identical Molecules *348*
9.3.4	Reaction Networks *348*
9.4	Classical Lotka–Volterra Differential Equation *350*
9.4.1	Model Formation and History *350*
9.4.2	Phase-Plane Analysis and Equilibria *351*
9.4.3	Constant of Motion *352*
9.4.4	Average Population Densities *353*
9.4.5	Effect of Fishing on the Population Densities *353*
9.5	Model of Killer T-Cell and Cancer Cell Dynamics *354*
9.5.1	Model Definition *354*
9.5.1.1	Resource Dynamics *354*
9.5.1.2	Cancer Cell Dynamics *355*
9.5.1.3	Killer T-Cell Dynamics *356*
9.5.2	Model Dynamics Without Treatment *357*
9.5.3	Treatment Effects *358*
9.6	Conclusion *359*
	Acknowledgments *359*
	References *360*
10	**Network Modelling Methods for Precision Medicine** *363*
	Elio Nushi, Victor-Bogdan Popescu, Jose-Angel Sanchez Martin, Sergiu Ivanov, Eugen Czeizler, and Ion Petre
10.1	Introduction *363*
10.2	Network Modelling Methods *364*
10.2.1	Network Centrality Methods *364*
10.2.1.1	Running Example *366*

10.2.1.2	Degree Centralities	*366*
10.2.1.3	Proximity Centralities	*368*
10.2.1.4	Path Centrality: Betweenness	*373*
10.2.1.5	Spectral Centralities	*377*
10.2.2	System Controllability Methods	*383*
10.2.2.1	Network Controllability	*384*
10.2.2.2	Minimum Dominating Sets	*387*
10.2.3	Software	*388*
10.2.3.1	NetworkX	*389*
10.2.3.2	Cytoscape	*390*
10.2.3.3	NetControl4BioMed	*390*
10.3	Applications of Network Modelling in Personalized Medicine	*392*
10.3.1	Constructing Personalized Disease Networks	*392*
10.3.2	Analysis Methods	*393*
10.3.3	Results	*398*
10.3.3.1	Structural Controllability Analysis	*398*
10.3.3.2	Minimum Dominating Set Analysis	*406*
10.4	Conclusion	*412*
	References	*413*

11 **Conclusion** *425*
Elisabetta De Maria

Index *427*

List of Contributors

Emna Ben Abdallah
Independent Researcher
Nantes
France

Matthieu Bouguéon
Inria, CNRS, IRISA, UMR 6074
University of Rennes
Rennes
France

Inserm, EHESP, Irset, UMR S1085
University of Rennes
Rennes
France

Pierre Boutillier
Nomadic Labs
Paris
France

Alberto Casagrande
Department Mathematics and
Geosciences
University of Trieste
Trieste
Italy

Eugen Czeizler
Department of Information
Technology
Åbo Akademi University
Turuku
Finland

National Institute of Research and
Development in Biological Sciences
Bucharest
Romania

Elisabetta De Maria
Université Côte d'Azur
CNRS, I3S
Sophia Antipolis
France

Jérôme Feret
Team Antique, Inria
Paris
France

École normale supérieure
DI-ENS (ÉNS, CNRS, PSL
University)
Paris
France

Maxime Folschette
Univ. Lille, CNRS, Centrale Lille
UMR 9189 CRIStAL
Lille
France

Raffaella Gentilini
Department of Mathematics and Computer Science
University of Perugia
Perugia
Italy

Octave Hazard
Team Antique, Inria
Paris
France

École normale supérieure
DI-ENS (ÉNS, CNRS, PSL University)
Paris
France

École Polytechnique
Palaiseau
France

Monika Heiner
Computer Science Department
Faculty of Mathematics, Natural Sciences and Computer Science
Brandenburg University of Technology Cottbus-Senftenberg
Cottbus
Germany

Sergiu Ivanov
IBISC Laboratory
Université Paris-Saclay
Université Évry
Paris
France

Merrill Knapp
Information and Computing Sciences
SRI International
Menlo Park, CA
USA

Keith Laderoute
Numentus Technologies, Inc.
Menlo Park, CA
USA

Fei Liu
School of Software Engineering
South China University of Technology
Guangzhou
China

Morgan Magnin
Centrale Nantes, Université de Nantes, CNRS, LS2N
Nantes
France

Hiroshi Matsuno
Graduate School of Science and Technology for Innovation
Yamaguchi University
Yamaguchi
Japan

Paolo Milazzo
Dipartimento di Informatica
Università di Pisa
Pisa
Italy

List of Contributors

Elio Nushi
Department of Computer Science
University of Helsinki
Helsinki
Finland

Kalle Parvinen
Department of Mathematics and Statistics
University of Turku
Turku
Finland

Advancing Systems Analysis Program, International Institute for Applied Systems Analysis
Laxenburg
Austria

Loïc Paulevé
Bordeaux INP, CNRS, LaBRI UMR5800
University of Bordeaux
Talence
France

Ion Petre
National Institute of Research and Development in Biological Sciences
Bucharest
Romania

Department of Mathematics and Statistics
University of Turku
Turku
Finland

Carla Piazza
Department of Mathematics
Computer Science, and Physics
University of Udine
Udine
Italy

Alberto Policriti
Department of Mathematics
Computer Science, and Physics
University of Udine
Udine
Italy

Victor-Bogdan Popescu
Department of Information Technology
Åbo Akademi University
Turku
Finland

Jose-Angel Sanchez Martin
Department of Computer Science
Technical University of Madrid
Madrid
Spain

Sylvain Sené
CNRS, LIS
Aix Marseille University
Marseille
France

Carolyn Talcott
Information and Computing Sciences
SRI International
Menlo Park, CA
USA

Nathalie Théret
Inria, CNRS, IRISA, UMR 6074
University of Rennes
Rennes
France

Inserm, EHESP, Irset, UMR S1085
University of Rennes
Rennes
France

Preface

Overview

Formal methods of computer science are nowadays unavoidable to model, study, and make advanced analysis of biological systems. Several formalisms are suitable to model biological systems: Petri nets, Boolean networks, reaction rules, process algebras, ordinary differential equations, timed and hybrid automata, etc. Once a biological system is encoded using one of these formalisms, some formal techniques such as model checking can be used to specify some expected properties of the system and verify whether they hold or not in the model at issue. This greatly helps in validating/refuting biological hypothesis, making predictions, and associating parameters with biological phenomena. In this book, we present and compare the main formalisms used in systems biology to model biological networks.

Organization and Features

Some crucial formal approaches used in systems biology are presented in detail, along with their advantages/drawbacks and main applications. Apart from Chapters 1 (Introduction) and 11 (Conclusion), each chapter of the book is devoted to one of the key formalisms used in the literature to model (and verify) biological systems. Each chapter includes an intuitive presentation of the targeted formalism, a brief history of the formalism and of its applications in systems biology, a formal description of the formalism and its variants, at least one realistic case study, some applications of formal techniques to validate and make deep analysis of models encoded with the formalism, and a discussion on the kind of biological systems for which the formalism is suited, along with concrete ideas on its possible evolutions.

Some chapters also include the description of a tool implementing the formalism and a sort of how-to practical guide about using the tool. The networks chosen to serve as case studies span the field of systems biology in a large way (they range from gene regulatory networks to prey-predatory networks).

Some chapters are quite technical and make use of an involved formal notation, but other chapters are more focused on the biological applications (in particular, the last chapter before the conclusion opens to some applications in precision medicine). For each chapter, the notation has been carefully chosen so that it looks the most natural and suited one to represent the formalism at issue. Please note that the authors of some chapters are the ones who first introduced the corresponding formalisms and/or tools in the literature. Also, note that all the chapters are thought to be self-contained: the reader will find in each chapter all the elements that are useful to understand and learn, without having to read other works. Some chapters contain references to other chapters to make comparisons, but there are no strong dependencies among chapters, and the reader can decide to read chapters in a different order than the one chosen in this book.

Some approaches appeared some decades ago, others are quite recent, but all of them are presented from a current and groundbreaking point of view. We describe how each formalism answers today needs in systems biology, which makes a real contribution to the scientific community.

The book is organized as follows.

Chapter 1 focuses on the necessity of using formal methods in the domain of systems biology. It introduces the main formal approaches to the modelling of biological systems and the steps to follow to improve and validate the obtained models.

Chapter 2 is devoted to Petri nets, an important tool for studying different aspects of biological systems, ranging from simple signaling pathways, metabolic networks, and genetic networks to tissues and organs. To explore such varieties of biological systems, many variants of Petri nets have been proposed. This chapter explains how these different net classes are applied to modelling and analysis of these different types of biological systems with the illustrative example of the yeast polarization model describing the pheromone-induced G-protein cycle in *Saccharomyces cerevisiae.*

Chapter 3 describes the development of Process Algebras and related analysis methods in the context of systems biology. It presents concepts that are at the basis of the application of this class of formalisms in the biological context, providing the relevant notions of biochemistry and cell biology, and discussing both qualitative and quantitative approaches. The π-calculus is chosen as representative Process Algebra in order to give modelling examples and clarify the relationships of the process algebraic approach with the traditional modelling of cell pathways as sets of chemical reactions.

Chapter 4 describes Kappa, a site graph rewriting language. As a realistic case study, a population of hepatic stellate cells under the effect of the *tgfb* protein is modeled. Kappa offers a rule-centric approach, inspired from chemistry, where interaction rules locally modify the state of a system that is defined as a graph of components, connected or not. In this case study, the components are occurrences of hepatic stellate cells in different states and occurrences of the protein *tgfb*. The protein *tgfb* induces different behaviors of hepatic stellate cells, thereby contributing either to tissue repair or to fibrosis. Better understanding the overall behavior of the mechanisms that are involved in these processes is a key issue to identify markers and therapeutic targets likely to promote the resolution of fibrosis at the expense of its progression.

Chapter 5 introduces Pathway Logic, a formal, rule-based system and interactive viewer for developing executable models of cellular processes. It includes a curated evidence knowledge base and a diverse collection of models for evaluation by users. This chapter presents the Pathway Logic representation system and the algorithms used by the Pathway Logic Assistant. The overview discusses rewriting logic and its implementation in the Maude system, the formal basis of Pathway Logic. Other sections in the chapter present the STM8 collection of signaling response networks, provide overviews of the curation process and how rules are inferred, and illustrate the utility of Pathway Logic using the *Lps* (lipopolysaccharide) and Cell Death models.

Chapter 6 presents Boolean networks, a mathematical model that has been widely used since decades in the context of biological regulation networks qualitative modelling. They consist in collections of entities, each having two possible local states (1 – active and 0 – inactive), which interact with each other over discrete time. The simplicity of their setting together with their high abstraction level are especially convenient to focus on foundations of information transmission in genetic regulation, and on mathematical explanation and prediction of phenomenological observations. This chapter aims to present the Boolean modelling framework by developing its theoretical bases and emphasizing its usefulness for capturing biological regulation phenomena. But it goes beyond that by covering their ability to capture the information transmission and its consequences depending on the ways the entities update their local state over time.

Chapter 7 deals with Answer Set Programming (ASP), which has proven to be a strong logic programming paradigm to deal with the inherent complexity of the biological models, allowing us to quickly investigate a wide range of configurations. ASP can efficiently enumerate a large number of answer sets, as well as easily filter the results thanks to constraints based on certain properties. This chapter first motivates the merits of ASP in biological studies based on the state of the art. Then, it introduces the basic concepts about ASP and its use in systems biology. After having given an overview of the different issues that can be tackled using ASP, it then focuses on one problem that is of critical importance: model-checking with ASP, and more specifically, the identification of attractors. The merits of this study are illustrated using case studies.

Chapter 8 focuses on hybrid automata, a formalism introduced and developed with the aim of integrating discrete and continuous ingredients in a single simulation tool. This chapter introduces some key logic formalisms for systems biology; illustrates some automata-based simulation tools; discusses the role, potential, and complexity of the notion of time in automata; and presents several methodologies to integrate discrete and continuous, time-oriented, formal instruments for systems biology. Several realistic case studies are treated.

Chapter 10 discusses several network modelling methods and their applicability to precision medicine. The chapter introduces a certain number of network centrality methods (degree centrality, closeness centrality, eccentricity centrality, betweenness centrality, and eigenvector-based prestige) and two systems controllability methods (minimum dominating sets and network structural controllability). Their applicability to precision medicine on three multiple myeloma patient disease networks is demonstrated. Each network consists of protein–protein interactions built around a specific patient's mutated genes, around the targets of the drugs used in the standard of care in multiple myeloma, and around multiple myeloma-specific essential genes. For each network, it is demonstrated how the discussed network methods can be used to identify personalized, targeted drug combinations uniquely suited to that patient.

Finally, Chapter 11 underlines some key concepts of the book and opens on possible evolutions in the domain of formal methods for systems biology.

Audience

This book can be of interest to scientists at all levels, from master students to senior scientists that want to learn about formal methods for systems biology. The book is mainly addressed to bioinformaticians but can also be appreciated by biologists, medical doctors, computer scientists, and mathematicians interested in this highly interdisciplinary area.

Nice, August 2022   *Elisabetta De Maria*

Acknowledgments

I would like to thank all the authors of this book. They are highly qualified and very busy researchers and teachers, and I'm honored they accepted to contribute to this book.

I'm deeply grateful to François Fages, who in 2008 introduced me to the field of systems biology and transmitted to me his passion for this research domain. I would like to express thanks to all the members of the team "Discrete Models for Complex Systems" at I3S research laboratory in Sophia Antipolis. At I3S, I could find a stimulating scientific environment which was very propitious to my researches and to the writing of this book. I would like to express my gratitude to all my students, from bachelor's to Ph.D., for the extraordinary inputs they have provided me during the last years.

I'm deeply indebted to my companion and my parents who supported my efforts in this endeavor. Finally a special thanks goes to my children Arthur and Juliette, who look at my work with enthusiasm, and are convinced that my profession is the most beautiful in the world.

1

Introduction

Elisabetta De Maria

Université Côte d'Azur, CNRS, I3S, Sophia Antipolis, France

This book is devoted to the use of *formal methods* of computer science in the domain of *systems biology*, which is a field that brings together researchers from biological, mathematical, computational, and physical sciences in order to study, conceive, simulate, and make advanced analysis of biological systems (Ideker et al., 2001). To this aim, biological knowledge is often extracted from high-throughput "omics" (genomics, transcriptomics, proteomics, metabonomics, etc.) data generated thanks to next-generation molecular technologies. The obtained pieces of information are then integrated into *interaction maps* or *networks*, which represent the interactions among the involved biological compounds. In these networks, nodes represent the modeled entities, and edges stand for their interactions. Several kinds of biological networks are in the scope of systems biology: gene regulatory networks (which represent genes, their regulators, and the regulatory relationships between them), protein–protein interaction networks (which model the physical contacts between proteins in a cell), metabolic networks (which represent the biochemical reactions catalyzed by enzymes in a cell), biological neural networks (which describe how neurons in the brain communicate through their synapses), ecological networks (which model the biological interactions of an ecosystem), etc. These networks help in having a *global view on data*, turning data from individual pieces to pieces that connect to form a system. As stated by Kitano in Kitano (2002), to understand biology at the system level, we must examine the structure and dynamics of cellular and organismal functions rather than the characteristics of isolated parts of a cell or an organism. The core of systems biology actually consists in turning data into networks to which we want to give a *dynamics*. As a matter of fact, these networks

Systems Biology Modelling and Analysis: Formal Bioinformatics Methods and Tools, First Edition. Edited by Elisabetta De Maria.
© 2023 John Wiley & Sons, Inc. Published 2023 by John Wiley & Sons, Inc.

are considered as dynamical systems, and we want to study the evolution of their components according to the time evolution. The *time dimension* is thus crucial: it is relevant to link the topology of these networks to their dynamics (Richard, 2010). It behoves us to underline that the dynamical systems we treat in systems biology are quite different from the ones studied in physics. The main complication derives from the fact that the large-scale data we get are often incomplete, heterogeneous, and interdependent (dependencies are usually hidden). Furthermore, the data in our possession are too few with respect to the search space. In fact, the search space grows exponentially with the number of measured compounds, which gives an explosion of the number of parameters. As a consequence, it is almost impossible to obtain a unique model from a given data set, and a system is generally described by a family of abstract models (Videla et al., 2017). The use of formal methods of computer science is then helpful to reason over these families of models and to discriminate models according to the biological properties they satisfy (Gilbert and Heiner, 2015).

1.1 Why Writing Models

Systems biology seeks at pointing out the emergence of biological phenomena (a system is more than the sum of its parts), and it would be reductive to consider the modeling task as a way to store as more information as possible concerning a biological system. Models are rarely an end in themselves, and they are very often written in order to answer some specific questions. We can identify the following main reasons for writing formal models of biological systems.

Validating/refuting a biological hypothesis: Biologists often get in touch with computer scientists, mathematicians, and/or physicists because they conjecture a hypothesis concerning a biological system and they need help in validating/refuting this hypothesis. In this case, we need to formalize not only the biological system at issue but also the associated (set of) hypothesis. Tools such as model checkers (Clarke et al., 1999) are then employed to test whether the hypothesis holds or not in the system.

Making predictions: To avoid to perform wet experiments in the laboratory, models can be exploited for their predictive power. In this regard, irrespective of whether experiments can be expensive, time-consuming, and intrusive for living creatures, we should consider that wet experiments are sometimes unsatisfactory for lack of *operability* (some compounds cannot be manipulated) and *observability* (the expression of some compounds cannot be observed). Models can thus help in predicting the values of some entities that cannot be observed. They can also predict the reactions the system will have under some special conditions, for example, when confronted with external factors such as disease, medicine, and environmental changes (Talcott and Knapp, 2017). The predicted behaviors need to be formally specified to automate reasoning on them.

Associating parameters with biological phenomena: Biologists often expect biological systems to display some known behaviors, but they do not always know under which conditions these behaviors can be observed. To tackle this problem, it is needed to write a model formalizing the biological knowledge concerning the system at issue and to look for parameters such that the expected dynamical properties are true in the model (these properties are often encoded using formal methods). *Parameter search* is crucial in artificial intelligence, and techniques from this domain are employed in systems biology to infer the parameters of biological systems. Observe that parameter search can also be exploited to find parameters such that a given model can reproduce some given execution traces (*data-fitting*). Again, these traces can be formally encoded.

Whatever the reason for writing a model is, formal methods are thus important for having deep insights into the system(s) involved. Of course, formal methods can also be exploited to *validate models* with respect to some already acquired biological knowledge, i.e. to verify whether the encoded models can reproduce some behaviors they are supposed to display (and do not display some unwanted behaviors). Also in this case, the behaviors to be checked should be formally encoded. In systems biology, models are not static, and they are often the object of modifications to incorporate new biological knowledge. Discovery passes through unceasing rounds between modelling and experiments. Therefore, it is like a cycle in which experiments lead to the proposition of new improved models and in turn these models suggest enhanced experiments and so on. Formal methods are again necessary to verify that the new information is correctly encoded and consistent with the previous knowledge. To summarize, formal methods are unavoidable to understand and control biological systems and, more generally, to automate reasoning on the different behaviors of biological systems.

1.2 Modelling and Validating Biological Systems: Three Steps

As stated before, formal methods greatly help in understanding how complex biological systems work. Formal approaches for systems biology usually follow the next three steps:

1. Describing the system at issue using a formal language;
2. Formalizing the biological properties to be verified using a formal language, which need not be the language used at step 1;
3. Using a tool to (automatically) test whether the encoded properties are verified by the modeled system, or to learn the conditions allowing the property satisfaction.

As far as the first step is concerned, a biological system can be modeled as a graph (*transition system*) whose nodes represent the different possible configurations of the system and whose edges encode meaningful configuration changes. Most of the formal languages employed in systems biology allow to (directly or indirectly) represent a biological system as a transition system. These formalisms, which are at the heart of this book, will be introduced in Section 1.2.1. Concerning the second step, a biological property concerning the temporal evolution of the biological species involved in the system can be encoded using formal languages such as temporal logics. These formalisms are introduced in Section 1.2.2. In these first two steps, the quality of interactions between biologists and computer scientists/mathematicians is very important. A gap between different terminologies often arises, and deep discussions are needed to find a common language. As far as the third step is concerned, tools called model checkers are largely used to verify that specific properties of a system hold at particular states. These tools are presented in Section 1.2.3.

1.2.1 Modelling Biological Systems

As far as the modelling of biological systems is concerned, in the literature, we can find both qualitative and quantitative approaches. Whereas qualitative methods are conceptual (e.g. they model the presence or absence of a biological entity), quantitative methods aim at counting, measuring, and representing data using numbers. To express the *qualitative* nature of dynamics, the most used formalisms are the following ones: Petri nets, Boolean networks and Thomas' networks, reaction rules, process algebras, logic programming languages, and pure logics.

Petri nets: They are directed-bipartite graphs with two different types of nodes: places and transitions. Places represent the resources of the system, while transitions correspond to events that can change the state of resources. Thanks to their graph-based structure, Petri nets are a mathematical formalism allowing an intuitive representation of biochemical networks (Reddy et al., 1993, Chaouiya, 2007). They are based on synchronous updating techniques.

Boolean networks and Thomas' networks: They are regulatory graphs, where nodes represent regulatory components (e.g. regulatory genes or proteins) and signed arcs (positive or negative) stand for regulatory interactions (activations or inhibitions). This graph representation is further associated with logical rules (or logical parameters), which specify how each node is affected by different combinations of regulatory inputs (Thomas et al., 1995). They can be other synchronous (Boolean networks (Kauffman, 1969, Ruz et al., 2018)) or asynchronous (Thomas'networks (Thomas, 1973)).

Reaction rules: Dedicated rule-based languages allow us to model biochemical reactions, defining how (sets of) reactants can be transformed into (sets of) products, and associating corresponding rate-laws (Chabrier-Rivier et al., 2004). The Systems Biology Markup Language (SBML) is the most common representation format for models of biological processes. It is based on XML, and it stores models as chemical reaction-like processes that act on entities (Hucka, 2014). The main rule-based modelling tools, such as BioNetGen (Blinov et al., 2004) or Biocham (Fages et al., 2004), provide both a textual and graphical format and are compatible with SBML.

Process algebras: They allow us to specify the communication and interactions of concurrent processes without ambiguities. There is a strong correspondence between concurrent systems described by process algebras and the biological ones: biological entities may be abstracted as processes that can interact with each other and reactions may be modeled as actions. The most widely used process algebras in systems biology are *pi-calculus* (Regev et al., 2001), where processes communicate on complementary channels identified by specific names, *bio-ambients* (Regev et al., 2004), which are based on bounded places where processes are contained and where communications take place, and *process-hitting* (Folschette et al., 2015), where biological components are abstracted as sorts divided into different processes, interactions between components are represented as a hit from one process of a sort to another process of another sort, and some cooperative sorts allow us to represent the combined influences of multiple components on a single target.

Logic programming languages: Declarative problem-solving languages belonging to the family of logic programming languages, such as Answer Set Programming (ASP), allow us to model biological systems with inherent tolerance of incomplete knowledge and to generate hypotheses about required expansions of biological models (Gebser et al., 2010, Videla et al., 2015, Fayruzov et al., 2011).

Pure logics: Different extensions of Linear Logic, such as HyLL (Despeyroux and Chaudhuri, 2013) and SELL (Olarte et al., 2015), allow us to specify systems that exhibit modalities such as the temporal or spatial ones. In these logics, propositions are called resources, and rules can be viewed as rewrite rules from a set of resources into another set of resources, where a set of resources describes a state of the system. Thus, biological systems can be modeled by a set of rules of the above form. Higher-order logic (HOL (Rashid et al., 2017)) can also be conveniently exploited to formalize reaction kinetics. Pathway Logic uses rewrite theories to formalize biological entities and processes (Talcott, 2016).

Some recent qualitative approaches evoke the use of languages for reactive systems. These are systems that constantly interact with the environment and which may have an infinite duration. As many biological systems can be seen as reactive systems continuously reacting to some stimuli, languages for reactive systems such as Nusmv (Goldfeder and Kugler, 2018) or Lustre (De Maria et al., 2017, De Maria et al., 2020) are suited to model them. To capture the dynamics of a biological system from a *quantitative* point of view, the most used approaches are the following ones: ordinary and stochastic differential equations, hybrid Petri nets, timed and hybrid automata, rule-based languages with continuous/stochastic dynamics, and stochastic process algebras.

Ordinary or stochastic differential equations: The most classical quantitative models resort to ordinary differential equations. The interaction between components is captured by sigmoid expressions embedded in differential equations. Both positive and negative regulations can be considered. Models based on ordinary differential equations can hardly be studied analytically but are often employed to simulate and predict the answer of biological systems. To track not only individuals but total populations, stochastic differential equations are often used (Székely and Burrage, 2014).

Hybrid Petri nets: They are characterized by the presence of two kinds of places (discrete and continuous) and two kinds of transitions (discrete and continuous) (Hofestädt and Thelen, 1998). A continuous place can hold a non-negative real number as its content, and a continuous transition fires continuously at the speed of an assigned parameter. Biological pathways can be observed as hybrid systems, showing both discrete and continuous evolution. For instance, protein concentration dynamics behave continuously when coupled with discrete switches.

Timed automata and hybrid automata: Timed automata (Maler and Batt, 2008) are finite state automata extended with timed behaviors: constraints allow us to limit the amount of time spent within particular states and the time intervals in which transitions are enabled. They are suited to model the time aspects of biological systems, such as time durations of activities. Hybrid automata (Campagna and Piazza, 2010) combine finite state automata with continuously evolving variables. A hybrid automaton exhibits two kinds of state changes: discrete jump transitions occur instantaneously and continuous flow transitions occur when time elapses. The presence of both discrete and continuous dynamics makes this formalism appealing to model biological systems (Bortolussi and Policriti, 2008).

Rule-based languages with continuous/stochastic dynamics: They allow to write mechanistic models of complex reaction systems, associating continuous or stochastic dynamics to rules (Kim and Gelenbe, 2012). In the popular tool Kappa (Boutillier et al., 2018), entities are graphical structures, rules are graph-rewrite directives, and rules can fire stochastically, as determined by standard continuous-time Monte Carlo algorithms.

Stochastic process algebras: They are process algebras where models are decorated with quantitative information used to generate stochastic processes. The most exploited formalisms are stochastic *pi*-calculus (Phillips and Cardelli, 2007), where each channel is associated with a stochastic rate, and process algebras such as Bio-PEPA (Ciocchetta and Hillston, 2009), which allows the specification of complex kinetic formulae.

The choice of the modelling approach must be suited to the biological phenomenon/system at issue: it is important to choose a formalism whose features and granularity of representation are sufficient to explain the phenomenon and to answer to the questions raised by biologists. The chosen formalism should be expressive enough without being too expressive (to avoid combinatorial explosion).

1.2.2 Specifying Biological Systems

The most widespread formal languages to specify properties concerning the dynamical evolution of biological systems are *temporal logics*. They are formalisms for describing sequences of transitions between states (Emerson, 1990). The *Computation Tree Logic* CTL* (Clarke et al., 1999) allows one to describe properties of computation trees. Its formulas are obtained by (repeatedly) applying Boolean connectives, *path quantifiers*, and *state quantifiers* to atomic formulas. The path quantifier **A** (resp., **E**) can be used to state that all paths (resp., some path) starting from a given state have some property. The state quantifiers are the next time operator **X**, which can be used to impose that a property holds at the next state of a path, the operator **F** (sometimes in the future), that requires that a property holds at some state on the path, the operator **G** (always in the future), that specifies that a property is true at every state on the path, and the until binary operator **U**, which holds if there is a state on the path where the second of its argument properties holds and, at every preceding state on the path, the first of its two argument properties holds. The *Computation Tree Logic* CTL (Clarke et al., 1986) is a fragment of CTL* that allows quantification over the paths starting from a given state. Unlike CTL*, it constrains every state quantifier to be immediately preceded by a path quantifier. The *Linear Time Logic* LTL (Sistla and Clarke, 1985) is another known fragment of CTL* where one may only describe events along a single computation path. Its formulas are of the form **A**φ, where φ does not contain path quantifiers, but it allows the nesting of state quantifiers. The *Probabilistic Computation Tree Logic* PCTL (Hansson and Jonsson, 1994) quantifies the different paths by replacing the **E** and **A** modalities of CTL by probabilities. Other formalisms, such as languages for reactive systems or logics, provide a unified framework to encode not only biological systems but also temporal properties of their dynamic behavior. For instance, the language Lustre (Halbwachs, 1998) offers an original means to express properties as *observers*. An observer of a property is a program, taking as inputs the inputs/outputs of the program under verification, and deciding at each instant whether the property is violated or not.

1.2.3 Validating Biological Systems

Until 1980s, the most common validation techniques for software and hardware systems were simulation and testing, which consist in injecting some signals in a given system at some given times, let the system evolve, and observe the output signals at some given times. The problem is that controlling all the possible interactions and all the bugs of a system using these techniques is rarely possible (too many executions should be considered) (Clarke et al., 1999). *Formal verification* was initially introduced in 1980s to prove that a piece of software or hardware is free of errors (Clarke et al., 1999) with respect to a given model. The field of systems biology is a more recent application area for formal verification, and the most common approaches to the formal verification of biological systems are based on *model checking*

(Clarke et al., 1999). In order to apply model checking, the biological system at issue should be encoded as a finite transition system, and relevant system properties should be specified using temporal logic. Formally, a transition system over a set AP of atomic propositions is a tuple $M = (Q, T, L)$, where Q is a finite set of states, $T \subseteq Q \times Q$ is a total transition relation (that is, for every state $q \in Q$, there is a state $q' \in Q$ such that $T(q, q')$), and $L : Q \to 2^{AP}$ is a labeling function that maps every state into the set of atomic propositions that hold at that state. Given a transition system $M = (Q, T, L)$, a state $q \in Q$, and a temporal logic formula φ expressing some desirable property of the system, the *model checking problem* consists of establishing whether φ holds at q or not, namely, whether $M, q \vDash \varphi$. Another formulation of the model checking problem consists of finding all the states $q \in Q$ such that $M, q \vDash \varphi$. Observe that the second formulation is more general than the first one. There exist several tools to automatically check whether a finite transition system verifies a given CTL, LTL, or PCTL formula, e.g., NuSMV (Cimatti et al., 1999), SPIN (Holzmann, 2004), and PRISM (Kwiatkowska et al., 2011). Probabilistic model checkers such as PRISM allow us to directly compute the probabilities for some given temporal logic formulae to hold in the system (provided that some probabilities are associated with the modeled transitions). Other approaches to the formal verification of biological systems are based on the use of theorem provers. These are formal proof systems providing a formal language to write mathematical definitions, executable algorithms, and theorems, together with an environment for the development of machine-checked proofs. Formulas are proved in a logical calculus, and some tactics can be exploited to perform backward reasoning and replace the formula to be proved (conclusion or goal) with the formulas that are needed to prove it (premises or subgoals). To be able to prove the properties of biological systems using theorem provers, both systems and specifications should be encoded in the logic implemented by the theorem prover. State-of-the art theorem provers are Coq (Bertot and Castéran, 2004) and Isabelle/HOL (Nipkow et al., 2002). While model checking is a press-button methodology (the user just has to encode the system and the temporal logic formula she wants to test and to press a button to query the tool), theorem provers often do not make proofs automatically (the presence of an expert is needed, and this can be time-consuming).

References

Yves Bertot and Pierre Castéran. *Interactive Theorem Proving and Program Development - Coq'Art: The Calculus of Inductive Constructions*. Texts in

Theoretical Computer Science. An EATCS Series. Springer, 2004. ISBN 978-3-642-05880-6. doi: https://doi.org/10.1007/978-3-662-07964-5.

Michael L. Blinov, James R. Faeder, Byron Goldstein, and William S. Hlavacek. BioNetGen: software for rule-based modeling of signal transduction based on the interactions of molecular domains. *Bioinformatics*, 20 (17): 3289–3291, 2004. doi: https://doi.org/10.1093/bioinformatics/bth378.

Luca Bortolussi and Alberto Policriti. Hybrid systems and biology: continuous and discrete modeling for systems biology. In *Proceedings of the Formal Methods for the Design of Computer, Communication, and Software Systems 8th International Conference on Formal Methods for Computational Systems Biology*, SFM'08, pages 424–448. Springer-Verlag, Berlin, Heidelberg, 2008. ISBN 3-540-68892-7, 978-3-540-68892-1. URL http://dl.acm.org/citation.cfm?id=1786698.1786712.

Pierre Boutillier, Mutaamba Maasha, Xing Li, Héctor F. Medina-Abarca, Jean Krivine, Jérôme Feret, Ioana Cristescu, Angus G. Forbes, and Walter Fontana. The Kappa platform for rule-based modeling. *Bioinformatics*, 34 (13): i583–i592, 2018. doi: https://doi.org/10.1093/bioinformatics/bty272.

Dario Campagna and Carla Piazza. Hybrid automata, reachability, and systems biology. *Theoretical Computer Science*, 411 (20): 2037–2051, 2010. doi: https://doi.org/10.1016/j.tcs.2009.12.015.

Nathalie Chabrier-Rivier, Marc Chiaverini, Vincent Danos, François Fages, and Vincent Schächter. Modeling and querying biomolecular interaction networks. *Theoretical Computer Science*, 325 (1): 25–44, 2004. doi: https://doi.org/10.1016/j.tcs.2004.03.063.

Claudine Chaouiya. Petri net modelling of biological networks. *Briefings in Bioinformatics*, 8 (4): 210–219, 2007. ISSN 1477-4054. doi: https://doi.org/10.1093/bib/bbm029.

Alessandro Cimatti, Edmund M. Clarke, Fausto Giunchiglia, and Marco Roveri. NUSMV: A new symbolic model verifier. In *Computer Aided Verification, 11th International Conference, CAV '99, Trento, Italy, July 6–10, 1999, Proceedings*, pages 495–499, 1999. doi: https://doi.org/10.1007/3-540-48683-6_44.

Federica Ciocchetta and Jane Hillston. Bio-PEPA: A framework for the modelling and analysis of biological systems. *Theoretical Computer Science*, 410 (33–34): 3065–3084, August 2009. ISSN 0304-3975. doi: https://doi.org/10.1016/j.tcs.2009.02.037.

Edmund M. Clarke, E. Allen Emerson, and A. Prasad Sistla. Automatic verification of finite-state concurrent systems using temporal logic specifications. *ACM Transactions on Programming Languages and Systems*, 8 (2): 244–263, 1986. doi: https://doi.org/10.1145/5397.5399.

E. M. Clarke, O. Grumberg, and D. A. Peled. *Model Checking*. MIT Press, Cambridge, MA, USA, 1999. ISBN 0-262-03270-8.

Vincent Danos and Cosimo Laneve. Formal molecular biology. *Theoretical Computer Science*, 325 (1): 69–110, 2004. doi: https://doi.org/10.1016/j.tcs.2004.03.065.

Elisabetta De Maria, Thibaud L'Yvonnet, Daniel Gaffé, Annie Ressouche, and Franck Grammont. Modelling and formal verification of neuronal archetypes coupling. In *Proceedings of the 8th International Conference on Computational Systems-Biology and Bioinformatics, Nha Trang City, Viet Nam, December 7-8, 2017*, pages 3–10, 2017. doi: https://doi.org/10.1145/3156346.3156348.

Elisabetta De Maria, Abdorrahim Bahrami, Thibaud L'yvonnet, Amy Felty, Daniel Gaffé, Annie Ressouche, and Franck Grammont. On the use of formal methods to model and verify neuronal archetypes. *Frontiers of Computer Science*, 40, December 2020. URL https://hal.archives-ouvertes.fr/hal-03053930.

Joëlle Despeyroux and Kaustuv Chaudhuri. A hybrid linear logic for constrained transition systems. In *19th International Conference on Types for Proofs and Programs, TYPES 2013, April 22-26, 2013, Toulouse, France*, pages 150–168, 2013. doi: https://doi.org/10.4230/LIPIcs.TYPES.2013.150.

E. Allen Emerson. Temporal and modal logic. In E. Allen Emerson and Jan van Leeuwen, editors, *Handbook of Theoretical Computer Science*, pages 995–1072. Elsevier and MIT Press, 1990. URL https://doi.org/10.1016/b978-0-444-88074-1.50021-4.

François Fages, Sylvain Soliman, and Nathalie Chabrier-Rivier. Modelling and querying interaction networks in the biochemical abstract machine Biocham. *Journal of Biological Physics and Chemistry*, 4: 64–73, 2004.

Timur Fayruzov, Jeroen Janssen, Dirk Vermeir, Chris Cornelis, and Martine De Cock. Modelling gene and protein regulatory networks with answer set programming. *IJDMB*, 5 (2): 209–229, 2011. doi: https://doi.org/10.1504/IJDMB.2011.039178.

Maxime Folschette, Loïc Paulevé, Katsumi Inoue, Morgan Magnin, and Olivier F. Roux. Identification of biological regulatory networks from process hitting models. *Theoretical Computer Science*, 568: 49–71, 2015. doi: https://doi.org/10.1016/j.tcs.2014.12.002.

M. Gebser, A. Konig, T. Schaub, S. Thiele, and P. Veber. The BioASP library: ASP solutions for systems biology. In *2010 22nd IEEE International Conference on Tools with Artificial Intelligence*, volume 1, pages 383–389, October 2010. doi: https://doi.org/10.1109/ICTAI.2010.62.

David R. Gilbert and Monika Heiner. Advances in computational methods in systems biology. *Theoretical Computer Science*, 599: 2–3, 2015. doi: https://doi.org/10.1016/j.tcs.2015.08.013.

Judah Goldfeder and Hillel Kugler. Temporal logic based synthesis of experimentally constrained interaction networks. In *Molecular Logic and Computational Synthetic Biology - First International Symposium, MLCSB 2018*, Santiago, Chile, December 17–18, 2018, Revised Selected Papers, pages 89–104, 2018. doi: https://doi.org/10.1007/978-3-030-19432-1_6.

N. Halbwachs. Synchronous programming of reactive systems. In Alan J. Hu and Moshe Y. Vardi, editors, *Computer Aided Verification, 10th International Conference, CAV '98, Vancouver, BC, Canada, June 28 - July 2, 1998, Proceedings*, volume 1427 of *Lecture Notes in Computer Science*, pages 1–16. Springer, 1998. doi: https://doi.org/10.1007/BFb0028726.

Hans Hansson and Bengt Jonsson. A logic for reasoning about time and reliability. *Formal Aspects of Computing*, 6 (5): 512–535, 1994. doi: https://doi.org/10.1007/BF01211866.

R. Hofestädt and S. Thelen. Quantitative modeling of biochemical networks. *In Silico Biology*, 1: 39–53, 1998.

Gerard J. Holzmann. *The SPIN Model Checker - Primer and Reference Manual*. Addison-Wesley, 2004. ISBN 978-0-321-22862-8.

Michael Hucka. Systems biology markup language (SBML). In *Encyclopedia of Computational Neuroscience*. 2014. doi: https://doi.org/10.1007/978-1-4614-7320-6_376-4.

Trey Ideker, Timothy Galitski, and Leroy Hood. A new approach to decoding life: systems biology. *Annual Review of Genomics and Human Genetics*, 2 (1): 343–372, 2001. doi: https://doi.org/10.1146/annurev.genom.2.1.343. PMID: 11701654.

Stuart Kauffman. Homeostasis and differentiation in random genetic control networks. *Nature*, 224: 177–178, 1969.

Haseong Kim and Erol Gelenbe. Stochastic gene expression modeling with hill function for switch-like gene responses. *IEEE/ACM Transactions on Computational Biology and Bioinformatics*, 9 (4): 973–979, 2012. doi: https://doi.org/10.1109/TCBB.2011.153.

Hiroaki Kitano. Systems biology: a brief overview. *Science*, 295 (5560): 1662–1664, 2002. ISSN 0036-8075. doi: https://doi.org/10.1126/science.1069492.

Marta Z. Kwiatkowska, Gethin Norman, and David Parker. PRISM 4.0: Verification of probabilistic real-time systems. In *Computer Aided Verification - 23rd International Conference, CAV 2011, Snowbird, UT, USA, July 14–20, 2011. Proceedings*, pages 585–591, 2011. doi: https://doi.org/10.1007/978-3-642-22110-1_47.

Oded Maler and Grégory Batt. Approximating continuous systems by timed automata. In *Formal Methods in Systems Biology, 1st International Workshop, FMSB 2008, Cambridge, UK, June 4–5, 2008. Proceedings*, pages 77–89, 2008. doi: https://doi.org/10.1007/978-3-540-68413-8_6.

Tobias Nipkow, Markus Wenzel, and Lawrence C. Paulson. *Isabelle/HOL: A Proof Assistant for Higher-order Logic*. Springer-Verlag, Berlin, Heidelberg, 2002. ISBN 3-540-43376-7.

Carlos Olarte, Elaine Pimentel, and Vivek Nigam. Subexponential concurrent constraint programming. *Theoretical Computer Science*, 606: 98–120, 2015. doi: https://doi.org/10.1016/j.tcs.2015.06.031.

Andrew Phillips and Luca Cardelli. Efficient, correct simulation of biological processes in the stochastic pi-calculus. In *Computational Methods in Systems Biology, International Conference, CMSB 2007, Edinburgh, Scotland, September 20–21, 2007, Proceedings*, pages 184–199, 2007. doi: https://doi.org/10.1007/978-3-540-75140-3_13.

Adnan Rashid, Osman Hasan, Umair Siddique, and Sofiane Tahar. Formal reasoning about systems biology using theorem proving. *PLOS ONE*, 12 (7): 1–27, 2017. doi: https://doi.org/10.1371/journal.pone.0180179.

Venkatramana N. Reddy, Michael L. Mavrovouniotis, and Michael N. Liebman. Petri net representations in metabolic pathways. In *Proceedings of the 1st International Conference on Intelligent Systems for Molecular Biology, Bethesda, MD, USA, July 1993*, pages 328–336, 1993. URL http://www.aaai.org/Library/ISMB/1993/ismb93-038.php.

Aviv Regev, William Silverman, and Ehud Shapiro. Representation and simulation of biochemical processes using the pi-calculus process algebra. In *Proceedings of the 6th Pacific Symposium on Biocomputing, PSB 2001, Hawaii, USA, January 3–7, 2001*, pages 459–470, 2001. URL http://psb.stanford.edu/psb-online/proceedings/psb01/regev.pdf.

Aviv Regev, Ekaterina M. Panina, William Silverman, Luca Cardelli, and Ehud Shapiro. BioAmbients: an abstraction for biological compartments. *Theoretical Computer Science*, 325 (1): 141–167, 2004. doi: https://doi.org/10.1016/j.tcs.2004.03.061.

A. Richard. Negative circuits and sustained oscillations in asynchronous automata networks. *Advances in Applied Mathematics*, 44 (4): 378–392, 2010.

Gonzalo A. Ruz, Ana Zúniga, and Eric Goles. A Boolean network model of bacterial quorum-sensing systems. *IJDMB*, 21 (2): 123–144, 2018. doi: https://doi.org/10.1504/IJDMB.2018.096405.

A. Prasad Sistla and Edmund M. Clarke. The complexity of propositional linear temporal logics. *Journal of the ACM*, 32 (3): 733–749, 1985. doi: https://doi.org/10.1145/3828.3837.

Tamàs Székely and Kevin Burrage. Stochastic simulation in systems biology. *Computational and Structural Biotechnology Journal*, 12 (20): 14–25, 2014. ISSN 2001-0370. doi: https://doi.org/https://doi.org/10.1016/j.csbj.2014.10.003.

Carolyn L. Talcott. The pathway logic formal modeling system: diverse views of a formal representation of signal transduction. In *IEEE International Conference on Bioinformatics and Biomedicine, BIBM 2016, Shenzhen, China, December 15-18, 2016*, pages 1468–1476, 2016. doi: https://doi.org/10.1109/BIBM.2016.7822740.

Carolyn L. Talcott and Merrill Knapp. Explaining response to drugs using pathway logic. In *Computational Methods in Systems Biology - 15th International Conference, CMSB 2017, Darmstadt, Germany, September 27-29, 2017, Proceedings*, pages 249–264, 2017. doi: https://doi.org/10.1007/978-3-319-67471-1_15.

René Thomas. Boolean formalization of genetic control circuits. *Journal of Theoretical Biology*, 42 (3): 563–585, December 1973.

René Thomas, Denis Thieffry, and Marcelle Kaufman. Dynamical behaviour of biological regulatory networks—I. Biological role of feedback loops and practical use of the concept of the loop-characteristic state. *Bulletin of Mathematical Biology*, 57 (2): 247–276, March 1995. ISSN 1522-9602. doi: https://doi.org/10.1007/BF02460618.

Santiago Videla, Carito Guziolowski, Federica Eduati, Sven Thiele, Martin Gebser, Jacques Nicolas, Julio Saez-Rodriguez, Torsten Schaub, and Anne Siegel. Learning Boolean logic models of signaling networks with ASP. *Theoretical Computer Science*, 599: 79–101, 2015. doi: https://doi.org/10.1016/j.tcs.2014.06.022.

Santiago Videla, Julio Saez-Rodriguez, Carito Guziolowski, and Anne Siegel. Caspo: a toolbox for automated reasoning on the response of logical signaling networks families. *Bioinformatics*, 33 (6): 947–950, 2017. doi: https://doi.org/10.1093/bioinformatics/btw738.

2

Petri Nets for Systems Biology Modelling and Analysis

Fei Liu[1], Hiroshi Matsuno[2], and Monika Heiner[3]

[1] School of Software Engineering, South China University of Technology, Guangzhou, China
[2] Graduate School of Science and Technology for Innovation, Yamaguchi University, Yamaguchi, Japan
[3] Computer Science Department, Faculty of Mathematics, Natural Sciences and Computer Science, Brandenburg University of Technology Cottbus-Senftenberg, Cottbus, Germany

2.1 Introduction

Systems biology aims to study a biological system as a whole to explore its overall behavior and interactions. Modelling and simulation play an important role in achieving this aim. Among many modelling methods, Petri nets have become an important tool for studying different aspects of biological systems, ranging from simple signaling pathways, metabolic networks, and genetic networks to tissues and organs, since Reddy et al. (1993) introduced qualitative Petri nets to represent metabolic pathways. Petri nets offer a number of advantages for representing and exploring biological systems (Heiner et al., 2008, Chaouiya, 2007). For example, Petri nets are intuitive graphical modelling formalisms, which can be executed via animation and simulation. Petri nets offer rich mathematically founded analysis techniques, covering both structural and behavioral properties. Petri nets also integrate qualitative and quantitative analysis techniques and methods, supported by a wealth of computer tools. For representing different concurrent, asynchronous, and dynamic behaviors of biological systems, many variants of Petri nets (Baldan et al., 2010, Gilbert and Heiner, 2006) have been proposed, such as stochastic Petri nets, timed Petri nets, continuous Petri nets (CPNs), fuzzy Petri nets, hybrid Petri nets, and colored Petri nets, each net class aiming at a specific aspect of biological systems. In this chapter, we will briefly introduce how these different net classes are applied to modelling and analysis of biological systems with an illustrative example.

Systems Biology Modelling and Analysis: Formal Bioinformatics Methods and Tools,
First Edition. Edited by Elisabetta De Maria.
© 2023 John Wiley & Sons, Inc. Published 2023 by John Wiley & Sons, Inc.

2.2 A Running Example

Throughout the chapter, we use the yeast polarization model describing the pheromone-induced G-protein cycle in *Saccharomyces cerevisiae* (Daigle Jr. et al., 2011, Drawert et al., 2010) to illustrate these different net classes. G proteins, also known as guanine nucleotide-binding proteins, are a family of proteins that act as molecular switches inside cells and are involved in transmitting signals from a variety of stimuli outside a cell to its interior. They also receive the ligand for the pheromone receptors to G protein activation cycle. The binding of ligands to pheromone receptors leads to conformational changes that promote the exchange of guanosine diphosphate (GDP) for guanosine triphosphate (GTP). Then, the GTP dimer released by GTP continues to stimulate many downstream effectors. When the GTP is hydrolyzed to GDP, the original receptor is reduced. This model consists of the following eight reactions.

$$r_1 : \quad \emptyset \xrightarrow{k_1} R$$

$$r_2 : \quad R \xrightarrow{k_2} \emptyset$$

$$r_3 : \quad L + R \xrightarrow{k_3} L + RL$$

$$r_4 : \quad RL \xrightarrow{k_4} R$$

$$r_5 : \quad RL + G \xrightarrow{k_5} G_a + G_{bg}$$

$$r_6 : \quad G_a \xrightarrow{k_6} G_d$$

$$r_7 : \quad G_d + G_{bg} \xrightarrow{k_7} G$$

$$r_8 : \quad \emptyset \xrightarrow{k_8} RL$$

In this model, R, L, and RL represent the pheromone receptors, ligands, and receptorligand complexes, respectively. G denotes the G-protein, and G_a, G_d, and G_{bg} represent its three separate units. The ligands L bind with the receptors R to form complexes RL, which activate the G-proteins G to separate its two subunits G_a and G_{bg}. G_a, acting as an autophosphotase, can be dephosphorylated to G_d, which then rebinds with G_{bg} to return to G. This model basically describes the whole pheromone-induced G-protein cycle.

2.3 Petri Nets

Petri nets (Murata, 1989) are weighted, directed bipartite graphs consisting of two types of nodes, i.e. places and transitions, and arcs that connect them.

Places usually represent passive system components such as conditions or resources, while transitions represent active system components such as actions or events. In systems biology, places may represent species or any kind of chemical compounds, e.g. genes, proteins, or protein complexes, while transitions represent any kind of chemical reactions, e.g. association, disassociation, translation, or transcription. For a chemical reaction, its precursors correspond to the preplaces of a transition while its products to the postplaces of the transition. The arcs lead from places to transitions or from transitions to places. Their weights indicate the multiplicity of each arc, reflecting, e.g., stoichiometries for chemical reactions. The arc weight 1 is usually not labeled explicitly. A place may contain an arbitrary (natural) number of tokens, represented as black dots or a natural number. A distribution of tokens over all places of a Petri net represents a state of the net, which is called a marking. A transition is called enabled if each of its preplaces contains at least the number of tokens specified by the weight of the corresponding arc. An enabled transition may fire and the firing of a transition transfers tokens from its preplaces to its postplaces according to the corresponding weight. The firing of a transition updates the current marking to a new reachable one. The repeated firing of transitions establishes the behavior of a net. The set of markings reachable from the initial marking constitutes the state space of the net. These reachable markings and transitions between them constitute the reachability graph of the net.

2.3.1 Modelling

Constructing a Petri net model usually has the following steps:

1. Collect knowledge and data about the biological problem to be studied.
2. Determine the modelling requirements by analyzing the biological problem.
3. Determine the species and reactions involved in the biological system.
4. Represent species as places and the reactions between species as transitions with arcs to connect places and transitions.
5. Adjust the layout of the constructed model for the display or further analysis.

According to the above steps, we create a Petri net model of the G-protein cycle based on the reactions given above, which is shown in Figure 2.1. Obviously, each chemical reaction can be directly mapped to a transition and a set of places and arcs.

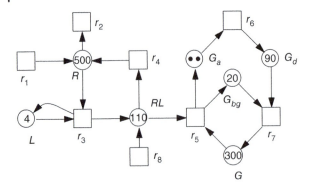

Figure 2.1 A Petri net model for yeast polarization based on the eight reactions given in Section 2.2. Source: Liu et al. (2016) / PLOS One / Licensed under CC BY 4.0.

2.3.2 Analysis

When a Petri net model is constructed, we can check it first using animation, which gives a first impression about the behavior of the model. We can then analyze its structural properties using a tool such as Charlie (Franzke, 2009). Structural analysis (Marsan et al., 1995, Heiner et al., 2008) means to investigate the properties of Petri nets from their structure without constructing the reachability graph. If a property is proved structurally for a Petri net, it holds for this Petri net in any initial marking. The important structural properties can be classified as follows (Heiner et al., 2008):

1. *Elementary graph properties*, e.g. connectedness and strongly connectedness. All of them are decided only by the graph structure using graph algorithms. They can be used for preliminary checks of the design of a net.
2. *P- and T-invariants*. A place invariant (P-invariant) represents a set of places over which the weighted token count keeps constant, and a transition invariant (T-invariant) describes how often the transitions contained by it have to fire to return to the original marking. Both of them can be obtained by solving a linear equation system that describes a net and are also independent of the initial marking.
3. *Deadlock and trap*: A deadlock is a set of places whose set of its pretransitions is contained in the set of its posttransitions, while a trap is a set of places whose set of its posttransitions is contained in the set of its pretransitions. Both of them contribute to the study of the liveness of Petri nets.

Among them, P- and T-invariants play crucial roles in analyzing biological systems because of their biological interpretations. For example,

2.3 Petri Nets

in a metabolic network, P-invariants correspond to the conservation law in chemistry, reflecting substrate conservations, and T-invariants reflect the steady-state behavior (Baldan et al., 2010, Heiner et al., 2008). Charlie (Franzke, 2009) is a software tool to analyze standard Petri nets, e.g. analyzing the structural properties given above. In order to do that, we have to export a Petri net file in Snoopy (Heiner et al., 2012) to a file in the APNN format, which will be inputted to Charlie to check those properties. See Franzke (2009) for more details about how to use Charlie. Figure 2.2 gives some analysis results for the model given in Figure 2.1. From this figure, we can obtain the following conclusions.

- This net is connected (CON) but not strongly connected (SC).
- It is live (LIV) without any dead transition (DTr). A dead transition is never enabled.
- It is covered by T-variants (CTI) but not covered by P-invariants (CPI).
- It is neither structurally bounded (SB) nor k-bounded (k-B).
- Every place has a pretransition (FP0) and a posttransition (PF0).

These properties can help us to deeply understand the model to be studied. Besides, Charlie offers many other analysis functions such as reachability/

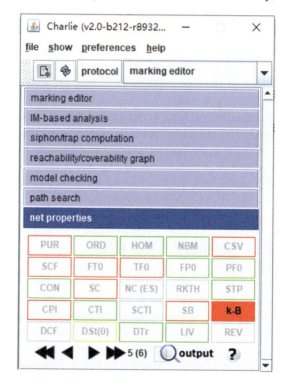

Figure 2.2 Structural analysis results by Charlie for the model in Figure 2.1.

coverability graph construction and computational tree logic/linear time logic (CTL/LTL) model checking (Liu, 2012).

2.3.3 Applications

Petri nets have been widely used for representing many biological systems. For example, Heiner et al. (2003) constructed a Petri net model of apoptosis and demonstrated how to validate this model with Petri net analysis methods. Gilbert et al. (2007b) built a Petri net model of the ERK pathway and deeply discussed many analysis techniques applicable for Petri nets, e.g. stochastic simulation, deterministic simulation, and simulative model checking. See Chaouiya (2007) for a review of some other applications.

2.4 Extended Petri Nets

There have been many various extensions based on the basic Petri nets (Heiner et al., 2008) so far. One of these extensions is to consider special arcs, e.g. inhibitor arcs, read arcs (also called test arcs), equal arcs, and reset arcs. The inhibitor arc reverses the logic of the enabling condition of a place, i.e. it imposes a precondition that a transition may only fire if the place contains less tokens than the weight the arc indicates. The read arc allows to model that some resource is read but does not consume the tokens of the source place. The equal arc imposes a precondition that a transition may only fire if the number of tokens of the place connected by the equal arc is equal to the arc weight. The reset arc makes it possible to empty the place connected by this arc once the transition connected by it is fired. These special arcs either make the model representation more compact or extend the modelling power of the Petri net formalism. They are particularly helpful for biological modelling. See Liu (2012) for the discussion about these kinds of special arcs.

2.5 Stochastic Petri Nets

Stochastic Petri nets (SPNs) are an extension of qualitative Petri nets by associating random delay firing rates with transitions, which provide a possibility to model (biological) systems with stochastics (Heiner et al., 2008). Because of inherent stochastics in biological processes, SPNs have recently become a popular modelling paradigm for capturing their complex stochastic dynamics, which offer a suitable way to understand the behavior of complex biological systems by integrating detailed biochemical data and providing quantitative analysis results, see e.g. Goss and Peccoud (1998) and Peleg et al. (2005). Contrary to qualitative Petri nets, in SPNs, a firing delay rate is introduced and associated with each transition t of a Petri net, which is a random variable X_t, defined by the following exponential probability distribution:

$$F_{X_t}(\tau) = 1 - e^{-\lambda_t \bullet \tau}, \tau \geq 0$$

Here, λ represents the parameter of the probability distribution and τ denotes the time. The semantics of a stochastic Petri net is equivalent to a continuous time Markov chain (CTMC), which is constructed from the reachability graph of the underlying qualitative Petri net by labeling the arcs between the states with transition rates. For more details, see Heiner et al. (2009).

2.5.1 Modelling

The steps to construct SPN models are similar to the steps for Petri nets except that we have to assign to each transition. For the model given in Figure 2.1, if we assign the rate functions given in Table 2.1, we can obtain an SPN model.

2.5.2 Stochastic Simulation

Gillespie simulation (Gillespie, 1977) offers a possibility for approximately analyzing models with huge state space. Gillespie simulation usually follows these steps:

Table 2.1 Rate functions of the model.

Transition	Rate function	Rate constant value
r_1	MassAction(k_1)	$k_1 = 0.38$
r_2	MassAction(k_2)	$k_2 = 0.04$
r_3	MassAction(k_3)	$k_3 = 0.082$
r_4	MassAction(k_4)	$k_4 = 0.12$
r_5	MassAction(k_5)	$k_5 = 0.021$
r_6	MassAction(k_6)	$k_6 = 0.1$
r_7	MassAction(k_7)	$k_7 = 0.005$
r_8	MassAction(k_8)	$k_8 = 13.21$

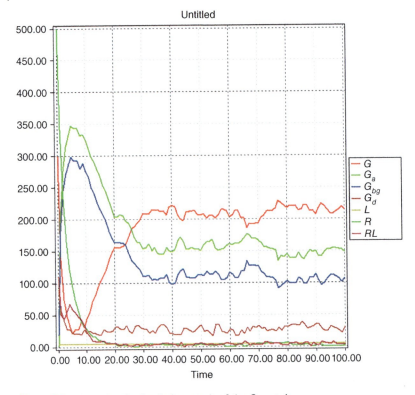

Figure 2.3 A stochastic simulation result of the G-protein.

1. Set the initial state of the chemical species involved in a model.
2. Calculate the propensity of each transition.
3. Compute the time that it will take for a transition to take place by sampling an exponential distribution.
4. Choose which transition takes place.
5. Update the time step by the chosen time.
6. Update the states after a transition fires.
7. Repeat this procedure until simulation reaches the end time.

For example, Figure 2.3 gives a stochastic simulation result of the G-protein cycle model.

2.5.3 CSL Model Checking

The underlying semantics of a stochastic Petri net is described by a CTMC, so it can be quantitatively analyzed by numerical model checking based on the Continuous Stochastic Logic (CSL) (Aziz et al., 2000, Baier et al., 2003), which is very useful for the validation of biological systems. Some tools for CSL model checking have been developed. For example, PRISM (Kwiatkowska et al., 2009) has been used to analyze probabilistic systems in a variety of application areas. Marcie (Schwarick et al., 2011) particularly aims to quantitative analysis of generalized stochastic Petri nets. CSL is based on the temporal logics CTL (Clarke et al., 1986) and PCTL (Probabilistic Computation Tree Logic) (Hansson and Jonsson, 1994). CSL provides a powerful means to specify both path-based and state-based properties on a CTMC. CSL is composed of not only the standard propositional logic operators but also two probabilistic operators, P and S, where the P operator checks the probability of a path formula, whereas the S operator checks the steady-state behavior of a CTMC. The syntax of CSL (Aziz et al., 2000, Baier et al., 2003) is defined as follows:

$$\Phi ::= P_{\trianglelefteq p}[\phi] \mid S_{\trianglelefteq p}[\Phi]$$
$$::= \neg \Phi \mid \Phi \wedge \Phi \mid \Phi \wedge \Phi \mid \Phi \rightarrow \Phi$$
$$::= true \mid false \mid ap.$$
$$\phi ::= X^I \Phi \mid F^I \Phi \mid G^I \Phi \mid \Phi U^I \Phi$$

where $ap \in AP$ is an atomic proposition, $\trianglelefteq \in \{<, \leq, \geq, >\}, p \in [0,1]$, and I is an interval of $\mathbb{R}^+$. CSL contains state formulas, Φ, interpreted over states of a CTMC, and path formulas, ϕ, interpreted over paths of a CTMC. It also has two probabilistic operators. $P \trianglelefteq p[\phi]$ asserts that the probability of the path formula ϕ being satisfied from a given state meets the bound given by $\trianglelefteq p$, and $S \trianglelefteq p[\Phi]$ indicates that the steady probability of being in a state satisfying Φ meets the bound given by $\trianglelefteq p$. For example, for the stochastic yeast polarization model, we can write the following property using CSL:

$$P_{>p}[F_{[t,t]}L = 1]$$

It checks if the probability of the species L being 1 at some time instance t is greater than a specified value p.

2.5.4 Applications

SPNs were successfully applied to simulate random fluctuations on the level of gene expression in Goss and Peccoud (1998). An SPN model of the RKIP-inhibited ERK pathway was illustrated in Heiner et al. (2010). Besides, in Rohr (2017), there were several SPN biological models that were discussed, e.g. the RKIP-inhibited ERK pathway, mitogen-activated protein kinase, and E. coli K-12 metabolic model.

2.6 Continuous Petri Nets

In a CPN, the discrete token values of places are replaced with continuous values, which describe the overall behavior of species represented by places via concentrations. A deterministic rate is associated with each transition, which makes a CPN model represent a set of ordinary differential equations (ODEs). Contrary to discrete Petri nets, the state space for a CPN is continuous and linear (Gilbert et al., 2007a). The underlying semantics of a CPN is a system of ODEs, where each equation describes the continuous token flow of a given place, i.e. continuously increasing its pretransitions' flow and decreasing its posttransitions' flow. An equation for a place p has the following form:

$$\frac{dm(p)}{dt} = \sum_{t \in {}^\bullet p} f(t,p)v(t) - \sum_{t \in p^\bullet} f(p,t)v(t)$$

Here, $f(t,p)$ is the weight of the arc from t to p, and $v(t)$ is the rate function of transition t. See Heiner et al. (2009) for more explanations about CPNs.

2.6.1 Modelling

The steps to construct a CPN model are similar to SPNs except the following two points:

- mapping the molecule number to the corresponding concentration of the species;
- adjusting the kinetic parameters if possible.

For the model given in Figure 2.1, if we assign the rate function given in Table 2.1 and consider it deterministic, we can obtain a CPN model.

2.6.2 Deterministic Simulation

There have been many deterministic simulation algorithms proposed for simulating a set of ODEs, such as the famous Runge–Kutta method, which has usually the following four steps:

1. Calculate the reaction rate of each related reaction, get the initial value.

$$y' = f(t,y), y(t_0) = y_0$$

2. Give the initial step length, rate of change, and initial value of the substance. The RK4 (the fourth-order Runge–Kutta) equation is as follows:

$$y_{n+1} = y_n + \frac{h}{6}(k_1 + 2k_2 + 2k_3 + k_4)$$
$$k_1 = f(t_n, y_n)$$
$$k_2 = f\left(t_n + \frac{h}{2}, y_n + \frac{h}{2}k_1\right)$$
$$k_3 = f\left(t_n + \frac{h}{2}, y_n + \frac{h}{2}k_2\right)$$
$$k_4 = f(t_n + h, y_n + hk_3)$$

The next value (y_{n+1}) is determined by the product of the current value (y_n) plus the time interval (h) and an estimated slope.

3. Adjust step size by precision. The local error of the formula is $O(h^5)$, and the estimated formula is obtained by comparing the step length of $\frac{h}{2}$ with h.

$$y(t_{n+1}) - y_{n+1}^{\left(\frac{h}{2}\right)} \approx \frac{1}{15}\left[y_{n+1}^{\left(\frac{h}{2}\right)} - y_{n+1}^{(h)}\right]$$

$$\Delta = \left\| y_{n+1}^{\left(\frac{h}{2}\right)} - y_{n+1}^{(h)} \right\|$$

Consider two scenarios:
1. For a given precision ϵ, if $\Delta > \epsilon$, repeatedly fold the step length into half until $\Delta < \epsilon$ and get $y_{n+1}^{\left(\frac{h}{2}\right)}$.
2. If $\Delta < \epsilon$, double the step again and again until $\Delta > \epsilon$, then fold in half to get the result.

4. Get a more accurate calculation

By determining the new step size, repeat step 2 to get the result.

Following these steps, Figure 2.4 gives a deterministic simulation result of the G-protein model.

2.6.3 Simulative Model Checking

Numerical model checking adopts numerical techniques for transient analysis of CTMC models (Aziz et al., 2000), which is usually highly accurate. However, it suffers from the problem of state space explosion because of its intensive computation and is often limited to models with the Markovian behavior (Younes et al., 2006). In contrast, simulative model checking follows the idea of Monte Carlo sampling and analyzes only a subset of the state space to obtain an approximate result. Therefore, it is not limited to specified formal models and not subject to the state space explosion; however,

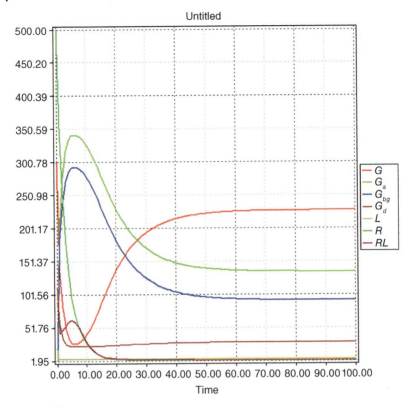

Figure 2.4 Deterministic simulation result of the G-protein model.

it is less accurate (Donaldson and Gilbert, 2008). PLTLc is the logic used in simulative model checking. The semantics of PLTLc is defined over a finite set of finite linear traces of temporal behavior, coming from, e.g., stochastic simulation runs. The temporal operators, X, G, F, and U, follow the standard LTL semantics. For example, for $P_{\trianglelefteq p}[\phi]$, each trace is evaluated to a Boolean value in terms of a property ϕ being true in this trace, and the probability of the property ϕ holding true in a set of traces is computed by the fraction of the set that is evaluated to true over the whole set. We can use MC2 (Donaldson and Gilbert, 2008), a model checker by Monte Carlo sampling, for simulative PLTLc model checking. MC2 can read sets of simulation traces as generated by Snoopy and expects additionally a file with the temporal-logical formulas. Therefore, in order to use MC2 for model checking of stochastic Petri nets, we only need to run simulation to obtain simulation traces and feed them to MC2. For example, we can use the following query to check the maximum on the trace of the place, G.

$$P_{=?}[G([G] < 250)]$$

2.6.4 Applications

Herajy and Heiner (2018) discussed two semantics of CPNs, adaptive semantics and bio-semantics, and illustrated how to apply these two semantics. Gilbert and Heiner (2006) used a RKIP model to illustrate how to construct a CPN model for biological systems.

2.7 Fuzzy Stochastic Petri Nets

Fuzzy stochastic Petri nets (FSPNs) (Liu et al., 2016) were proposed to model and analyze biological systems with uncertain kinetic information. FSPNs combine SPNs and fuzzy sets, thereby taking into account both randomness and fuzziness of biological systems. For a biological system, SPNs model the randomness, while fuzzy sets model the kinetic parameters with fuzzy uncertainty or variability by associating each kinetic parameter with a fuzzy number instead of a crisp real value. A fuzzy set defines a class of objects, each object having a continuum of grades of membership, while an object in a (crisp) set only has two membership grades, either 0 or 1. A fuzzy number is a special fuzzy set defined on the set of real numbers. A triangular fuzzy number (TFN) has a membership function of the triangular form. See Liu et al. (2016) for more explanations. In an FSPN model, a kinetic parameter is assigned either a crisp value or a fuzzy value, usually a TFN.

2.7.1 Modelling

The construction of an FSPN model is similar to SPN except determining fuzzy parameters. To do this, we may adopt the following way. First, ask the professional experts of the field to determine a rough range of the fuzzy parameter to be studied and then to specify the pessimistic value, the most possible value, and the optimistic value for the fuzzy parameter, which then form a TFN. This TFN is then used as the value of the parameter.

2.7.2 Fuzzy Stochastic Simulation

The idea to simulate an FSPN model is as follows. Firstly, decompose all fuzzy parameters into its α-cuts and then discretize each α-cut to obtain a set of crisp values for each parameter. For a fuzzy set, a α-cut defines a crisp set whose element has a membership grade equal to or more than the given α level. Secondly, for each sample of fuzzy parameters, run stochastic simulations and obtain the α-cut for each output of interest. Thirdly, compose all the α-cuts and obtain the membership function for each output, which reflects the effect of the uncertainties of the input parameters. For example, Figure 2.5 gives a plot of the model by setting k_4 to a FTN (0.1, 0.12, 0.14); see Table 2.2 for the rate functions of the model. The top subfigure gives

Table 2.2 Rate functions of the model.

Transition	Rate function	Rate constant value
t_1	MassAction(k_1)	$k_1 = 0.38$
t_2	MassAction(k_2)	$k_2 = 0.04$
t_3	MassAction(k_3)	$k_3 = 0.082$
t_4	MassAction(k_4)	$k_4 = (0.1, 0.12, 0.14)$
t_5	MassAction(k_5)	$k_5 = 0.021$
t_6	MassAction(k_6)	$k_6 = 0.1$
t_7	MassAction(k_7)	$k_7 = 0.005$
t_8	MassAction(k_8)	$k_8 = 13.21$

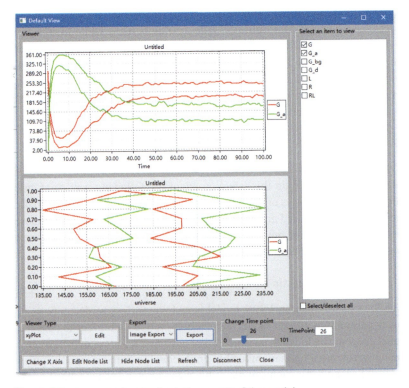

Figure 2.5 Fuzzy stochastic simulation result of the model.

the uncertain bands of species G and G_a, and the bottom subfigure shows the member functions of the two species at time point 26. We can of course check the member function at each time point.

2.7.3 Applications

A yeast polarization FSPN model was given in Liu et al. (2016), which illustrates the power of this new type of Petri nets. In Assaf et al. (2019), they also discussed some small case studies of FSPNs, e.g. the repressilator model.

2.8 Fuzzy Continuous Petri Nets

Fuzzy continuous Petri nets (FCPNs) (Liu et al., 2018) were proposed by combining CPNs with fuzzy numbers. For a biological model, if some of its kinetic parameters are not available, thus precluding ODE simulation, we can still use FCPNs to quantitatively analyze the model by giving an uncertain band of any output, rather than crisp values.

2.8.1 Modelling

The construction of an FCPN model is similar to CPN except determining fuzzy parameters, which are also discussed above.

2.8.2 Fuzzy Deterministic Simulation

For achieving numerical simulation of an FCPN model, we adopt the following idea. We first represent each fuzzy number as a union of its α-cuts according to Zadeh's extension principle (Zadeh, 1965), which is a fuzzy version of united extension, namely, it extends a point-to-point function to a set-to-set function. By sampling the α-cut at each α level, we obtain a combination of samples for all parameters. For each combination, we run numerical simulation on the corresponding CPN model at an α level. After running simulations for all considered α levels, we compose all the α-cuts to obtain the membership function of each output at each simulation time point. That is, we obtain the uncertainties of outputs caused by the uncertainties of kinetic parameters. For example, Figure 2.6 gives a plot of the model by setting k_6 to a FTN (0.05, 0.1, and 0.15). The top subfigure gives the uncertain bands of species G and G_a, and the bottom subfigure shows the member functions of the two species at time point 75. We can of course check the member function at each time point.

2 Petri Nets for Systems Biology Modelling and Analysis

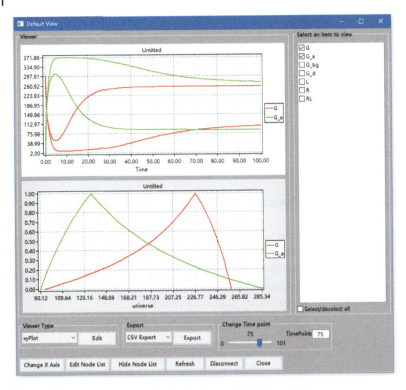

Figure 2.6 Fuzzy deterministic simulation result of the model.

2.8.3 Applications

A heat shock response FCPN model was given in Liu et al. (2018) to illustrate the power of FCPNs, e.g. modelling a biological system in which some of its kinetic parameters are not available or imprecise. In Assaf et al. (2019), they also discussed some small case studies of FCPNs.

2.9 Conclusions

In this chapter, we discussed some Petri net classes that have been widely used in the modelling and analysis of biological systems. This chapter is intended to guide modelers in the use of Petri nets for biological modelling.

Acknowledgment

This work has been supported by the National Natural Science Foundation of China (61873094), the Science and Technology Program of Guangzhou, China (201804010246), and the Natural Science Foundation of Guangdong Province of China (2018A030313338).

References

G. Assaf, M. Heiner, and F. Liu. Biochemical reaction networks with fuzzy kinetic parameters in Snoopy. In L. Bortolussi and G. Sanguinetti, editors, *Proceedings of the CMSB 2019*, volume 11773 of *LNCS/LNBI*, pages 302–307. Springer, September 2019. doi: https://doi.org/10.1007/978-3-030-31304-3_17. First Online: 17 September 2019.

A. Aziz, K. Sanwal, V. Singhal, and R. Brayton. Model checking continuous time Markov chains. *ACM Transactions on Computational Logic*, 1 (1): 162–170, 2000.

C. Baier, B. Haverkort, H. Hermanns, and J. P. Katoen. Model-checking algorithms for continuous-time Markov chains. *IEEE Transactions on Software Engineering*, 29 (6): 524–541, 2003.

P. Baldan, N. Cocco, A. Marin, and M. Simeoni. Petri nets for modelling metabolic pathways: a survey. *Natural Computing*, 9 (4): 955–989, 2010.

Claudine Chaouiya. Petri net modelling of biological networks. *Briefings in Bioinformatics*, 8 (4): 210–219, 2007. ISSN 1467-5463. doi: https://doi.org/10.1093/bib/bbm029. _eprint: https://academic.oup.com/bib/article-pdf/8/4/210/591466/bbm029.pdf.

E. Clarke, E. Emerson, and A. Sistla. Automatic verification of finite-state concurrent systems using temporal logics. *ACM Transactions on Programming Languages and Systems*, 8 (2): 244–263, 1986.

Bernie J. Daigle Jr., Min K. Roh, Dan T. Gillespie, and Linda R. Petzold. Automated estimation of rare event probabilities in biochemical systems. *The Journal of Chemical Physics*, 134: 044110–1–13, 2011.

R. Donaldson and D. Gilbert. A model checking approach to the parameter estimation of biochemical pathways. In *Proceedings of the 6th International Conference on Computational Methods in Systems Biology*, LNCS 5307, pages 269–287. Springer, 2008.

Brian Drawert, Michael J. Lawson, Linda Petzold, and Mustafa Khammash. The diffusive finite state projection algorithm for efficient simulation of the stochastic reaction-diffusion master equation. *The Journal of Chemical Physics*, 132: 074101-1–12, 2010.

A. Franzke. *Charlie 2.0 – A Multi-Threaded Petri Net Analyzer*. Master's thesis, Computer Science Department, Brandenburg University of Technology Cottbus, December 2009.

D. Gilbert and M. Heiner. From Petri nets to differential equations - an integrative approach for biochemical network analysis. In *Proceedings of the 27th International Conference on Applications and Theory of Petri Nets and Other Models of Concurrency*, LNCS 4024, pages 181–200. Springer, 2006.

D. Gilbert, M. Heiner, and S. Lehrack. A unifying framework for modelling and analysing biochemical pathways using Petri nets. In *Proceedings of the 5th International Conference on Computational Methods in Systems Biology*, LNCS 4695, pages 200–216. Springer, 2007a.

D. Gilbert, M. Heiner, and S. Lehrack. A unifying framework for modelling and analysing biochemical pathways using Petri nets. In *Proceedings of the 5th International Conference on Computational Methods in Systems Biology (CMSB 2007), Edinburgh*, volume 4695 of *LNCS/LNBI*, pages 200–216. Springer, 2007b. URL http://www.springerlink.com/content/d88056wn51n37255/?p=4c67a9984be7494681f0f84b0784b89c&pi=2. This paper is a short version of Technical Report I-02/2007, BTU Cottbus.

D. T. Gillespie. Exact stochastic simulation of coupled chemical reactions. *Journal of Physical Chemistry*, 81 (25): 2340–2361, 1977.

P. J. E. Goss and J. Peccoud. Quantitative modeling of stochastic systems in molecular biology by using stochastic Petri nets. *The Proceedings of the National Academy of Sciences of the United States of America*, 95 (12): 6750–6755, 1998.

H. Hansson and B. Jonsson. A logic for reasoning about time and reliability. *Formal Aspects of Computing*, 6 (5): 512–535, 1994.

M. Heiner, I. Koch, and J. Will. Model validation of biological pathways using Petri nets - demonstrated for apoptosis. In *Proceedings of the 1st International Workshop on Computational Methods in Systems Biology (CMSB 2003), Rovereto, February 2003*, volume 2602 of *LNCS*, page 173. Springer, 2003. doi: https://doi.org/10.1016/j.biosystems.2004.03.003.

M. Heiner, D. Gilbert, and R. Donaldson. Petri nets for systems and synthetic biology. In *Proceedings of the 8th International Conference on Formal Methods for Computational Systems Biology*, LNCS 5016, pages 215–264. Springer, 2008.

M. Heiner, S. Lehrack, D. Gilbert, and W. Marwan. Extended stochastic Petri nets for model-based design of wetlab experiments. In Corrado Priami,

Ralph-Johan Back, and Ion Petre, editors, *Transaction on Computational Systems Biology XI*, LNBI 5750, pages 138–163, 2009.

M. Heiner, R. Donaldson, and D. Gilbert. *Petri Nets for Systems Biology*, Chapter 3, pages 61–97. Jones & Bartlett Learning, LCC, 2010. URL http://www.jblearning.com/catalog/9780763753702/.

M. Heiner, M. Herajy, F. Liu, C. Rohr, and M. Schwarick. Snoopy — a unifying Petri net tool. In *Proceedings of PETRI NETS 2012*, volume 7347 of *LNCS*, page 398–407. Springer, June 2012. doi: https://doi.org/10.1007/978-3-642-31131-4_22. URL http://www.springerlink.com/content/l27w488386w03m45/.

M. Herajy and M. Heiner. Adaptive and bio-semantics of continuous Petri nets: choosing the appropriate interpretation. *Fundamenta Informaticae*, 160 (1–2): 53–80, 2018. doi: https://doi.org/10.3233/FI-2018-1674. accepted for publication: May, 18 2017.

M. Kwiatkowska, G. Norman, and D. Parker. PRISM: Probabilistic model checking for performance and reliability analysis. *ACM SIGMETRICS Performance Evaluation Review*, 36 (4): 40–45, 2009.

Fei Liu. *Colored Petri Nets for Systems Biology*. PhD thesis, BTU Cottbus, Dep. of Computer Science, January 2012.

Fei Liu, Monika Heiner, and Ming Yang. Fuzzy stochastic Petri nets for modeling biological systems with uncertain kinetic parameters. *PLOS ONE*, 11 (2): 1–19, 2016. doi: https://doi.org/10.1371/journal.pone.0149674.

F. Liu, S. Chen, M. Heiner, and H. Song. Modeling biological systems with uncertain kinetic data using fuzzy continuous Petri nets. *BMC Systems Biology*, 12 (Suppl. 4): 42, 2018. doi: https://doi.org/10.1186/s12918-018-0568-8. accepted for publication: January 2018.

M. A. Marsan, G. Balbo, G. Conte, S. Donatelli, and G. Franceschinis. *Modelling with Generalized Stochastic Petri Nets*. Wiley Series in Parallel Computing. John Wiley and Sons, 1995.

T. Murata. Petri nets: properties, analysis and applications. *Proceedings of the IEEE*, 77 (4): 541–580, 1989.

M. Peleg, D. Rubin, and R. B. Altman. Using Petri net tools to study properties and dynamics of biological systems. *Journal of the American Medical Informatics Association*, 12 (2): 181–199, 2005.

V. N. Reddy, M. L. Mavrovouniotis, and M. N. Liebman. Petri net representations in metabolic pathways. In *Proceedings of the 1st International Conference on Intelligent Systems for Molecular Biology*, pages 328–336. AAAI Press, 1993.

Christian Rohr. *Simulative Analysis of Coloured Extended Stochastic Petri Nets*. PhD thesis, BTU Cottbus, Dep. of CS, January 2017.

M. Schwarick, C. Rohr, and M. Heiner. MARCIE - Model checking and reachability analysis done efficiently. In *Proceedings of the 8th International Conference on Quantitative Evaluation of SysTems*, pages 91–100. IEEE, 2011.

H. L. S. Younes, M. Kwiatkowska, G. Norman, and D. Parker. Numerical vs. statistical probabilistic model checking. *International Journal on Software Tools for Technology Transfer*, 8 (3): 216–228, 2006.

L. A. Zadeh. Fuzzy sets. *Information and Control*, 8: 338–353, 1965.

3

Process Algebras in Systems Biology
Paolo Milazzo

Dipartimento di Informatica, Università di Pisa, Pisa, Italy

3.1 Introduction

Process Algebras have been developed in the context of concurrency theory and have been widely applied to investigate the dynamical properties about correctness of execution, security threats, information flow, and so on (Baeten, 2004). They are essentially formal notations for the specification (or modelling) of concurrent entities able to interact with each other. In this sense, they are similar to agent-based modelling approaches (Abar et al., 2017).

In the early 2000s, with the increasing availability of biological data and the emerging need of computational techniques for the analysis of such data and the understanding of biological phenomena, researchers in concurrency theory started to propose applications of Process Algebras also in the biological application domain (Regev et al., 2001). This was motivated by the fact that there are some aspects of the cell functioning that are very similar to those studied in the context of concurrency (Regev and Shapiro, 2002). Moreover, the analysis of concurrent systems typically focuses on the *interaction* between the system components, and this was fitting very well with the aims of the *systems biology* approach emerging in the same years (Kitano, 2002). Indeed, the main aim of systems biology is to study cell processes at the *systems* level by describing the behavior of components (e.g. genes and proteins) in an abstract way and by emphasizing interactions.

After an initial period in which already available Process Algebras were applied to biological case studies, the research on Process Algebras in the context of systems biology led to the proposal of a number of new algebras specifically defined to capture peculiar aspects of biological systems.

Systems Biology Modelling and Analysis: Formal Bioinformatics Methods and Tools,
First Edition. Edited by Elisabetta De Maria.
© 2023 John Wiley & Sons, Inc. Published 2023 by John Wiley & Sons, Inc.

In this chapter, we describe the development of Process Algebras and related analysis methods in the context of systems biology. Surveys of Process Algebras for biological applications are already available (Bernini et al., 2018, Fisher and Henzinger, 2007, Machado et al., 2011). Here, we choose to describe in more detail the concepts that are at the basis of the adoption of Process Algebras in the context of systems biology.

This chapter is structured as follows. In Section 3.2, we will describe some of the main features of Process Algebras as formalisms for concurrent systems, and we will introduce the definition of the π-calculus. In Section 3.3, we will give some background notions of biochemistry and cell biology, and we will describe the analogies between cell pathways and concurrent systems. In Section 3.4, we will present the concepts behind qualitative modelling and analysis of biological systems with Process Algebras, while in Section 3.5, we will describe the quantitative aspects of cell pathways related with chemical kinetics and show how they can be dealt with by using stochastic extensions of Process Algebras. Finally, in Section 3.6, we draw some conclusions.

3.2 Process Algebras in Concurrency Theory

Process algebras (or process calculi) are a family of formal modelling notations for concurrent and distributed systems whose development started in the 1980s. Before the proposal of the first Process Algebras, formal models developed in the context of concurrency theory were based mainly on automata-like notations such as Petri nets (Reisig, 2012). Petri nets have been very successful (and are still very widely used) because they capture essential aspects of concurrency in a simple way and with an effective graphical notation (see Chapter 2). It is indeed quite easy to use Petri nets to model systems consisting of several concurrent components competing for (limited) shared resources. The dynamics of Petri nets is governed by formally specified rules, and this allows exploring the concurrent system behavior and the reachable configurations in a rigorous way in order, for instance, to detect potential deadlock situations. Moreover, Petri nets enable the application of formal analysis methods by applying both static and dynamic approaches. Static approaches allow assessing dynamical properties of the net by exploring only its structure. For instance, techniques based on the computation of invariants can be used to assess the properties of *liveness* (persistent availability of resources that make the net always able to evolve), *boundedness* of produced resources (that imply a finite set of reachable configurations), etc. On the other hand, dynamic techniques make it possible to determine whether the system can reach a specific configuration (*reachability*), or a configuration with *at least* a given amount of resources (*coverability*). Dynamical properties of this kind can only partially be assessed using static approaches but can be investigated using dynamic techniques based on the computation of the *reachability graph*, which represents all the possible system behaviors in an exhaustive way. Determining whether a particular configuration can be reached is a decidable problem (Petri nets are not Turing-equivalent), although it is in general a problem of exponential complexity in the size of the net (Esparza and Nielsen, 1994).

Process Algebras provide alternative modelling and analysis approaches for concurrent systems. Their main added value is *compositionality* (Valmari, 1996). In a Petri net, a concurrent system consisting of many components and resources has to be modeled as a single huge graph. Considering a variant of the system that differs in only one of the components may require a refactoring of the whole graph. In Process Algebras, a concurrent system is described as a composition of terms (or processes) through a set of specific operators representing relationships between system components. The dynamics of a model is defined by a formal semantics that typically follows either an *operational*, *denotational*, or *equational* approach (Winskel, 1993). Probably, the most successful approach among the three is the operational one, in particular following the *structural operational semantics (SOS)* style (Aceto et al., 2001), originally proposed by Plotkin (1981). Operational semantics describe the dynamics of processes in terms of a *transition system (TS)* that is the analogous of the reachability graph of Petri nets. The SOS style requires the TS to be defined by a set of inference rules that are in a strict relationship with the syntax of the Process Algebra.

Semantics of Process Algebras are usually defined in a compositional way; hence, the semantics of a composition of processes can be computed from the semantics of the individual processes. Compositionality often requires the TS to be defined in order to represent not only the actual behavior of the process but also the potential behavior that the process may exhibit by interacting with other processes to be composed with. In technical terms, potential behaviors are represented in the TS as additional transitions distinguished by special labels describing the expected interaction. As a consequence, compositional semantics are usually expressed as labeled transition systems (LTS) rather than standard TS. A LTS is simply a TS in which transitions can be labeled.

Compositionality implies that in the model of a big concurrent system, it is often possible to study the properties of individual components on the semantics of only the corresponding processes. Moreover, considering a variant of the system in which only a few of the components are changed requires only to redefine the processes describing those components, without changing the rest of the model.

Formal semantics of Process Algebras provide descriptions of the system behaviors in mathematical terms. This opens to the possibility of proving the behavioral properties of the systems under study by using techniques such as model checking (Clarke Jr. et al., 2018, Clarke, 1997) and abstract interpretation (Cousot, 1996). Moreover, compositionality of semantics favors the application of modular analysis methods. In particular, approaches based on *behavioral equivalences* have been particularly successful in the context of concurrency. A behavioral equivalence is an equivalence relation on processes that essentially compares their LTSs. Many behavioral equivalences of many Process Algebras have been proved to be congruences, with the effect that they can be used to compare the behavior of two concurrent systems component-wise.

Among the most successful Process Algebras in concurrency theory, we mention the *Calculus of Communicating Systems (CCS)* (Milner et al., 1980), the formalism of *communicating sequential processes (CSP)* (Hoare, 1978), the π-calculus (Milner, 1999), and the *Mobile Ambients* calculus (Cardelli and Gordon, 2000).

Both CCS and CSP aim at modelling concurrent processes able to synchronize either over synchronization channels (CCS) or on events (CSP). The π-calculus can be seen as an extension of CCS in which processes can use channels also to communicate with each other. Messages in inter-process communications of the π-calculus are channel names, and this makes processes reconfigurable as if they were moving from one environment to another (where they receive different channels to be used). Also Mobile Ambients focus on mobility of processes. It uses an explicit notion of ambient (as a pair of brackets) representing the environment where the process is placed, and a process can perform actions to move from one ambient to another.

3.2.1 π-Calculus

As a representative Process Algebra, we present in more detail the π-calculus (Milner, 1999, Sangiorgi and Walker, 2003). We introduce here its syntax and semantics, and we will use it in the rest of the chapter to give examples and discuss different aspects of biological modelling. We choose the π-calculus since it is has been successfully applied in both concurrency theory and systems biology. Actually, the first steps toward the application of Process Algebras in the biological domain have been done using such a formalism.

3.2 Process Algebras in Concurrency Theory

The π-calculus is a formalism that allows describing concurrent processes able to communicate over channels through message passing. Channel names can be used as messages, and a process that receives a channel name can communicate over it. This makes processes reconfigurable and able to adapt their behavior depending on (what received from) the environment. For this reason, the π-calculus has been originally proposed as a formalism for concurrent and *mobile* systems.

We recall syntax and semantics of the π-calculus from the definition given in Quaglia and BRICS Lecture Series (1999) with slight modifications.

Definition 3.1 *(Syntax of the π-Calculus)* Let $\mathcal{N}$ be an enumerable infinite set of *channel names* ranged over by $x, y, z, \ldots$. The *syntax* of π-calculus processes, ranged over by $P, Q, Z, \ldots$, is defined by the following grammar:

$$
\begin{array}{lll}
P ::= & 0 & \text{inaction} \\
  | & \alpha.P & \text{prefix} \\
  | & [x = y]P & \text{match} \\
  | & P + P & \text{non-deterministic choice} \\
  | & P \mid P & \text{parallel composition} \\
  | & \nu x\, P & \text{restriction} \\
  | & A(x_1, \ldots, x_n) & \text{macro expansion}
\end{array}
$$

Prefixes describe actions that can be performed by a process P, defined by

$$\alpha ::= \tau \;\mid\; \bar{x}y \;\mid\; x(y)$$

where τ represents an *internal (silent) action* performed autonomously by P; $\bar{x}y$ represents an *output action*, corresponding to sending name y over channel x to a process external to P; and $x(y)$ represents an *input action*, corresponding to receiving name y over channel x from a process external to P.

We denote with $\mathcal{P}$ the set of all syntactically correct π-calculus processes.

Informally, 0 describes a terminated process; $\alpha.P$ a process ready to perform action α and then continue as P; $[x = y]P$ expresses a condition: if x and y are the same channel, the process continues behaving as P; $P + Q$ means that the process can non-deterministically behave either as P or as Q, and once one of the two alternatives is chosen (the first action of either P or Q is performed), the other is discarded; $P \mid Q$ instead means that the process consists of two parallel components that can be interleaved and can communicate with each other; $\nu x\, P$ restricts the scope of channel name x to process P: any occurrence of x outside of $\nu x\, P$ actually refers to a different channel; finally, $A(x_1, \ldots, x_n)$ recalls a process previously defined as A with

n channel names as formal parameters (like parametric macro expansion). We omit technical details about process definition and assume them to be always available and specified as in the following example:

$$A(x_1,\ldots,x_n) := P_1$$
$$B(y_1,\ldots,y_m) := P_2$$
$$\vdots \qquad \vdots$$

where $P_1, P_2, \ldots$ are π-calculus processes. For the sake of simplicity, in the case of process definitions without parameters, we omit the empty parentheses after the process name, that is, we write A instead of $A()$.

We remark that process definitions can be *recursive*, with the consequence of enabling infinite executions and Turing-completeness. Moreover, in the π-calculus literature, process definitions are often replaced by the simpler replication operator $!P$. Intuitively, $!P$ means that there exist infinitely many copies of P ready to be executed (hence, $!P \equiv P \mid !P$). In this chapter, we prefer to opt for process definitions instead of replication since this will make it clearer the correspondence between biological entities and process examples of the following sections.

Before giving the formal semantics of the calculus, we have to introduce a couple of auxiliary concepts. First of all, when considering channel names used in a process, we have to distinguish *bound names* from *free names*. In the language, we have two operators that limit the scope of a channel name: the first one is $vx\ P'$, that restricts the scope of x to P', and the second one is the input prefix $x(y).P'$ that restricts the scope of y to P'. Given a process P, let us denote with $n(P)$ the set of all channel names syntactically present in it. Among them, we define the set of free names $fn(P)$ as follows:

$$fn(0) = \emptyset$$
$$fn(\tau.P') = fn(P')$$
$$fn(\overline{x}y.P') = \{x,y\} \cup fn(P')$$
$$fn(x(y).P') = (fn(P')\setminus\{y\}) \cup \{x\}$$
$$fn(vx\ P') = fn(P')\setminus\{x\}$$
$$fn(Q+Z) = fn(Q \mid Z) = fn(Q) \cup fn(Z)$$
$$fn(A(y_1,\ldots,y_n)) = (fn(Q)\setminus\{x_1,\ldots,x_n\}) \cup \{y_1,\ldots,y_n\}$$
$$\text{if } A(x_1,\ldots,x_n) := Q\ .$$

This definition of fn does not work for mutually recursive process definitions (e.g. as in $A := \overline{x}y.B$ and $B := v(z).A$). We assume fn to compute a fixpoint

$$
\begin{array}{ll}
\text{(Tau)} & \tau.P \xrightarrow{\tau} P \qquad\qquad\qquad \text{(Out)} \quad \overline{x}y.P \xrightarrow{\overline{x}y} P \\[4pt]
\text{(Inp)} & x(y).P \xrightarrow{x(w)} P\{w/y\} \quad w \notin \text{fn}(\nu y\, P) \\[6pt]
\text{(Sum1)} & \dfrac{P \xrightarrow{\ell} P'}{P+Q \xrightarrow{\ell} P'} \qquad\qquad \text{(Sum2)} \quad \dfrac{Q \xrightarrow{\ell} Q'}{P+Q \xrightarrow{\ell} Q'} \\[10pt]
\text{(Par1)} & \dfrac{P \xrightarrow{\ell} P'}{P\,|\,Q \xrightarrow{\ell} P'\,|\,Q} \quad \text{bn}(\ell) \cap \text{fn}(Q) = \emptyset \\[10pt]
\text{(Par2)} & \dfrac{Q \xrightarrow{\ell} Q'}{P\,|\,Q \xrightarrow{\ell} P\,|\,Q'} \quad \text{bn}(\ell) \cap \text{fn}(P) = \emptyset \\[10pt]
\text{(Com)} & \dfrac{P \xrightarrow{\overline{x}y} P' \quad Q \xrightarrow{x(w)} Q'}{P\,|\,Q \xrightarrow{\tau} P'\,|\,Q'\{y/w\}} \qquad \text{(Close)} \quad \dfrac{P \xrightarrow{\overline{x}(w)} P' \quad Q \xrightarrow{x(w)} Q'}{P\,|\,Q \xrightarrow{\tau} \nu w\,(P'\,|\,Q')} \\[14pt]
\text{(Open)} & \dfrac{P \xrightarrow{\overline{x}y} P'}{\nu y\, P \xrightarrow{\overline{x}(w)} P'\{w/y\}} \quad y \neq x,\, w \notin \text{fn}(\nu y\, P') \\[10pt]
\text{(Res)} & \dfrac{P \xrightarrow{\ell} P'}{\nu y\, P \xrightarrow{\ell} \nu y\, P'} \quad x \notin \text{n}(\ell) \\[10pt]
\text{(Def)} & \dfrac{P\{y_1/x_1\}\ldots\{y_n/x_n\} \xrightarrow{\ell} P'}{A(y_1,\ldots,y_n) \xrightarrow{\ell} P'} \quad A(x_1,\ldots,x_n) := P
\end{array}
$$

Figure 3.1 Semantic rules of the π-calculus.

in such a case. Finally, bound names of a process P are channel names that are not bound, namely, $\text{n}(P) = \text{bn}(P) \cup \text{fn}(P)$ and $\text{bn}(P) \cap \text{fn}(P) = \emptyset$.

We now introduce the definition of the formal semantics of the calculus. It is given in terms of a LTS in a compositional way and adopting the SOS style (Aceto et al., 2001).

Definition 3.2 *(Semantics of the π-Calculus)* The *semantics* of the pi-calculus is defined as a LTS in $\mathcal{P} \times \mathcal{L} \times \mathcal{P}$, where $\mathcal{P}$ is the set of all π-calculus processes (Definition 3.1) and $\mathcal{L}$ is the set of all possible labels defined as follows:

$$\mathcal{L} = \{\tau\} \cup \{\overline{x}y \mid x,y \in \mathcal{N}\} \cup \{x(y) \mid x,y \in \mathcal{N}\} \cup \{\overline{x}(y) \mid x,y \in \mathcal{N}\}$$

The semantics is then defined as the least LTS obtained from the inference rules in Figure 3.1, with $P \xrightarrow{\ell} Q$ denoting a generic transition.

Inference rules of the formal semantics describe how each operator can cause the process to perform one or more transitions. Rules (Tau), (Out), and (Inp) define transitions performed by a process starting with a prefix. The prefix itself becomes the label of the transition, for composition purposes. In the case of the input prefix $x(y)$, the rule defines infinitely many transitions, each with a different (free) name w replacing y. This is because in $x(y).P$, the prefix $x(y)$ causes y to be a bound name in P. Hence, allowing y to be replaced by any free name w will permit to solve name clashes when composing with other processes. Rules (Sum1) and (Sum2) for non-deterministic choices are straightforward. Rules (Par1) and (Par2) define transitions of parallel processes in which one of the two performs an action without interacting with the other. The side condition $bn(\ell) \cap fn(Q) = \emptyset$ ensures that if ℓ corresponds to an input prefix, the transition generated by (Inp) that is considered for the composition is one of those in which the name of the received channel does not clash with free names in Q. Rules (Com) and (Close) define transitions performed by two processes that communicate with each other. In these cases, it is required that the two processes can perform an output and an input transition, respectively, on the same channel. If so, the parallel composition translates such transitions in a single transition with label τ describing an internal action (since now the composition cannot interact with any other process in the environment). The difference between (Com) and (Close) is that the latter considers the cases in which the channel w sent by the process P performing the output action was bound in P. This is denoted by the parenthesis in the label $\bar{x}(w)$ that is produced by Rule (Open) for the restriction operator vx. In such a case, we have that the scope of the channel name has to be enlarged to include also Q. This is obtained by removing the original restriction vx, as specified by (Open), and by adding it before the whole parallel composition, as specified by (Close). This change of scope for the local channel name is called *scope extrusion* and is one of the distinctive features of the π-calculus. Finally, rule (Res) describes the preservation of the restriction operator when no scope extrusion is necessary, and (Def) describes the instantiation of a previously defined process.

3.3 Analogies between Biology and Concurrent Systems

In this section, we start by providing some basic notions of cell biology and cell pathways, and then, we describe the analogies between cells and concurrent systems that led to the application of Process Algebras in the biological domain.

3.3.1 Elements of Cell Biology

There are two basic classifications of cell: *prokaryotic* and *eukaryotic*. Traditionally, the distinguishing feature between the two types is that an eukaryotic cell possesses a membrane-enclosed nucleus and a prokaryotic cell does not. Prokaryotic cells are usually small and relatively simple, and they are considered representative of the first types of cell to arise in biological evolution. Prokaryotes include, for instance, almost all bacteria. Eukaryotic cells, on the other hand, are generally larger and more complex, reflecting an advanced evolution, and include multicellular plants and animals.

In eukaryotic cells, different biological functions are segregated in discrete regions within the cell, often in membrane-limited structures. Subcellular structures that have distinct organizational features are called *organelles*. As an organelle, for example, the *nucleus* contains chromosomal deoxyribonucleic acid (DNA) and the enzymatic machinery for its expression and replication, and the *nuclear membrane* separates it from the rest of the cell, which is called *cytoplasm*. There are organelles within the cytoplasm, e.g. *mitochondria*, sites of respiration, and (in some cells) *chloroplasts*, sites of photosynthesis.

The most abundant class of biomolecules in cells are *proteins*. Proteins are synthesized according to the genetic information contained in the cell chromosomes. They exhibit a very high functional versatility and are therefore utilized in a variety of biological roles. A few examples of biological functions of proteins are enzymatic activity (catalysis of chemical reactions), transport, and storage of molecule and cellular structuring.

Although biologically active proteins are macromolecules that may be very different in size and shape, they are all polymers consisting of a chain of amino acids. The number and order of the amino acids determine the distinctive three-dimensional structure of the protein, which, in turn, determines its functionality. In this three-dimensional structure, often very complex and involving more than one chain of amino acids, it is sometimes possible to identify places where chemical interaction with other molecules can occur. These places are called *interaction sites* and are usually the basic entities in the abstract description of the behavior of a protein.

Similar to proteins, *nucleic acids* are polymers with a sequential structure. More precisely, they are chains of nucleotides. Two types of nucleic acid exist: the *DNA* and the *ribonucleic acid* (RNA). The former contains the genetic instructions for the biological development of a cellular form of life. In eukaryotic cells, it is placed in the nucleus and it is shaped as a double helix, while in prokaryotic cells, it is placed directly in the cytoplasm and it is circular. DNA contains the genetic information that is inherited by the offspring of an organism. A strand of DNA contains genes, areas that regulate genes, and areas that either have no function, or a function yet unknown. Genes are the units of heredity and can be loosely viewed as the organism's "cookbook."

Like DNA, most biologically active RNAs are chains of nucleotides. Various types of RNA exist, among which we mention the *messenger ribonucleic acid* (mRNA) which carries information from DNA to sites of protein synthesis in the cell, and the *transfer ribonucleic acid* (tRNA) which transfers a specific amino acid to a growing protein chain.

The description of proteins and nucleic acids we have given suggests a route for the flow of biological information in cells. In fact, we have seen that DNA contains instructions for the biological development of a cellular form of life, RNA carries information from DNA to sites of protein synthesis in the cell and provides amino acids for the development of new proteins, and proteins perform activities of several kinds in the cell. Schematically, we have this flux of information:

$$\text{DNA} \xrightarrow{\text{transcription}} \text{RNA} \xrightarrow{\text{translation}} \text{Protein}$$

in which *transcription* and *translation* are the activities of performing a "copy" of a portion of DNA into a mRNA molecule and of building a new protein by following the information found on the mRNA and by using the amino acids provided by tRNA molecules. This process is known as the *Central Dogma of Molecular Biology*.

3.3.2 Cell Pathways

Very roughly speaking, a cell is essentially a container of a huge number of molecules floating in a fluid medium organized into membrane-delimited compartments. Molecules interact with each other by performing chemical reactions. Different groups of molecules can interact at the same time in different parts of the cell. The interaction of two or more molecules causes some changes in their configuration (the creation or the removal of a binding, a conformational change, etc.) that, in turn, enables other chemical reactions. Consequently, cell functioning is based on big sequences (or networks) of chemical reactions that can be performed concurrently inside the cell itself. In biological terms, such networks are called *pathways*, and they mostly involve proteins. Pathways are often activated by environmental stimuli, such as the availability of some nutrients to be used to produce energy (*metabolic pathways*), or the presence of some signaling molecules produced by neighbor cells (*signaling pathways*). The detection of an environmental stimulus may require the cell to activate new functionalities (i.e. new pathways) by starting the production of the relevant proteins. This involves the genes of the DNA (from which proteins are synthesized) whose activity is controlled by complex regulation processes (called *gene regulation networks*) that are again networks of chemical reactions working as circuits that activate the synthesis of proteins in accordance with the whole cell configuration and the environmental stimuli.

Cell pathways can be described as sets of *chemical reactions*, specifying that a group of molecules (the *reactants*) can be transformed into another group of molecules (the *products*). As an example of chemical reaction, we can think of the classical example of transformation of water production:

$$2H_2 + O_2 \rightarrow 2H_2O$$

This reaction specifies that two molecules of hydrogen (H_2) together with one molecule of oxygen (O_2) can produce two molecules of water (H_2O).

In chemistry, chemical reactions are often considered as equations, but from the viewpoint of a computer scientist, they can be more naturally seen as rewrite rules: the reactants are somehow rewritten into the products.

We now give a toy example useful to acquire confidence with chemical reactions and pathways. It will also be used as a running example in the presentation of Process Algebras in the next sections.

Example 3.1 *(Toy Pathway)*
Let us consider a minimal signaling pathway consisting of the following four chemical reactions:

$(R_a) \quad S + R \rightleftharpoons SR$
$(R_b) \quad SR + P \rightarrow SR + P'$
$(R_c) \quad P' \rightarrow P$
$(R_d) \quad P' + G_i \rightleftharpoons G_a$

Reaction (R_a) describes the binding of a *signal* protein S present in the cell environment with a *receptor* protein R located on the cell surface. This gives rise to protein complex SR. Since the reaction is *reversible* (denoted $\rightleftharpoons$), the complex SR could also disassemble into S and R. A reversible reaction is simply a shorter way of denoting two separate reactions in which reactants and products are swapped, namely, $S + R \rightarrow SR$ and $SR \rightarrow S + R$.

Reaction (R_b) shows that the complex SR is able to catalyze the transformation of P into P', essentially acting as an *enzyme*. We can imagine that P' is the *active* form of P: the complex SR caused a conformational change to P that enabled one or more interaction sites on it. This reaction is not reversible, but P' can take again its inactive form P autonomously through reaction (R_c). Finally, the reversible reaction (R_d) describes the binding of P' with gene G by moving it from an inactive state (denoted G_i) to an active state (denoted G_a). We can imagine that now G_a can cause the synthesis of some new proteins that correspond to the activation of some new functionalities in response to the detection of the initial signal S. The synthesis of such new proteins is not modeled in this toy example.

The toy pathway we presented is an extremely (and non-realistically) simplified example of cell pathway. A real cell pathway typically involves dozens or hundreds of proteins. A classical example of real signaling pathway, very often considered in the literature of formal methods with biological applications, is the epidermal growth factor receptor (EGFR) pathway (Wiley et al., 2003, Schoeberl et al., 2002). It is widely studied from the biological viewpoint and has become particularly well known for its involvement in many types of cancer development.

The dynamics of a pathway is governed by the occurrence of reactions. The chemical solution contained in the cell can be seen as a multiset of molecules that is "updated" every time a reaction takes place. A number of quantitative aspects should be taken into account at this point. For example, one should consider the time required by each reaction to be completed, the propensity of each reaction to take place in a given configuration, the density of the molecules in the cell, etc. We will deal with these quantitative aspects in the following. For the moment, we describe the dynamics of pathways in a *qualitative* way. We assume the reactions to be equiprobable and instantaneous (hence, it is impossible that two reactions take place at the same time). Under this assumption, starting from an initial multiset of molecules, the dynamics can be described simply as a sequence of reactions with the corresponding state updates. More precisely, since in the same state several reactions can take place, the qualitative dynamics can be properly described in terms of a TS.

Definition 3.3 *(Qualitative Pathway dynamics)* Given a set of reactions $\mathcal{R}$ (a *pathway*), let S be the set of all the molecules being reactants or products of reactions in $\mathcal{R}$ and $\mathcal{M}_S$ be the set of all possible multisets over S. The *qualitative dynamics* of $\mathcal{R}$ is the TS $(\mathcal{M}_S, \mapsto)$, where $\mapsto \subseteq \mathcal{M}_S \times \mathcal{M}_S$ is the least transition relation defined by the following inference rule:

$$\frac{R \rightarrow P \in \mathcal{R}}{M \cup R \mapsto M \cup P}$$

where R and P are multisets representing reactants and products, respectively, of the considered reaction.

Note that a reaction cannot take place if the corresponding reactants are not present in the current state.

Example 3.2 As an example of evolution of the toy pathway presented in Example 3.1, let us assume an initial cell configuration with 10 receptor proteins R, 10 inactive proteins P, and 1 inactive gene G_i. This is described by the following multiset (represented as a set of pairs):

$$\{(R, 10), (P, 10), (G_i, 1)\} \tag{3.1}$$

In this state, no reaction can take place. Let us now add a signal protein S:

$$\{(S, 1), (R, 10), (P, 10), (G_i, 1)\} \tag{3.2}$$

Now, reaction (R_a) is enabled, causing the activation of the pathway. A possible evolution, according to the pathway dynamics given in Definition 3.3, is the one described by the following sequence of transitions in which reactions (R_a), (R_b), and (R_d) take place:

$$\{(S, 1), (R, 10), (P, 10), (G_i, 1)\} \mapsto \{(SR, 1), (R, 9), (P, 10), (G_i, 1)\} \mapsto$$
$$\{(SR, 1), (R, 9), (P, 9), (P', 1), (G_i, 1)\} \mapsto$$
$$\{(SR, 1), (R, 9), (P, 9), (G_a, 1)\}$$

This is actually the shorter sequence of transitions in the pathway dynamics that lead to the activation of G_a. Another one, in which also the reverse direction of (R_a) and reaction (R_c) takes place, is the following slightly longer sequence of transitions:

$$\{(S, 1), (R, 10), (P, 10), (G_i, 1)\} \mapsto \{(SR, 1), (R, 9), (P, 10), (G_i, 1)\} \mapsto$$
$$\{(S, 1), (R, 10), (P, 10), (G_i, 1)\} \mapsto \{(SR, 1), (R, 9), (P, 10), (G_i, 1)\} \mapsto$$
$$\{(SR, 1), (R, 9), (P, 9), (P', 1), (G_i, 1)\} \mapsto$$
$$\{(SR, 1), (R, 9), (P, 10), (G_i, 1)\} \mapsto$$
$$\{(SR, 1), (R, 9), (P, 9), (P', 1), (G_i, 1)\} \mapsto$$
$$\{(SR, 1), (R, 9), (P, 9), (G_a, 1)\}$$

The qualitative dynamics given in Definition 3.3 shows that chemical reactions can be seen essentially as *multiset rewriting rules* and pathways as *multiset rewriting systems*. Indeed, pathways can be trivially expressed using the MultiSet Rewriting formalism (Bistarelli et al., 2003). This natural representation of chemical reactions in terms of rewriting rules motivated the definition of several rule-based bio-inspired formalisms (see Chapter 4 and e.g. (Pedersen et al., 2015, Pérez-Jiménez and Romero-Campero, 2006, Barbuti et al., 2006, Danos and Laneve, 2004, Eker et al., 2004, Barbuti et al., 2005)).

3.3.3 "Molecules as Processes" Abstraction

From this very concise description of the cell functioning, a number of similarities with concurrent systems in computer science can be observed. First of all, molecules (in particular, proteins and genes) can be seen as active resources able to interact with each other. This makes them similar to processes that are active computational entities able to synchronize and communicate with each other. Moreover, interaction between different groups of molecules can happen at the same time, as well as processes can be executed concurrently or in parallel. Furthermore, the interaction between molecules leads to changes in their configuration that enable new interactions, and the same typically happens to processes. As a consequence, a cell can be abstracted as a big system of concurrent processes in which synchronizations and inter-process communications simulate chemical reactions.

In order to clarify the relationship between chemical reactions and process synchronizations, let us consider the toy pathway described in the previous section (Example 3.1). In π-calculus terms, we can translate it into a set of process definitions, one for each molecule involved in the pathway. Moreover, we assume each reaction to correspond to a channel name (in general, we assume channel x to correspond to reaction (R_x)). Then,

- each reaction (R_x) involving *a single reactant* is translated into a τ action of the corresponding process;
- each reaction (R_x) involving *two reactants* is translated into a synchronization between the two corresponding processes on channel x;
- reactions involving more than two reactants cannot be translated directly.

Example 3.3 The four reactions of the toy pathway can be translated into the following seven processes (one for each molecule type), in which we omit, since it is irrelevant, the channel name transmitted in each interprocess communication. Since we only need synchronizations (and not communications) for this example, we adopt the notations $\bar{x}$ and x instead of $\bar{x}y$ and $x(y)$, respectively.

These are the process definitions corresponding to the reactions of the toy pathway:

$$S := \bar{a}.SR$$
$$R := a.0$$
$$SR := \tau.(S \mid R) + \bar{b}.SR$$
$$P := b.P'$$
$$P' := \tau.P + \bar{d}.0$$

3.3 Analogies between Biology and Concurrent Systems | 49

$$G_i := d.G_a$$
$$G_a := \tau.(P' \mid G_i)$$

By inspecting the process definitions in the example, it can be observed that reaction $(R_a)S + R \rightleftharpoons SR$ has been translated into a synchronization on channel a between processes S and R and a τ performed by process SR. The roles (sender/receiver) of the processes involved in each synchronization are not relevant: by swapping a and $\bar{a}$ in the two processes, we obtain exactly the same dynamics. Moreover, since (R_a) has two reactants and only one product SR, the process representing SR is activated by S after performing the synchronization on a, while R terminates with 0. Again, by swapping SR and 0 in the two processes, we would obtain the same dynamics. The other reactions are translated similarly. We only notice that reactions with a single reactant and two products are translated into a τ action followed by a parallel composition of the processes representing the products. Moreover, when the same reactant can be involved in more than one reaction, the corresponding synchronizations are composed using a non-deterministic choice. Process SR is an example in which both parallel composition and non-deterministic choice are used.

It can be trivially proved that there exists a correspondence between the TS describing the dynamics of a pathway and the LTS obtained the π-calculus semantics computed on the processes obtained from the translation of the pathway. We choose to omit technicalities and simply present an example of dynamics of the toy pathway.

Example 3.4 Let us consider again the evolution of the toy pathway described in Example 3.2. The initial multiset 3.1 corresponds to the following parallel composition of π-calculus processes:

$$Init = \overbrace{R \mid \ldots \mid R}^{10 \text{ times}} \mid \overbrace{P \mid \ldots \mid P}^{10 \text{ times}} \mid G_i$$

where R, P, and G_i are processes defined as in Example 3.3. It is immediate to see that the processes in this parallel composition cannot interact with each other since they are ready to synchronize on different channels. Consequently, all the sequences of transitions they can perform start with a potential interaction with another process in the environment (with a non-τ label) as in the following example:

$$Init \xrightarrow{\bar{a}} SR \mid \overbrace{R \mid \ldots \mid R}^{9 \text{ times}} \mid \overbrace{P \mid \ldots \mid P}^{10 \text{ times}} \mid G_i \xrightarrow{\tau}$$

$$SR \mid \overbrace{R \mid \ldots \mid R}^{9 \text{ times}} \mid \overbrace{P \mid \ldots \mid P}^{9 \text{ times}} \mid P' \mid G_i \xrightarrow{\tau}$$

$$SR \mid \overbrace{R \mid \ldots \mid R}^{9 \text{ times}} \mid \overbrace{P \mid \ldots \mid P}^{9 \text{ times}} \mid G_a \qquad (3.3)$$

The evolution of the initial multiset 3.1 in which S is present can be computed (compositionally) from the evolution of the previous case as follows:

$$S \mid Init \xrightarrow{\tau} SR \mid \overbrace{R \mid \ldots \mid R}^{9 \text{ times}} \mid \overbrace{P \mid \ldots \mid P}^{10 \text{ times}} \mid G_i \xrightarrow{\tau}$$

$$SR \mid \overbrace{R \mid \ldots \mid R}^{9 \text{ times}} \mid \overbrace{P \mid \ldots \mid P}^{9 \text{ times}} \mid P' \mid G_i \xrightarrow{\tau}$$

$$SR \mid \overbrace{R \mid \ldots \mid R}^{9 \text{ times}} \mid \overbrace{P \mid \ldots \mid P}^{9 \text{ times}} \mid G_a \qquad (3.4)$$

The example shows that transitions included in the qualitative dynamics of a pathway correspond to τ transitions in the semantics of the corresponding π-calculus processes. Moreover, the π-calculus semantics can describe potential evolutions that could take place in a proper context. This can be seen in the case of Eq. (3.3), which shows that in a context that can provide an interaction on channel a, gene G_a is activated after a few internal steps. Finally, the semantics of a composition can be computed from the semantics of the components, as we did to compute the evolution of $S \mid Init$ from that of $Init$ in Eq. (3.4). Compositionality is also an important property for analysis purposes: it enables component-wise reasoning on the behavior of the system.

In our example, we did not use the main feature of the π-calculus, that is, the ability to communicate channel names over channels. In the biological context, such a feature can be used to model the aspects of interactions between molecules that are not easily captured by traditional chemical reactions. In particular, processes representing two molecules could share a private channel that allows them to perform reactions in isolation with respect to the rest of the system. This allows, for instance, to model reactions that take place between molecules in a separate compartment of the cell (e.g. an organelle or a membrane vesicle). Moreover, this feature can also be useful to encode reactions with more than two reactants into sequences of binary synchronizations on a number of private channels to be shared among processes representing molecules to be involved.

3.4 Process Algebras for Qualitative Modelling

Among Process Algebras originally proposed for the modelling of concurrent processes and then applied also in the context of systems biology, the most successful one is for sure the π-calculus. This because it was the formalism used in some of the most relevant papers introducing biological applications in the concurrency theory community.

In Regev et al. (2001) and Regev and Shapiro (2002), Regev and Shapiro proposed to describe metabolic pathways as π-calculus processes, and in Priami et al. (2001), they showed how the stochastic variant of the model, defined by Priami (1995), can be used to represent both qualitative and quantitative aspects of the systems described. Moreover, Regev et al. (2004) defined the BioAmbients calculus, a model inspired by both the π-calculus and the Mobile Ambients calculus (Cardelli and Gordon, 2000), which can be used to describe biochemical systems with a notion of compartments (as, for instance, membranes). More details of membrane interactions have been considered by Cardelli in the definition of Brane Calculi (Cardelli, 2004), which are elegant formalisms for describing intricate biological processes involving membranes. Another process algebraic formalism that aims at capturing modularity of biological systems is Beta Binders (Priami and Quaglia, 2004) that then evolved into the Blenx modelling language (Dematté et al., 2008).

Another pioneering formalism for the description of cell pathways is the κ-calculus (Danos and Laneve, 2004). It is a formal language for protein interactions, it is enriched with a very intuitive visual notation, and it has been encoded into the π-calculus. Essentially, the κ-calculus idealizes protein–protein interactions as a particular restricted kind of graph-rewriting operating on graphs with sites. A formal protein is a node with a fixed number of sites, and a complex (i.e. a bundle of proteins bound together) is a connected graph built over such nodes, in which connections are established between sites.

3.4.1 Formal Analysis Techniques

Formal notations enable unambiguous description of the systems under study. Moreover, they allow verification of properties of these systems by means of methods, such as model checking or abstract interpretation, that are well established in theoretical computer science and could find applications also in biology. In the specific context of Process Algebras, methods based on *behavioral equivalences* have been proposed and widely applied to reason about concurrent systems and verify their behavioral properties.

LTSs may describe the behavior of the modeled system in great detail. Relations on states of a LTS can be defined to compare the behavior of two modeled systems. In particular, behavioral equivalences are reflexive, transitive, and symmetric relations that relate systems that are not distinguished by any external observer, according to a given notion of observation. We recall here the notion of (strong) *bisimulation equivalence* (Sangiorgi, 1998), which relates two states in a LTS when they are step by step able to perform transitions with the same labels.

Definition 3.4 *(Strong bisimulation)* Given an LTS $(S, L, \rightarrow)$, a relation $R \subseteq S \times S$ is a *strong bisimulation* if whenever $(s_0, s_2) \in R$, the following two conditions hold:

$s_0 \rightarrow s_1 \implies \exists s_3 \in S$ such that $s_2 \rightarrow s_3$ and $(s_1, s_3) \in R$;

$s_2 \rightarrow s_3 \implies \exists s_1 \in S$ such that $s_0 \rightarrow s_1$ and $(s_1, s_3) \in R$.

The *strong bisimilarity* $\sim$ is the largest of such relations.

Comparing the behaviors of two systems, typically internal (hidden) actions can be ignored. For this reason, a different notion of bisimulation equivalence, called *weak bisimulation*, is often considered. In the definition of weak bisimulation, we denote with $\overset{l}{\Longrightarrow}$ a sequence of transitions $\overset{\tau}{\rightarrow} \cdots \overset{\tau}{\rightarrow} \overset{l}{\rightarrow} \overset{\tau}{\rightarrow} \cdots \overset{\tau}{\rightarrow}$, where τ is a label describing unobservable actions (as in the π-calculus semantics).

Definition 3.5 *(Weak bisimulation)* Given an LTS $(S, L, \rightarrow)$ in which $\tau \in L$ is a special label describing hidden actions, a relation $R \subseteq S \times S$ is a *weak bisimulation* if whenever $(s_0, s_2) \in R$, the following two conditions hold:

$s_0 \overset{l}{\rightarrow} s_1 \implies \exists s_3 \in S$ such that $s_2 \overset{l}{\Longrightarrow} s_3$ and $(s_1, s_3) \in R$;

$s_2 \overset{l}{\rightarrow} s_3 \implies \exists s_1 \in S$ such that $s_0 \overset{l}{\Longrightarrow} s_1$ and $(s_1, s_3) \in R$.

The *weak bisimilarity* $\approx$ is the largest of such relations.

Bisimulations are well-established behavioral equivalences that are now widely used for the verification of properties of computer science systems. In principle, properties can be verified by assessing the bisimilarity of the considered system with a system one knows to enjoy them. Moreover, given the model of a system, bisimulations can be used to consider equivalent simplified models. The notion of congruence is very important for a compositional account of behavioral equivalence. This is true, in particular, for complex systems such as the biological ones. Being defined for LTSs, bisimilarities can be naturally adopted for comparing the behavior of processes of a Process Algebra.

In Antoniotti et al. (2004), bisimulations are used to perform qualitative analysis of biological systems described in terms of hybrid automata. In Barbuti et al. (2008b), bisimulations are defined specifically for Process Algebras making an explicit modelling of cell membranes. In Cardelli et al. (2015), bisimulations are proposed as a tool to minimize models of chemical reactions.

Apart from bisimulations, other formal reasoning and analysis methods that are well established in concurrency theory and that have been applied also in the biological context are mostly based on static analysis techniques (Nielson et al., 2003, Bodei et al., 2015), such as abstract interpretation (Danos et al., 2008, Gori and Levi, 2010) and type systems (Aman et al., 2009), and causal semantics (Curti et al., 2003, Busi, 2007). Model checking techniques have instead been applied mostly in the quantitative setting, which we will describe in the next section.

3.5 Process Algebras for Quantitative Modelling

Quantitative aspects of the dynamics of cell pathways deal essentially with the *frequency* or, better, the *rate* of occurrence of chemical reactions. How often a certain reaction takes places inside a cell depends on many factors: the abundancy of reactants, the presence of specific enzymes, the affinity between the reactants (how easily they can bind with each other), the temperature, etc. Pathways fit perfectly the definition of *complex systems*: they are *dynamical systems* consisting of many "simple" components (the molecules) that are able to interact with each other, with an emerging collective dynamics that is sometimes surprising.

Traditional approaches for dynamical systems (e.g. those based on differential equations or Markov chains) can be used to perform simulations or study properties of cell pathways. The literature on Process Algebras already contained some formalisms able to deal with *stochastic* rates, developed originally for applications in the context of performance analysis. Some of these formalisms, such as the stochastic π-calculus and the PEPA Process Algebra, have then started to be applied in the biological domain. Moreover, some new quantitative formalisms have been proposed with the specific aim of modelling biological systems.

In this section, we start from recalling some notions of chemical kinetics necessary to understand how reaction rates can be determined. Then, we describe how these notions have been adopted in stochastic Process Algebras to apply them for the quantitative modelling of pathways.

3.5.1 Chemical Kinetics

The fundamental empirical law governing reaction rates in biochemistry is the *law of mass action*. This states that for a reaction in a homogeneous medium, the reaction rate will be proportional to the concentrations of the individual reactants involved. A chemical reaction following a mass action kinetics is usually represented with the following notation:

$$\ell_1 S_1 + \ell_2 S_2 \underset{k_{-1}}{\overset{k}{\rightleftharpoons}} \ell_3 S_3 + \ell_4 S_4$$

where $S_1, \ldots, S_4$ are molecules, $\ell_1, \ldots, \ell_4$ are their stoichiometric coefficients, and k, k_{-1} are the kinetic constants. Irreversible reactions are denoted by the single arrow $\overset{k}{\longrightarrow}$.

For example, given the simple reaction

$$2A \underset{k_{-1}}{\overset{k}{\rightleftharpoons}} B$$

the rate of the production of molecule B for the law of mass action is

$$k[A]^2$$

and the rate of destruction of B is

$$k_{-1}[B]$$

where $[A], [B]$ are the *concentrations* (i.e. moles over volume unit) of the respective molecules. In general, the rate of a reaction is

$$k[S_1]^{\ell_1} \cdots [S_p]^{\ell_p}$$

where $S_1, \ldots, S_p$ are all the distinct molecular reactants of the reaction.

In Gillespie (1977), Gillespie gives a stochastic formulation of chemical kinetics that is based on the theory of collisions and that assumes a stochastic reaction constant c_μ for each considered chemical reaction R_μ. The reaction constant c_μ is such that $c_\mu dt$ is the probability that a particular combination of reactant molecules of R_μ will react in an infinitesimal time interval dt and can be derived with some approximations from the kinetic constant of the chemical reaction.

The probability that a reaction R_μ will occur in the whole solution in the time interval dt is given by $c_\mu dt$ multiplied by the number of distinct R_μ molecular reactant combinations. For instance, the reaction

$$R_1 : S_1 + S_2 \rightarrow 2S_1 \tag{3.5}$$

will occur in a solution with X_1 molecules S_1 and X_2 molecules S_2 with probability $X_1 X_2 c_1 dt$. Instead, the inverse reaction

$$R_2 : 2S_1 \rightarrow S_1 + S_2 \tag{3.6}$$

will occur with probability $\frac{X_1(X_1-1)}{2!} c_2 dt$. The number of distinct R_μ molecular reactant combinations is denoted by Gillespie with h_μ; hence, the probability of R_μ to occur in dt (denoted with $a_\mu dt$) is

$$a_\mu dt = h_\mu c_\mu dt$$

Rates computed using the law of mass action can be used to define systems of ordinary differential equations (ODEs) (see Chapter 9). On the other hand, stochastic rates computed by following Gillespie's approach are parameters of exponential distributions and hence can be used to describe the system dynamics in terms of continuous time Markov chains (CTMCs).

From the viewpoint of the biological modeler, the approach based on stochastic rates and CTMCs offers several advantages compared to ODEs. First of all, CMTC models of pathways are more accurate: they take into account the stochastic variability of the dynamics, and they are more precise in counting molecules. These two advantages become crucial when the modeled pathway may have molecules present in very small quantities, such as genes. A gene is a single molecule that can have two (or a few) different states (e.g. active/inactive). In a CTMC, we would typically have one integer variable for each gene state, and reactions could cause such variables to take only values 0 and 1, where 1 means that the gene is in such a state. ODEs, on the other hand, are based on real-valued variables that are changed in a continuous way. As a consequence, in the modelling of a gene state, the variable modelling of such a state would typically take a value in [0,1] (say, 0.5) that is not realistic for a single molecule. The same problem also occurs when it is necessary to model proteins that are present in small quantities. Approximating them with a continuous variable would introduce a significant approximation error. In addition to this, being deterministic, ODEs cannot capture stochastic fluctuations.

On the other hand, stochastic approaches have scalability problems. When quantities of molecules become high, rates increase as well and make reactions very frequent. A CTMC describes reaction occurrences one by one, and this immediately causes the model to become difficult to be analyzed. ODEs instead do not suffer from this problem since they assume that reactions happen continuously, and an increase of the rates due to higher quantities has a significantly smaller impact. Also, ODEs can have scalability problems in the case of stiff systems, but in general, they can be considered as much more scalable and efficient to be analyzed than CTMCs.

Since the two approaches based on ODEs and CTMCs offer complementary advantages, they are both rather common in the systems biology literature. The most common analysis techniques applied to them are numerical integration for ODEs and stochastic simulation for CTMCs. They often provide traces of the dynamics of the studied system in relatively short times. In the stochastic setting, CTMCs are actually often not mentioned, but they are implicitly under the hood of the simulation algorithms applied (as in the case of Gillespie's algorithm).

In order to see how quantitative aspects of chemical kinetics can be dealt with, let us consider a version of our toy pathway extended with reaction rates.

Example 3.5 Let us extend the toy pathway introduced in Example 3.1 by adding kinetic constants to reactions:

$(R_a) \quad S + R \underset{1}{\overset{1}{\rightleftharpoons}} SR$

$(R_b) \quad SR + P \xrightarrow{0.5} SR + P'$

$(R_c) \quad P' \xrightarrow{1} P$

$(R_d) \quad P' + G_i \underset{10}{\overset{1}{\rightleftharpoons}} G_a$

We associate each reaction with one (or two, if reversible) kinetic constants of our choice. In real pathways, these constants are either measured through wet lab experiments or estimated by fitting the available data from lab observations of the pathway dynamics.

The results of numerical integration of ODEs and stochastic simulation of the pathway are depicted in Figures 3.2 and 3.3, respectively. Numerical integration gives smooth curves, while stochastic simulation shows the effect of stochastic noise. In this case, both approaches give qualitatively the same results: protein P gets activated (P' is denoted as Pp in the figures) and the same holds for gene G, which takes mostly its G_a form. However, the difference between the two approaches can be observed by inspecting more in detail the dynamics of G_i and G_a. In the deterministic case (Figure 3.2), the system reaches an equilibrium with a concentration of 0.9 for G_a and 0.1 for G_i. On the other hand, in the stochastic case (Figure 3.3), the gene more realistically oscillates between the G_a and G_i states.

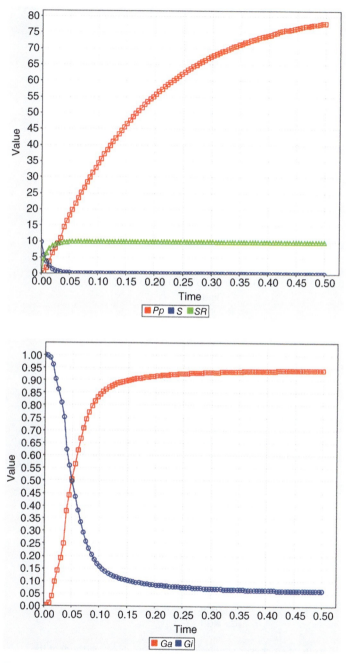

Figure 3.2 Results of numerical integration (adaptive Runge–Kutta method) of the quantitative variant of the toy pathway presented in Example 3.5.

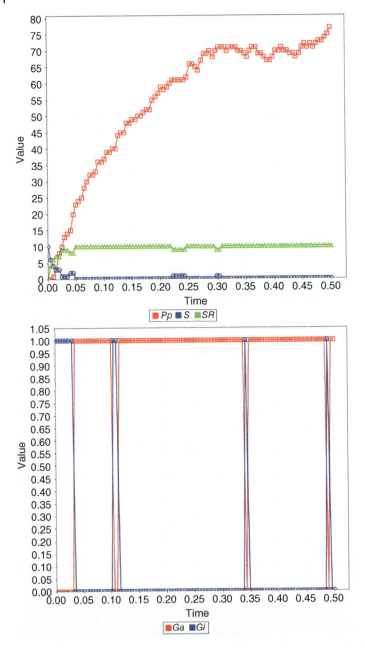

Figure 3.3 Results of stochastic simulation (Gillespie's direct method) of the quantitative variant of the toy pathway presented in Example 3.5.

3.5.2 Stochastic Process Algebras

Quantitative aspects of pathway modelling can be taken into account by applying a Stochastic Process Algebra, namely a Process Algebra in which actions are associated with stochastic rates describing how frequently they are performed by the process. This small syntactic extension has a significant impact on the semantics of the formalism. In particular, the definition of the semantics of a Stochastic Process Algebra typically uses rates of actions as stochastic rates of transitions, making the obtained LTS essentially the same as a CTMC.

As already done in the previous section, let us continue using the π-calculus as our reference Process Algebra. Indeed, its stochastic extension, stochastic π-calculus, proposed by Priami (1995), has been the first one used in the context of systems biology (Priami et al., 2001), and it has been successfully used to develop modelling and analysis tools (Phillips and Cardelli, 2004, 2007, Cardelli and Mardare, 2013) and to investigate biological case studies (Kuttler and Niehren, 2006, Segata et al., 2007, Palma-Orozco et al., 2009).

Without recalling the formal definition of the stochastic π-calculus, let us directly show how such a Process Algebra can be used to define a quantitative model of a pathway. We resort again on our toy pathway to give an example.

Example 3.6 We translate the quantitative model of the toy pathway presented in Example 3.5 into stochastic π-calculus. This simply corresponds to extend the π-calculus process we already define in Example 3.3 with the kinetic constants we introduced in Example 3.5. The kinetic constants are to be associated with each single action representing a reaction by paying attention, in the case of synchronizations, to associate the same constant with both actions involved in the synchronization.

$$S := (\overline{a}, 1).SR$$
$$R := (a, 1).0$$
$$SR := (\tau, 1).(S \mid R) + (\overline{b}, 0.5).SR$$
$$P := (b, 0.5).P'$$
$$P' := (\tau, 1).P + (\overline{d}, 1).0$$
$$G_i := (d, 1).G_a$$
$$G_a := (\tau, 10).(P' \mid G_i)$$

In order to obtain a description of the dynamics that takes into account reactants' multiplicities in the computation of reaction rates (as in Gillespie's approach), it is necessary to consider the semantics of the stochastic π-calculus described in Priami et al. (2001). In this definition, processes ready to perform an action on a given channel are considered as reactants of the corresponding reaction, and the rate to be associated with the transition in the LTS counts and multiplies them by the kinetic constant. This gives, as a result, a LTS equivalent to the CTMC obtained by following Gillespie's approach.

The stochastic π-calculus is not the only stochastic Process Algebraic formalism that has been applied to biological systems. We chose it as a representative Process Algebra, but there are several alternatives available in the literature. Among them, we mention, first of all, Performance Evaluation Process Algebra (PEPA) (Gilmore and Hillston, 1994, Calder et al., 2006) and its derivative Bio-PEPA (Ciocchetta and Hillston, 2008). Moreover, we can mention formalisms based on stochastic concurrent constraint programming (Bortolussi and Policriti, 2007, Chiarugi et al., 2010) and stochastic rule-based calculi (Barbuti et al., 2008a).

Although also in the stochastic setting the Process Algebraic approach opens to the possibility of applying formal analysis methods, the most common analysis approach that is applied to them when they are used for biological applications is still stochastic simulation. The real motivation for this is that in many cases, the ability of obtaining few traces of the quantitative dynamics of the modeled pathway is simply enough. As we already said, quantitative dynamics are influenced by many environmental or experimental factors, such as temperature and availability of substances. As a consequence, quantitative models of pathway naturally contain a number of approximations of the real system under study that make very accurate analysis techniques such as the formal ones an overkill. It is often not necessary to investigate in great detail the dynamics of a model that is inherently not accurate.

In the quantitative context, formal methods such as static analysis or bisimulations have been used not to prove the properties of models, but, for example, to find equivalent simplified models (Cardelli et al., 2015) or to deal with uncertainty in models parameters (Barbuti et al., 2012, Coletta et al., 2009). Maybe, the formal analysis techniques that more than the others has been applied to investigate the properties of biological case studies is stochastic model checking (Kwiatkowska et al., 2008, Chabrier and Fages, 2003), in particular with the PRISM tool (Kwiatkowska et al., 2011). Also in this case, however, in order to face the efficiency problems of model checking techniques in the case of very complex systems such as the biological ones, the choice of modelers is often to apply *statistical* or *Bayesian* model checking techniques (Legay et al., 2010, Clarke et al., 2008, Jha et al., 2009) that can provide answers to exactly the same questions that could be faced with the stochastic approach but not by considering all possible behaviors exhaustively. Statistical model checking actually performs its analyses on a number of traces obtained by stochastic simulation. Hence, such a technique can be seen as a systematic way of performing simulations and analyzing their results and, although nonexhaustive, in many cases, it turns out to be a very effective way of studying pathway dynamics.

In addition to stochastic aspects, several formalisms have been proposed to deal with other kinds of quantitative factors, such as delays (Caravagna and Hillston, 2010, 2012, Barbuti et al., 2011) and spatiality (Cardelli and Gardner, 2010, Bartocci et al., 2010).

3.6 Conclusions

In this chapter, we described how Process Algebras can be applied to model biological systems. In particular, we focused on the modelling of cell pathways by following a systems biology approach. Rather than presenting a survey of algebras, we preferred to focus on the concepts that are behind the application of these classes of formalisms to the biological domains. Surveys of approaches are available in the literature (Bernini et al., 2018, Fisher and Henzinger, 2007, Machado et al., 2011).

The use of Process Algebras in the biological application area is motivated by analogies that can be observed between cell pathways and concurrent processes, and it opens to the application of analysis techniques that are common in concurrency theory but unknown in the biological domain. We have shown that both qualitative and quantitative modelling of pathways are of interest and lead to the application of different analysis techniques.

Acknowledgments

This work has been supported by the Università di Pisa under the "PRA – Progetti di Ricerca di Ateneo" (Institutional Research Grants) – Project no. PRA_2020-2021_26 "Metodi Informatici Integrati per la Biomedica".

References

Sameera Abar, Georgios K. Theodoropoulos, Pierre Lemarinier, and Gregory M. P. O'Hare. Agent based modelling and simulation tools: a review of the state-of-art software. *Computer Science Review*, 24: 13–33, 2017.

Luca Aceto, Wan Fokkink, and Chris Verhoef. Structural operational semantics. In Corrado Priami, A. Ponse, and S.A. Smolka, editors, *Handbook of Process Algebra*, pages 197–292. Elsevier, 2001.

Bogdan Aman, Mariangiola Dezani-Ciancaglini, and Angelo Troina. Type disciplines for analysing biologically relevant properties. *Electronic Notes in Theoretical Computer Science*, 227: 97–111, 2009.

Marco Antoniotti, Carla Piazza, Alberto Policriti, Marta Simeoni, and Bud Mishra. Taming the complexity of biochemical models through bisimulation and collapsing: theory and practice. *Theoretical Computer Science*, 325 (1): 45–67, 2004.

J. C. M. Baeten. *A brief history of process algebra*. Computer science reports. Technische Universiteit Eindhoven, 2004.

Roberto Barbuti, Stefano Cataudella, Andrea Maggiolo-Schettini, Paolo Milazzo, and Angelo Troina. A probabilistic model for molecular systems. *Fundamenta Informaticae*, 67 (1–3): 13–27, 2005.

Roberto Barbuti, Andrea Maggiolo-Schettini, Paolo Milazzo, and Angelo Troina. A calculus of looping sequences for modelling microbiological systems. *Fundamenta Informaticae*, 72 (1–3): 21–35, 2006.

Roberto Barbuti, Andrea Maggiolo-Schettini, Paolo Milazzo, Paolo Tiberi, and Angelo Troina. Stochastic calculus of looping sequences for the modelling and simulation of cellular pathways. In Corrado Priami, editor, *Transactions on Computational Systems Biology IX*, pages 86–113. Springer, 2008a.

Roberto Barbuti, Andrea Maggiolo-Schettini, Paolo Milazzo, and Angelo Troina. Bisimulations in calculi modelling membranes. *Formal Aspects of Computing*, 20 (4): 351–377, 2008b.

Roberto Barbuti, Giulio Caravagna, Andrea Maggiolo-Schettini, and Paolo Milazzo. Delay stochastic simulation of biological systems: a purely delayed approach. In Corrado Priami, Ralph-Johan Back, Ion Petre, and Erik Vink, editors, *Transactions on Computational Systems Biology XIII*, pages 61–84. Springer, 2011.

Roberto Barbuti, Francesca Levi, Paolo Milazzo, and Guido Scatena. Probabilistic model checking of biological systems with uncertain kinetic rates. *Theoretical Computer Science*, 419: 2–16, 2012.

Ezio Bartocci, Flavio Corradini, Maria Rita Di Berardini, Emanuela Merelli, and Luca Tesei. Shape calculus. a spatial mobile calculus for 3D shapes. *Scientific Annals of Computer Science*, 20, 2010.

Andrea Bernini, Linda Brodo, Pierpaolo Degano, Moreno Falaschi, and Diana Hermith. Process calculi for biological processes. *Natural Computing*, 17 (2): 345–373, 2018.

Stefano Bistarelli, Iliano Cervesato, Gabriele Lenzini, Roberto Marangoni, and Fabio Martinelli. On representing biological systems through multiset rewriting. In *International Conference on Computer Aided Systems Theory*, pages 415–426. Springer, 2003.

Chiara Bodei, Roberta Gori, and Francesca Levi. Causal static analysis for brane calculi. *Theoretical Computer Science*, 587: 73–103, 2015.

Luca Bortolussi and Alberto Policriti. Stochastic concurrent constraint programming and differential equations. *Electronic Notes in Theoretical Computer Science*, 190 (3): 27–42, 2007.

Nadia Busi. Causality in membrane systems. In *International Workshop on Membrane Computing*, pages 160–171. Springer, 2007.

Muffy Calder, Stephen Gilmore, and Jane Hillston. Modelling the influence of RKIP on the ERK signalling pathway using the stochastic process algebra PEPA. In Corrado Priami, editor, *Transactions on computational systems biology VII*, pages 1–23. Springer, 2006.

Giulio Caravagna and Jane Hillston. Modeling biological systems with delays in bio-PEPA. *arXiv preprint arXiv:1011.0493*, 2010.

Giulio Caravagna and Jane Hillston. Bio-PEPAd: a non-Markovian extension of Bio-PEPA. *Theoretical Computer Science*, 419: 26–49, 2012.

Luca Cardelli. Brane calculi. In *International Conference on Computational Methods in Systems Biology*, pages 257–278. Springer, 2004.

Luca Cardelli and Philippa Gardner. Processes in space. In *Conference on Computability in Europe*, pages 78–87. Springer, 2010.

Luca Cardelli and Andrew D. Gordon. Mobile ambients. *Theoretical Computer Science*, 240 (1): 177–213, 2000. Special issue on Coordination.

Luca Cardelli and Radu Mardare. Stochastic pi-calculus revisited. In *International Colloquium on Theoretical Aspects of Computing*, pages 1–21. Springer, 2013.

Luca Cardelli, Mirco Tribastone, Max Tschaikowski, and Andrea Vandin. Forward and backward bisimulations for chemical reaction networks. *arXiv preprint arXiv:1507.00163*, 2015.

Nathalie Chabrier and François Fages. Symbolic model checking of biochemical networks. In *International Conference on Computational Methods in Systems Biology*, pages 149–162. Springer, 2003.

Davide Chiarugi, Moreno Falaschi, Carlos Olarte, and Catuscia Palamidessi. Compositional modelling of signalling pathways in timed concurrent constraint programming. In *Proceedings of the 1st ACM International*

Conference on Bioinformatics and Computational Biology, pages 414–417, 2010.

Federica Ciocchetta and Jane Hillston. Bio-PEPA: An extension of the process algebra PEPA for biochemical networks. *Electronic Notes in Theoretical Computer Science*, 194 (3): 103–117, 2008.

Edmund M. Clarke. Model checking. In *International Conference on Foundations of Software Technology and Theoretical Computer Science*, pages 54–56. Springer, 1997.

Edmund M. Clarke, James R. Faeder, Christopher J. Langmead, Leonard A. Harris, Sumit Kumar Jha, and Axel Legay. Statistical model checking in BioLab: applications to the automated analysis of T-cell receptor signaling pathway. In *International Conference on Computational Methods in Systems Biology*, pages 231–250. Springer, 2008.

Edmund M. Clarke Jr., Orna Grumberg, Daniel Kroening, Doron Peled, and Helmut Veith. *Model Checking*. MIT Press, 2018.

Alessio Coletta, Roberta Gori, and Francesca Levi. Approximating probabilistic behaviors of biological systems using abstract interpretation. *Electronic Notes in Theoretical Computer Science*, 229 (1): 165–182, 2009.

Patrick Cousot. Abstract interpretation. *ACM Computing Surveys (CSUR)*, 28 (2): 324–328, 1996.

Michele Curti, Pierpaolo Degano, and Cosima Tatiana Baldari. Causal π-calculus for biochemical modelling. In *International Conference on Computational Methods in Systems Biology*, pages 21–34. Springer, 2003.

Vincent Danos and Cosimo Laneve. Formal molecular biology. *Theoretical Computer Science*, 325 (1): 69–110, 2004.

Vincent Danos, Jérôme Feret, Walter Fontana, and Jean Krivine. Abstract interpretation of cellular signalling networks. In Francesco Logozzo, Doron A. Peled, and Lenore D. Zuck, editors, *Verification, Model Checking, and Abstract Interpretation, 9th International Conference, VMCAI 2008, San Francisco, USA, January 7–9, 2008, Proceedings*, volume 4905 of *Lecture Notes in Computer Science*, pages 83–97. Springer, 2008.

Lorenzo Dematté, Corrado Priami, and Alessandro Romanel. The BlenX language: a tutorial. In *International School on Formal Methods for the Design of Computer, Communication and Software Systems*, pages 313–365. Springer, 2008.

Steven Eker, Merrill Knapp, Keith Laderoute, Patrick Lincoln, and Carolyn Talcott. Pathway logic: executable models of biological networks. *Electronic Notes in Theoretical Computer Science*, 71: 144–161, 2004.

Javier Esparza and Mogens Nielsen. Decidability issues for Petri nets. *Petri Nets Newsletter*, 94: 5–23, 1994.

Jasmin Fisher and Thomas A. Henzinger. Executable cell biology. *Nature Biotechnology*, 25 (11): 1239–1249, 2007.

D. Gillespie. Exact stochastic simulation of coupled chemical reactions. *Journal of Physical Chemistry*, 81: 2340–2361, 1977.

Stephen Gilmore and Jane Hillston. The PEPA workbench: a tool to support a process algebra-based approach to performance modelling. In *International Conference on Modelling Techniques and Tools for Computer Performance Evaluation*, pages 353–368. Springer, 1994.

Roberta Gori and Francesca Levi. Abstract interpretation based verification of temporal properties for bioambients. *Information and Computation*, 208 (8): 869–921, 2010.

Charles Antony Richard Hoare. Communicating sequential processes. *Communications of the ACM*, 21 (8): 666–677, 1978.

Sumit K. Jha, Edmund M. Clarke, Christopher J. Langmead, Axel Legay, André Platzer, and Paolo Zuliani. A bayesian approach to model checking biological systems. In *International conference on computational methods in systems biology*, pages 218–234. Springer, 2009.

Hiroaki Kitano. Systems biology: a brief overview. *Science*, 295 (5560): 1662–1664, 2002.

Céline Kuttler and Joachim Niehren. Gene regulation in the pi calculus: simulating cooperativity at the lambda switch. In Corrado Priami, editor, *Transactions on Computational Systems Biology VII*, pages 24–55. Springer, 2006.

Marta Kwiatkowska, Gethin Norman, and David Parker. Using probabilistic model checking in systems biology. *ACM SIGMETRICS Performance Evaluation Review*, 35 (4): 14–21, 2008.

Marta Kwiatkowska, Gethin Norman, and David Parker. Prism 4.0: verification of probabilistic real-time systems. In *International Conference on Computer Aided Verification*, pages 585–591. Springer, 2011.

Axel Legay, Benoît Delahaye, and Saddek Bensalem. Statistical model checking: an overview. In *International Conference on Runtime Verification*, pages 122–135. Springer, 2010.

Daniel Machado, Rafael S. Costa, Miguel Rocha, Eugénio C. Ferreira, Bruce Tidor, and Isabel Rocha. Modeling formalisms in systems biology. *AMB Express*, 1 (1): 1–14, 2011.

Robin Milner. *Communicating and Mobile Systems: The Pi Calculus*. Cambridge University Press, 1999.

Robin Milner. *A Calculus of Communicating Systems*, Springer-Verlag, 1980.

Flemming Nielson, Hanne Riis Nielson, Corrado Priami, and Debora Schuch da Rosa. Static analysis for systems biology, 2003.

Rosaura Palma-Orozco, Pablo Padilla-Longoria, and Pablo G. Padilla-Beltran. A stochastic pi-calculus model for the intrinsic apoptopic pathway. *Biophysical Journal*, 96 (3): 426a, 2009.

Michael Pedersen, Andrew Phillips, and Gordon D. Plotkin. A high-level language for rule-based modelling. *PLoS One*, 10 (6): e0114296, 2015.

Mario Jesús Pérez-Jiménez and Francisco José Romero-Campero. P systems, a new computational modelling tool for systems biology. In Corrado Priami and Gordon Plotkin, editors, *Transactions on Computational Systems Biology VI*, pages 176–197. Springer, 2006.

Andrew Phillips and Luca Cardelli. A correct abstract machine for the stochastic pi-calculus. *Electronic Notes in Theoretical Computer Science*, 10, 2004.

Andrew Phillips and Luca Cardelli. Efficient, correct simulation of biological processes in the stochastic pi-calculus. In *International Conference on Computational Methods in Systems Biology*, pages 184–199. Springer, 2007.

Gordon Plotkin. Structural operational semantics. *Aarhus University, Denmark*, pages 20–23, 1981.

Corrado Priami. Stochastic π–calculus. *The Computer Journal*, 38 (7): 578–589, 1995.

Corrado Priami and Paola Quaglia. Beta binders for biological interactions. In *International Conference on Computational Methods in Systems Biology*, pages 20–33. Springer, 2004.

C. Priami, A. Regev, E. Shapiro, and W. Silverman. Application of a stochastic name-passing calculus to representation and simulation of molecular processes. *Information Processing Letters*, 80 (1): 25–31, 2001.

Paola Quaglia and BRICS Lecture Series. The pi-calculus: notes on labelled semantic. *Bulletin of the EATCS*, 68: 104–114, 1999.

A. Regev and E. Shapiro. Cells as computation. *Nature*, 419: 343, 2002.

A. Regev, W. Silverman, and E. Y. Shapiro. Representation and simulation of biochemical processes using the pi-calculus process algebra. In *Pacific Symposium on Biocomputing*, pages 459–470, 2001.

A. Regev, E. M. Panina, W. Silverman, L. Cardelli, and E. Shapiro. Bioambients: an abstraction for biological compartments. *Theoretical Computer Science*, 325 (1): 141–167, 2004.

Wolfgang Reisig. *Petri Nets: An Introduction*, volume 4. Springer Science & Business Media, 2012.

Davide Sangiorgi. On the bisimulation proof method. *Mathematical Structures in Computer Science*, 8 (5): 447–479, 1998.

Davide Sangiorgi and David Walker. *The Pi-calculus: A Theory of Mobile Processes*. Cambridge University Press, 2003.

Birgit Schoeberl, Claudia Eichler-Jonsson, Ernst Dieter Gilles, and Gertraud Müller. Computational modeling of the dynamics of the map kinase cascade activated by surface and internalized egf receptors. *Nature Biotechnology*, 20 (4): 370–375, 2002.

Nicola Segata, Enrico Blanzieri, and Corrado Priami. Stochastic pi-calculus modelling of multisite phosphorylation based signaling: in silico analysis of the Pho4 transcription factor and the PHO pathway in saccharomyces cerevisiae, 2007.

Antti Valmari. Compositionality in state space verification methods. In *International Conference on Application and Theory of Petri Nets*, pages 29–56. Springer, 1996.

H. Steven Wiley, Stanislav Y. Shvartsman, and Douglas A. Lauffenburger. Computational modeling of the EGF-receptor system: a paradigm for systems biology. *Trends in Cell Biology*, 13 (1): 43–50, 2003.

Glynn Winskel. *The Formal Semantics of Programming Languages: An Introduction*. MIT Press, 1993.

4

The Rule-Based Model Approach: *A Kappa Model for Hepatic Stellate Cells Activation by TGFB1*

Matthieu Bouguéon[1,2], Pierre Boutillier[3], Jérôme Feret[4,5], Octave Hazard[4,5,6], and Nathalie Théret[1,2]

[1] Inria, CNRS, IRISA, UMR 6074, University of Rennes, Rennes, France
[2] Inserm, EHESP, Irset, UMR S1085, University of Rennes, Rennes, France
[3] Nomadic Labs, Paris, France
[4] Team Antique, Inria Paris, Paris, France
[5] École normale supérieure, DI-ENS (ÉNS, CNRS, PSL University), Paris, France
[6] École Polytechnique, Palaiseau, France

4.1 Introduction

4.1.1 Modelling Systems of Biochemical Interactions

The description and the analysis of the large-scale and highly combinatorial systems that emerge from some mechanistic models of systems biology are still out of scope of the state of the art. In such models, the individual behavior of proteins or other components, which may establish links and modify their capability of interaction, is driven by races against shared resources. Moreover, occurrences of proteins may form a large amount of distinct complexes. Concurrency between interactions at different time scale induces non-linear feedback loops that control the abundance of these complexes. Lastly, these systems involve interactions between very small molecules, as ions and ligands, and giant complexes as DNA strands, the ribosome, or the signalosome. Understanding how the collective behavior of population of proteins and other components emerges from interactions between individual proteins remains a crucial and mainly open challenge.

While technological progresses provide quickly an ever-increasing amount of details about the potential mechanistic interactions between the components of these systems, and at an affordable cost, the scientific community is far from a global understanding of how the macroscopic behavior of these systems emerges from these interactions. This is the holy

Systems Biology Modelling and Analysis: Formal Bioinformatics Methods and Tools,
First Edition. Edited by Elisabetta De Maria.
© 2023 John Wiley & Sons, Inc. Published 2023 by John Wiley & Sons, Inc.

grail of systems biology. Yet, this challenge is hopeless without the help of specific and innovative methods to describe these complex systems and analyze their properties. These methods must scale to the large amount of information that is published in the literature at an exponentially increasing rate.

4.1.2 Modelling Languages

Formal languages have been widely used to describe models of mechanistic interactions between occurrences of proteins. They provide mathematical tools to encode interactions and to define rigorously the behavior of the systems they represent by means of a choice of semantics, would they be qualitative, stochastic, or differential.

Languages as reaction networks and classical Petri nets Heiner and Koch (2004) are based on multi-set rewriting. Applying an interaction consists in consuming some reactants while producing some products. Kinetic constants specify, according to the choice of semantics, either the speed or the average frequencies of application of each kind of reactions. These languages are very convenient to model the behavior of small- or medium-sized interaction systems. Yet, they struggle to scale to large models because one name (or one placeholder in the case of Petri nets) is required for each distinct kind of complexes.

It is worth to make the distinction among agent-based and rule-based approaches. In agent-based approaches, each entity, would it be a process Ciocchetta and Hillston (2009) or an object Dematté et al. (2008), has to contain the specification of all its potential behaviors. The evolution of the configurations of the different entities is synchronized by means of communication rules that define the operational semantics of the model. There are usually very few rules. It is possible to restrict the behavior of an agent with respect to some conditions over the properties of some other agents to which this agent would be linked. Yet, some fictitious processes would then be required to fetch the necessary information. Such trick has been already used in the first models written in the π-calculus Regev et al. (2001). Nonetheless, in general, agent-based approaches lead to a network of finite state processes Kahramanogullari and Cardelli (2013). Thanks to this, the behavior of these models can be studied by means of symbolic model checking tools as PRISM Kwiatkowska et al. (2011).

Agent-based approaches fail to scale whenever occurrences of components admit too many distinct configurations or whenever their capabilities of interaction depend too much on the configurations of the components they are linked to. Such models cannot be described, and *a fortiori* their behavior cannot be computed with such approaches.

Rule-based approaches consist in defining models by means of interaction rules. Each rule specifies under which conditions over the configurations of the different occurrences of agents an interaction may happen and what is the impact of applying this interaction. This way, the state of an agent does not define once for all the capabilities of interaction of this agent. The capabilities of interaction are within the rules. This way, it is no longer necessary to itemize exhaustively the set of all the configurations agents may take. Rules only describe the parts that matter in the interactions that they describe. As a matter of fact, rule-based approaches scale better and ease the versioning of models. Moreover, as it is not necessary to describe every capability of interaction of the occurrences of the components, they ease unbiased modelling when the conception of the model is not influenced by a specific goal.

Ambient-calculus Cardelli and Gordon (1998, 2000), bioambient-calculus Regev et al. (2004), and brane-calculi Cardelli (2004) are particular cases of languages. They describe the behavior of hierarchies of compartments, which may be arbitrarily nested. Some agents in the compartments, or in the case of the brane-calculi, in the membranes of compartments, provide to their compartments some capabilities to move within the hierarchy of compartments and to fuse pairwise. Capabilities of interaction may depend on the relative localization of compartments within the hierarchy of compartments. Projective brane calculus Danos and Pradalier (2004) describes even more faithfully the organization of compartments within a cell, by making the description of the state of the system independent from the choice of the root of the hierarchy of compartments.

4.1.3 Kappa

Languages for site graph rewriting Danos and Laneve (2004), Faeder et al. (2005), Andrei and Kirchner (2008), John et al. (2011) aim at describing in a transparent way networks of interaction between occurrences of components, by means of a syntax that is inspired from chemistry.

4.1.3.1 Overview

In Kappa, each complex is described by a site graph. An example of site graph is given in Figure 4.1a. A site graph is made of some nodes that denote occurrences of some components. Each component is associated with a list of interaction sites. Sites may be free, or bound pairwise. Besides, some interaction sites may be tagged with a property, which may stand for an activation level. Interactions between occurrences of components may change the conformation of components. For instance, in the case of proteins, they may be

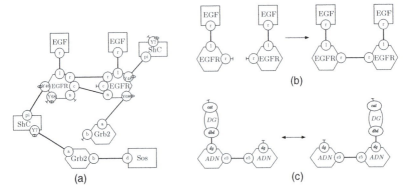

Figure 4.1 (a) A site graph. This is a biochemical complex made of two occurrences of the ligand protein *EGF*, two occurrences of the receptor *EGFR*, one occurrence of a scaffold protein (*Shc*), two occurrences of a transport protein (*Grb2*), and one occurrence of the protein *Sos*. (b) An example of binding rule. Two occurrences of the receptor *EGFR*, when both activated by a binding with some occurrences of the ligand *EGF*, may bind to each other to form a dimer. The other interaction sites are omitted because they play no rule in this interaction. (c) An example of movement rule. An occurrence of the enzyme glycolase (*DG*) may glide along both directions (according to a random walk process) along a DNA strand. (a) A site graph. (b) A rule to bind some occurrences of proteins. (c) A rule to make an occurrence of protein glide.

folded and/or unfolded, which may hide or reveal some interaction sites. In Kappa, there is no explicit notion of a three-dimensional structure. In contrast, the conditions for a site to be visible are specified in the description of the interactions themselves.

The behavior of a system that is written in Kappa is described by means of context-free rewriting rules. Figure 4.1(b) shows a rule of dimer formation. Two occurrences of the receptor *EGFR* that are activated by some occurrences of the ligand *EGF* may bind to each other and form a dimer. Figure 4.1(c) gives a second rule that is taken from a model of DNA repair Köhler et al. (2014). In this rule, an occurrence of an enzyme, the glycolase (*DG*), may glide randomly in both directions along a DNA strand.

4.1.3.2 Semantics of Kappa

A rule may be seen from an intensional point of view, as a local transformation of the state of the system, or extensionally as a potentially infinite set of reactions that may be obtained by fully specifying the context of application of these rules. From this set of reactions, several semantics may be induced. These semantics may be qualitative, stochastic, or differential, as for reaction networks and Petri nets (quantitive semantics – that is to say stochastic or differential – require the use of rate constant). Yet, the stochastic semantics of a model that is described in Kappa may be executed directly, without ever generating the underlying network of reactions. This execution is based on the iteration of the following event loop (which corresponds to the application of Gillespie's algorithm Gillespie (1977)). Given the current state of the system, denoted as a site graph, the set of all the potential events that may happen next is computed. An event consists in applying a rule in the site graph at an occurrence of the left-hand side of a rule. Each event has a propensity that is defined as the rate of the corresponding interaction rule. Then, the next event is drawn randomly with a probability that is defined proportionally to its propensity, while the time between two consecutive events is drawn according to the exponential law with the parameter equal to the sum of the propensity of all the available events. For the sake of scalability, it would not be reasonable to recompute the list of all the potential events after each rule application. This set can indeed be dynamically updated by accounting only the new potential events and the events that are no longer possible because of the application of the last event Danos et al. (2007). The actual simulator engine has been optimized thanks to the potential sharing between the patterns that occur in the left-hand side of rules Boutillier et al. (2017).

4.1.3.3 Kappa Ecosystem

4.1.3.3.1 *Modelling Platform*

Kappa models can be designed in a dedicated modelling platform Boutillier et al. (2018b). Rules can be specified in a text window while widgets provide access to most of the existing Kappa tools, including simulation, static analysis, and causal analysis.

The platform can be used online (for reasonable size models) or installed locally. Support for linking with python frontend/backend and exporting results in Jupiter notebooks is also provided.

4.1.3.3.2 *Stochastic Simulation*

The stochastic simulator KaSim samples the trajectories of Kappa models faithfully according to their probability density distribution. As explained in Section 4.1.3.2, it relies on a representation of the state of the system as a site graph. The set of events that may be applied in the current state is computed dynamically Danos et al. (2007). The use of a dedicated category-based data structure to describe and update this set while optimizing the benefit due to potential common regions among patterns Boutillier et al. (2017) has speed up the simulator.

Additionally, the Kappa platform provides support for end-user interactions during the execution of a model Boutillier (2019). In particular, the end-user can pause the simulation while observing the behavior of the model, specify modifications of the state of the model, and restart the simulation to observe the impact of his intervention.

4.1.3.3.3 Static Analysis

Static analysis enhances the confidence in models. The static analyzer KaSa Danos et al. (2008), Feret and Lý (2018), Boutillier et al. (2018a) computes and proves some structural properties about the complexes that may be formed in a given model, given an initial configuration. This tool uses the abstract interpretation framework Cousot and Cousot (1977) to approximate the computation of a least fix point over sets of patterns of interest. This way, it detects and proves which of these patterns may be reachable. As a matter of fact, any pattern of interest that is not in the result of the analysis will never occur in a state that may be reached from the initial state. Because of the approximation, we cannot conclude about the patterns that are discovered by the analysis: they may – or not – occur in a given reachable state. This analysis is particularly useful to detect some relationships among the states of the sites of some agents. It can also be proved that some rules will never be applied in a given model. More information about how static analysis can help the modelling process is provided in Boutillier et al. (2018a).

4.1.3.3.4 Causal Analysis

Causal analysis aims at extracting different scenarii of interest from a given simulation. While a simulation describes the behavior of a population of agents that may evolve back and forth, causal analysis aims at describing the evolution of some individuals while focusing on the computation steps that make progresses. Given an event of interest, causal analysis provides a set of minimal scenarii that are extracted from the simulation traces and that describe the events that were necessary to trigger an instance of the event of interest.

Causal analysis relies on two main ingredients. Firstly, scenarii are described as event structures. This way, concurrency between causally independent events is exploited and their interleaving orders is abstracted away. Secondly, operational research techniques are used to extract minimal sub-structures leading to the same ending event. This way, non-necessary events are discarded. More information about the formal background of causal analysis may be found in Danos et al. (2012).

4.1.3.3.5 Underlying Network and Model Reduction

As explained in Section 4.1.3.2, a Kappa model induces a (potentially infinite) network of reactions. The tool KaDe Camporesi et al. (2017) generates this network. Several export formats are available: the network may be exported in the DOTNET language Faeder et al. (2009), Blinov et al. (2004) or in SBML Hucka et al. (2010), or as a set of differential equations written in MAPLE Monagan et al. (2005), MATHEMATICA Wolfram Research (2017), MATLAB MATLAB (2017), or OCTAVE Eaton et al. (2015).

Model reduction may be used to simplify the underlying reaction network, the underlying system of differential equations, or even the underlying Markov chain. Exact model reduction techniques consist in discovering a change of variables. They find out sets of quantities that can be exactly described while discarding the others. Such changes of variables may be detected directly at the level of the site graph structure, hence without even generating the underlying network of reactions. These reductions are based on the detection of symmetries Feret (2015) and the static inspection of the flow of information among different regions of complexes Feret et al. (2009), Danos et al. (2010), Feret et al. (2013), Camporesi et al. (2013).

Conservative methods, based on tropicalization techniques, have also been proposed Beica et al. (2020). By exploiting separation between time and concentration scales, they permit to eliminate some variables, at the cost of numerical approximations. Here, the exact behavior of the variables of interest is lost, but it is safely approximated by means of intervals.

4.1.3.4 Main Limitations

Kappa suffers from several limitations. For instance, the name of the interaction sites of an agent shall be pairwise distinct; also, in regard to geometry, Kappa does not offer any support either for describing the tridimensional structure of the complexes or for describing their spatial distribution. Disallowing multiple occurrences of sites in a given agent greatly eases the detection of occurrences of patterns in graphs. Not only this is the cornerstone for an efficient stochastic simulation but also it is the root of some algebraic constructions widely used in static analysis and model reduction. Some languages get rid of this limitation either directly as in BNGL Faeder et al. (2005) and mød Andersen et al. (2016) or indirectly by encoding them by means of hyperlinks as in React(C) John et al. (2011). Nevertheless, it deeply impacts the efficiency of simulation engines. As far as geometry is concerned, some assumptions about the spatial conformation of agents may be implicitly encoded within rewriting rules. Some extensions of the language provide a syntax to describe the relative position of agents within complexes so as to restrict the potential events to those that satisfy some specific constraints Danos et al. (2015).

As far as the spatial distribution of complexes is concerned, the assumption is made that they are perfectly mixed. As a matter of fact, in the case of intracellular models, crowding effects that may result from the accumulation of proteins in some specific regions of a cell cannot be modeled. The same way, the gradients of proteins that may result from the action of a scaffold protein cannot be described (an occurrence of a scaffold protein holds some occurrences of proteins to maintain them in the same biochemical complex; in Kappa, no assumption is made about the position of these occurrences of proteins when they are released, even for a short amount of time). A partial alternative consists in encoding a grid of potential discrete positions. Then, some rewriting rules may be used to model the diffusion of proteins, which consists in making proteins glide from one position to an adjacent one. SpatialKappa Stewart (2010) offers a transparent syntax for discrete diffusion of agents based on this construction. Besides, the ML language Helms et al. (2017) provides support for describing models of interactions between proteins with continuous motion. In Kappa, it is also possible to define a static finite hierarchy of compartments. Yet, the transport of occurrences of proteins by means of vesicules cannot be modeled this way. The formal cell machinery Damgaard et al. (2012) addresses this issue but does not provide efficient simulation engines.

4.1.4 Modelling a Population of Hepatic Stellate Cells

In this chapter, we model in Kappa the behavior of a population of hepatic stellate cells (HSCs). This is an interesting case study as it illustrates the flexibility of the language. It is worth noting that the system that is modeled is not, as it is usually the case, a population of proteins but a population of cells interacting with some signaling proteins. Also, the abstraction level of the model is tailored to cope with the amount of information that is available about the interaction and the combinatorial complexity of the execution of the underlying mechanisms. We provide in this section some biological context about the model.

Chronic liver diseases (CLDs) are long-duration and slow progression pathologies that represent a major public health issue in terms of economic cost Byass (2014). CLDs are mainly associated with viral infections, alcoholic diseases, and more recently with the non-alcoholic fatty liver diseases (NAFLD) because of the increasing frequency of metabolic syndromes (insulin resistance, type 2 diabetes, and obesity). Chronic hepatitis is associated with the development of fibrosis, which results in the abnormal deposition of extracellular matrix rich in interstitial collagen and a severe dysfunction of liver functions. The terminal stage of fibrosis is cirrhosis, which constitutes the major risk of occurrence of hepatocellular carcinoma (HCC). The mortality linked to the complications of cirrhosis (hemorrhage, liver failure, and cancer) leads to the death of a little more than one million people per year in the world.

The matrix microenvironment is therefore the major regulator of events related to the fibrosis-cirrhosis-cancer progression, and HSC are the main actors for modifying the extracellular microenvironment (Figure 4.2). In response to hepatic insults, HSC undergo a process of activation from quiescent vitamin A-rich cells in normal liver to proliferating, fibrogenic, and contractile myofibroblasts Tsuchida and Friedman (2017). Among the molecules that drive HSC activation, the transforming growth factor TGFB1 plays the major role. In addition to the deposition of fibrillar matrix

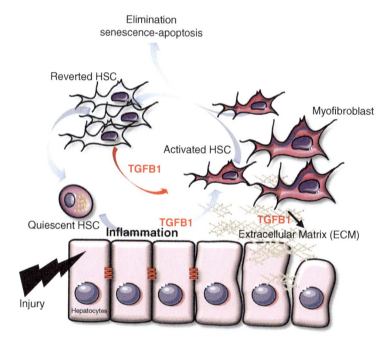

Figure 4.2 Dynamics of hepatic stellate cells. Upon injury, damaged hepatocytes produce signal to induce inflammation that in turn promotes TGFB1-dependent activation of HSC. Activated HSC orchestrate tissue repair and are either eliminated through senescence and apoptosis or deactivated toward a transient reverted state that can be reactivated more rapidly. Upon repeated injuries, activated HSC progress toward a myofibroblast state that escapes to control, leading to fibrosis.

components, activated HSC produce a wide variety of molecules involved in extracellular matrix remodeling, which in turn modulates the availability and signaling of TGFB1. Upon injury, HSC are activated to repair tissues and next are eliminated according to three mechanisms apoptosis, senescence, and reversion leading to return to the healthy situation Kisseleva and Brenner (2021). However, when injury persists, HSC remain activated with a myofibroblast phenotype, and extracellular matrix accumulates, leading to fibrosis, cirrhosis, and cancer. The understanding of the dynamics of HSC activation and regulation by TGFB1 is essential to identify markers and therapeutic targets likely to promote the resolution of fibrosis at the expense of its progression.

In this chapter, we developed a Kappa model to characterize the dynamics of HSC activation and the different states upon TGFB1 stimulation.

4.1.5 Outline

The rest of the chapter is organized in the following way. In Section 4.2, the main features of Kappa are informally explained in graphical representation (figures have been generated with the GKappa library Feret). In Section 4.3, the rules that model the behavior of the HSCs are given and explained. The kinetic parameters are also documented. Some references to the literature are provided to justify both rules and parameters. In Section 4.4, the model is checked and simulated. Static analysis is used so as to increase the confidence on the model. Then, the model is simulated under two scenarii. In the first one, the population of HSCs responds to an acute inflammatory aggression. In the second one, the case of chronic inflammation is considered. In Section 4.5, the chapter is concluded. The current state of the model is discussed as well as future extensions of it.

4.2 Kappa

We give the syntax and the semantics of Kappa.

4.2.1 Site Graphs

We introduce in this section the notion of site graphs. Site graphs will be used to describe not only the different states of the systems that we are modeling but also the different patterns that will be used in Section 4.2.2 to describe, by means of rewriting rules, the behavior of these systems.

4.2.1.1 Signature

The definition of a model starts with its signature. This signature specifies the alphabet, that is to say all the ingredients that may be involved in this model. It may be described graphically by means of a *contact map*, as the one that is given in Figure 4.3. A contact map is made of a list of nodes that specifies the different *kinds of agents* in the model. Each node has a name and is drawn with a specific shape. The notion of agent in Kappa is quite abstract. Agents can be used to encode not only instances of proteins but also individual cells, depending on the granularity of the model. Moreover, in order to tune the scaling of a model, an agent may also stand for a fixed amount of occurrences of a given kind of protein, or a given kind of cell, all in a same configuration. Each agent is also fitted with a set of *interaction sites*, which are depicted around it by means of named circles. In Kappa, a given kind of agent cannot bear two interaction sites with the same name. Lastly, each interaction site is fitted with a set of tags that may be used to denote its *activation level*, as the stage of differentiation of some cells, for instance. Activation states may also be used to describe the localization of

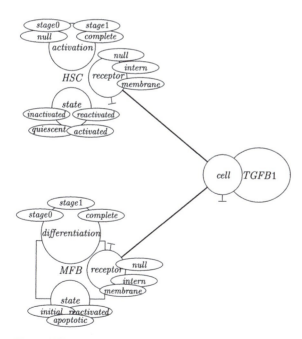

Figure 4.3 A contact map. This map specifies the signature of a model by itemizing the different kinds of agents, their interaction sites, the different activation states that each kind of sites may take, and the potential bindings between these interaction sites.

the occurrences of some agents or of some sites within a finite hierarchy of compartments. Interaction sites may also carry a *binding state*: sites with the symbol ⊣ may potentially remain free; the pairs of sites that may be bound pairwise are described in the contact map by means of undirected arcs. In the contact map, a site may be bound to several others (we will explain later that in such a case, there is a competition for the binding to this interaction site). Moreover, a site may be bound to itself in the contact map (in such a case, the corresponding interaction sites of two occurrences of the agent may be bound together).

Example 4.1 An example of contact map is given in Figure 4.3. It describes the components of our model. This contact map introduces three kinds of agents: HSCs *HSC*, myofibroblasts *MFB*, and occurrences of the transforming growth factor protein *TGFB1*. The model documents only one interaction site for the protein *TGFB1*. This site, which is called *cell*, enables the occurrences of the *TGFB1* protein to bind either a HSC or a myofibroblast. The protein *TGFB1* has also many other interaction sites and may exist in various forms (active, latent, degraded, …). Yet, these considerations do not matter in the scope of this model. They are useful only when considering the extracellular matrix molecular network. Thus, we omit them for the sake of simplicity. In the model, we are interested in the state of three interaction sites of the HSCs. Two of them describe the different forms of the cells and its different activation stages. We distinguish quiescent cells, activated cells, inactivated cells, and reactivated cells. Then, within each form, there may be different activation stages. When the notion of activation stage does not make sense, the stage *null* is used. For instance, quiescent cells have no intermediary stages; thus, their site *activation* is always in the state *null*. Otherwise, the cell may be in an intermediary stage *stage0* or *stage1*, or fully activated, which is written as *complete*. The third site, called *receptor*, is an abstraction of all the TGFB1 receptors in a HSC. This site *receptor* carries out a binding state. This site can be free or bound to an occurrence of the *TGFB1* agent. A given cell may fix many occurrences of the *TGFB1* protein. The sites of the occurrences of the *TGFB1* protein may also be free. For scalability issues, we abstract this by a single interaction site per cell. Hence, an agent *TGFB1* does not stand for a single occurrence of the *TGFB1* protein but for the average amount of occurrences of this protein that are bound to a HSC. The site *receptor* also carries out a localization, which ranges among *null*, *intern*, and *membrane*. The state *null* means that all the receptors of the cell are currently degraded. The state *intern* means that they all have been internalized. Lastly, the state *membrane* means that they are all available on the membrane of the cell.

As far as myofibroblasts are concerned, the model focuses on three interaction sites as well. Myofibroblasts carry out a state. We distinguish initial and apoptotic/senescent ones and also the ones coming from the differentiation of a reactivated hepatic cell. Besides, there exist several differentiation stages between different states, which is encoded within a site called *differentiation*. Myofibroblasts may be in an intermediary stage, which is written as *stage0* or *stage1*, or fully differentiated, which is written as *complete*. The third site, which is called *receptor*, copes for the receptors of an occurrence of myofibroblast. This site works exactly as the site *receptor* of the HSCs does. Lastly, the agent *TGFB1* has a unique interaction site. This site carries out a binding state. Indeed, a pack of occurrences of the TGFB1 protein may be either free or bound to the receptors of a hepatic stellate cell, or bound to the receptors of a myofibroblast.

4.2.1.2 Complexes

Kappa models describe the behavior of a soup of complexes. A complex is made of several occurrences of agents. Each occurrence of an agent is equipped with a set of interaction sites. Some sites carry out an activation state, but only one. Lastly, each occurrence of a site may be either free, or bound to exactly one other occurrence of a site. As opposed to the contact map, an occurrence of a site cannot be bound to itself in a complex. Additionally, an occurrence of a site cannot be bound to two distinct occurrences of sites. A complex forms a connected pattern, this means that it is always possible to go from a given occurrence of an agent to another one by following a potentially empty sequence of bonds.

Example 4.2 An example of complex is given in Figure 4.4. This complex is made of two occurrences of agents. The first one denotes an occurrence of HSC. The second one denotes a pack of TGFB1 proteins. The TGFB1 proteins are bound to the receptors of the HSC that is in its quiescent form (hence not activated yet). Lastly, the receptors of the HSC are located on the membrane of this cell.

Figure 4.4 A complex. It contains several occurrences of agents. Each occurrence documents the set of its interaction sites. The sites that may carry out an activation state, have one. Moreover, each site that may carry a binding state is either free, or bound to another site.

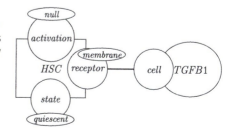

The signature constrains the set of complexes of a model. Not all the complexes that satisfy the syntax of Kappa are consistent with a signature. The contact map not only provides the list of all the potential kinds of agents with their interaction sites but also summarizes the potential states of the sites of each occurrence of agents. More precisely, each occurrence of an agent in a complex shall specify the same interaction sites as the corresponding agent in the contact map. Moreover, every site in a complex such that the corresponding site in the contact map admits at least one activation state shall bear an activation state as well. This is the same for the binding states. These constraints ensure that the state of each occurrence of agents in a complex is fully specified. Three additional constraints ensure that the state of sites matches with the contact map: firstly, an occurrence of site may carry an activation state only when the corresponding site in the contact map carries this activation state as well; secondly, an occurrence of site may be free only when the corresponding site in the contact map may be free as well; thirdly, two occurrences of sites may be bound together only when the two corresponding sites may be bound in the contact map. These constraints can be formalized by requiring that every complex can be projected onto the contact map, that is to say that the function that maps every occurrence of agents of a complex to the agent with the same name on the contact map is always a *homomorphism*. Said differently, the contact map may be understood as the folding of every complex of a model and every node of the contact map summarizes all the potential configurations of the occurrences of the corresponding agent.

Example 4.3 Figure 4.5 depicts the projection of the complex of Figure 4.4 onto the contact map of Figure 4.3.

4.2.1.3 Patterns

The behavior of complexes is defined by means of rewriting rules. These rules specify not only the necessary conditions to trigger an interaction but also the potential effects of these interactions. Before explaining more precisely what a rewriting rule is, it is necessary to introduce the notion of pattern. Indeed, patterns are used to specify under which conditions a rule may be applied.

We focus the presentation on the description of connected patterns. More sophisticated patterns may be obtained by putting several connected patterns side by side. A *connected pattern* is a contiguous part of a complex. This way, it may be made of zero, one, or several occurrences of each kind of agents. Each occurrence of an agent may be associated with a set of interaction sites. Each occurrence of an interaction site may potentially bear an

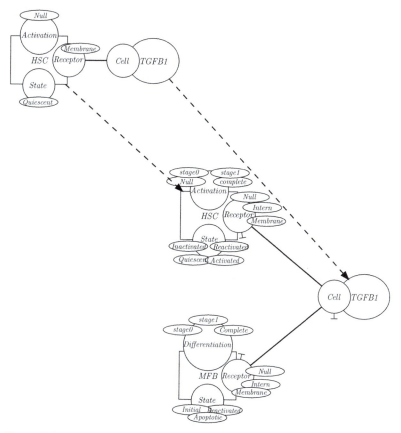

Figure 4.5 The unique projection from the complex of Figure 4.4 into the contact map of Figure 4.3. This projection is defined by mapping each occurrence of agents of the complex to the unique node corresponding to this agent in the contact map.

activation state. Lastly, each occurrence of an interaction site may be free or bound to exactly one other occurrence of an interaction site. The binding state of an occurrence of an interaction site may also remain unspecified.

Example 4.4 Two examples of connected patterns are given in Figure 4.6. The first pattern (see in Figure 4.6(a)) is made of a single occurrence of the agent *HSC* and documents the state of two sites. The site *state* is in the state *quiescent* whereas the site *activation* is in the state *null*. Neither the binding state nor the state of the interaction site *receptor* is specified.

The second pattern (see in Figure 4.6(b)) is made of one occurrence of the agent *HSC* and one occurrence of the agent *TGFB1* that are bound together via the site *receptor* of the first one and the site *cell* of the second one.

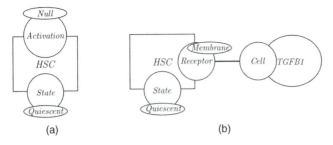

Figure 4.6 Two connected patterns. They are made of occurrences of agents. Each occurrence of agents documents a subset of its interaction sites. Each site may carry out an internal state and/or a binding state (while remaining consistent with the content of the contact map). An occurrence of an interaction site may be free or bound to a specific occurrence of an interaction site. (a) First pattern. (b) Second pattern.

The site *receptor* of the agent *HSC* is tagged with the string "*membrane*," which means that the receptors of this cell are on the membrane. It is also specified that the HSC must be in the quiescent state. The activation level of the cell is not documented.

As it was already the case for complexes, the contact map also contrains that the patterns may be used in a model. This way, the occurrence of an agent in a pattern can only document the interaction sites that are associated with the unique occurrence of this agent in the contact map. An occurrence of an interaction site may bear a given activation state only if the corresponding site in the contact map is tagged with this activation state. An occurrence of an interaction site may be free only when the corresponding site in the contact map may be free as well. Lastly, two occurrences of sites may be bound together in a pattern only when there is a link between the two corresponding sites in the contact map. Said differently, as it was already the case for complexes, it shall be possible to project the pattern onto the contact map. This means that the function mapping each agent of a pattern to the unique corresponding agent in the contact map shall be a homomorphism.

4.2.1.4 Embeddings Between Patterns

A pattern may specify more or less information. It is indeed possible to insert some interaction sites in the occurrence of an agent that does not document all its interaction sites. Moreover, it is also possible to insert a new binding or activation state to an interaction site that misses one. It is even possible to bound a given site to the fresh occurrence of an agent. In all these cases, we will say that the initial pattern occurs in the second one, or equivalently, that the second pattern contains an occurrence of the first one. That is to say that the relationship between the occurrences of agents in the first pattern and the occurrences of agents in the second pattern induces an embedding. An *embedding* from a pattern into another one is a function that maps each occurrence of agents in the first pattern to an occurrence of agents in the second pattern while preserving the structure of site graphs. This means that this mapping preserves the kinds of agents, the sites that are documented, the activation, and binding states.

It is worth noting that complexes are particular patterns. In a complex, each occurrence of an agent documents all its sites, with an activation state and a binding state whenever they have one. This way, it is not possible to insert more information in a complex. A complex is a connected pattern that cannot be embedded in any other connected pattern.

Example 4.5 Two examples of embeddings are given in Figure 4.7. These are the only embeddings between the patterns that were given in Figure 4.6 and the complex that was given in Figure 4.4. In both embeddings, each occurrence of an agent in the pattern is mapped into the unique occurrence of this agent in the complex.

It is worth noting that an embedding from a connected pattern into another pattern is fully defined by the image of one of its occurrences of agents. The other associations may be retrieved by following the links between interaction sites and by using the fact that embeddings shall preserve links. This property highly eases the research of occurrences of a given pattern inside another pattern. Site graphs are *rigid* Danos et al. (2010), Petrov et al. (2012).

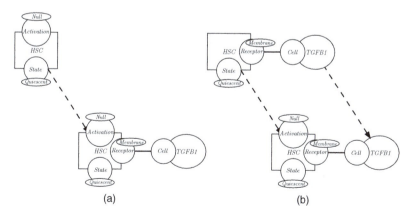

Figure 4.7 Two embeddings between the patterns that were given in Figure 4.6 and the complex that was given in Figure 4.4. (a) First embedding. (b) Second embedding.

4.2.2 Site Graph Rewriting

Patterns are used to speficy the potential behavior of Kappa models by means of rewriting rules. This is the objective of the current section.

4.2.2.1 Interaction Rules

Complexes may evolve by applying interaction rules. An *interaction rule* is defined as a pair of patterns with an implicit pairing relation between some occurrences of agents in the first pattern and some occurrences of agents in the second pattern. The first pattern specifies the local conditions under which an interaction may be triggered. The difference between both patterns specifies the transformation that is performed when this rule is applied. As a matter of fact, the second pattern in a rule shall be obtained from the first one by modifying the activation and/or the binding states of some interaction sites, inserting fresh occurrences of agents (in this case, the full interface of these occurrences shall be documented), and removing some occurrences of agents.

Example 4.6 In Figure 4.8, three examples of interaction rules are given.

In Figure 4.8(a), occurrences of the protein TGFB1 bind a HSC in its quiescent form. The receptors of this cell must be located on its membrane. The interaction consists in establishing a bond between the site *receptor* of the agent *HSC* and the site *cell* of the agent *TGFB1* while internalizing the receptors of the cell.

In Figure 4.8(b), an activated hepatic stellate cell (aHSC) in activation stage 0 and connected to some *TGFB1* proteins may proliferate. As a result, the *TGFB1* proteins are consumed, which frees the receptors of the cell, and another cell is created. This cell is in the same activation stage and is also activated. Moreover, its receptors are free and in the state *null*. We notice that the configuration of the newly produced cell is fully specified, as it is required when a fresh agent is created.

In Figure 4.8(c), a quiescent hepatic cell in the activation state *null* may be degraded. Whenever this interaction rule is applied, an occurrence of HSC in this configuration is removed. There is no requirement on the receptors of the cell. They may be null, on the membrane, or internalized. Additionally, when the receptors are bound, the corresponding binding is released before degrading the cell, which also frees the occurrences of TGFB1 proteins potentially bound to the cell. This update has not to be specified explicitly in the rule: this is called a side effect.

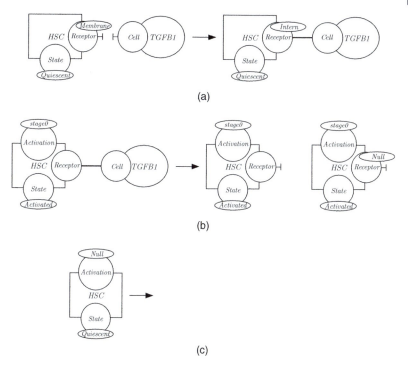

Figure 4.8 Three examples of rules: (a) a binding rule; (b) a proliferation rule; and (c) a degradation rule. (a) Binding of TGFB1 to quiescent hepatic stellate cells. (b) Proliferation of activated hepatic stellate cells at stage 0. (c) Degradation of quiescent hepatic stellate cells.

4.2.2.2 Reactions Induced by an Interaction Rule

As explained previously, the left-hand side of a rule specifies under which context a given interaction may happen. It is then possible to insert further constraints for the conditions under which a rule may be applied, by refining the patterns that occur on the left-hand side and on the right-hand side of a rule in exactly the same way. A rule that cannot be refined further (without creating a new connected component) is called a *reaction-rule* Harmer et al. (2010).

Special care has to be taken about agent degradation. On the first hand, when refining the state of an agent to be degraded in the left-hand side of a rule, the right-hand side is not impacted (since there is no corresponding agent). On the second hand, agent degradation may cause side effects. Indeed, the state of an occurrence of agent to be degraded may be refined by binding one of its sites to the site of another occurrence of agent. In such a case, if the latter occurrence is not degraded, then its site becomes free in the right hand side of the rule. There is no pending bound in Kappa.

Example 4.7 Figure 4.9 shows that some rules are refined into reaction-rules. Figure 4.9(a) shows a refinement of the rule that was given in Figure 4.8(a). It is additionally specified that the HSC shall be on the state *null* of activation. Figure 4.9(b) depicts the unique refinement of the rule that was given in Figure 4.8(b) (since this was already a reaction, no further information can be inserted). Figure 4.9(c) draws a refinement of the rule that was given in Figure 4.8(c). It is additionally specified that the receptors of the hepatic stelatte cell shall be internalized and bound to some occurrences of the *TGFB1* proteins. These occurrences are released when the HSC is degraded, as a result of side effects.

4.2.2.3 Underlying Reaction Network

A set of rules can be translated into a – potentially infinite – set of reaction-rules by replacing each interaction rule by the set of the reaction-rules that can be obtained as refinements of this rule. It is then enough to replace each complex by a name to get a proper (potentially infinite) reaction network, in which each rule is defined as a tuple of reactants and a tuple of products. This reaction network is defined uniquely up to the choice of the names of each complex. The behavior of a set of interaction rules may then be defined as the behavior of its underlying reaction network. Quantitative semantics require to assign rates to each rules, the rate of each reaction-rule being defined as the rate of the rule it has been generated from.

Example 4.8 We conclude this section by describing the compilation of a toy model written in Kappa into a reaction network. We consider a model with only one kind of agent, a protein. This protein has two sites, l and r. Each occurrence of these sites may be phosphorylated or not. The signature of the model is given in Figure 4.10(a) by means of a contact map. The phosphorylation and the dephosphorylation of each site of an occurrence of a protein is independent from the state of the other site in this occurrence, which is formalized in the four rules that are given in Figure 4.10(b). This way, neither the phosphorylation rules nor the dephosphorylation rules document the state of the other site.

The underlying reaction-rules are obtained by expanding the context of application of each interaction. This way, in our example, each rule gives birth to two reaction-rules according to the phosphorylation state of the site that is not specified in the initial rule. These reaction-rules are given in Figure 4.10(c).

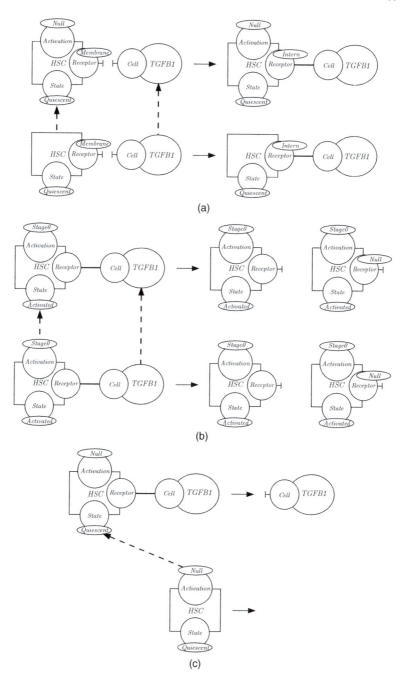

Figure 4.9 Three examples of reaction-rules for the rules of Figure 4.8.
(a) Binding of TGFB1 on quiescent *HSC*s in activation level *null*. (b) Proliferation of *HSC*s at stage 0. (c) Degradation of *HSC*s bound to some occurrences of the *TGFB1* proteins.

90 | *4 The Rule-Based Model Approach*

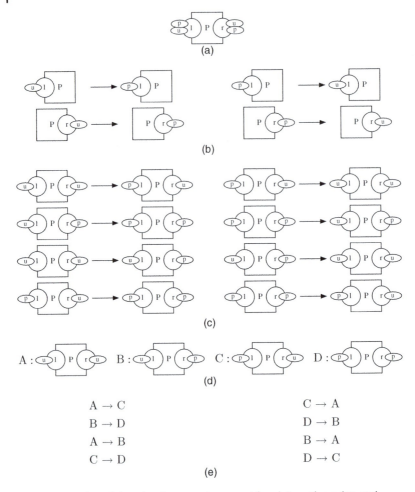

Figure 4.10 A model made of a contact map and four interaction rules, and its compilation into a reaction network. (a) Contact map. (b) Interaction rules. (c) Reaction-rules. (d) Dictionary. (e) Reactions.

The next step consists in naming the different kinds of complexes that are involved in the so-obtained reaction-rules. An occurrence of the protein with no phosphorylated site is called A, an occurrence of the protein with only the site r phosphorylated is called B, an occurrence of the protein with only the site l phosphorylated is called C, and an occurrence of the protein with both sites phosphorylated is called D. Named reactions are given in Figure 4.10(e). They have been obtained by replacing each occurrence of complex with its name in the reaction-rules.

Defining the behavior of a model by means of its underlying reaction networks has been done to ease the presentation. The semantics of the language BNGL was initially implemented this way Faeder et al. (2009). Yet, such a semantics is not so convenient in practice since a Kappa model usually induces too many reactions. The semantics may be formalized directly by means of a process calculus Danos et al. (2008), Feret et al. (2013) or in a categorical setting Danos et al. (2012), Feret (2015). The first approach provides a more operational perspective, whereas the second one abstracts away more computational details. It is worth noting that usual categorical rewriting frameworks (by single push-out Löwe (1993), double puch-out Corradini et al. (1997), or sesqui-pushout Corradini et al. (2006)) fail in modeling correctly side effects. Two known approaches solve this issue. It is possible to twist the definition of embeddings Danos et al. (2012), Feret (2015) or to enrich graphs with constraints Behr and Krivine (2019).

The simulator of Kappa directly applies rewriting rules in the graph that describes the state of the system Danos et al. (2007), Boutillier et al. (2017). The underlying reaction network is never computed explicitly.

4.3   Model of Activation of Stellate Cells

Now, we describe in Kappa a model of the behavior of a population of HSCs.

4.3.1   Overview of Model

Figure 4.11 shows the sketched potential behavior of an occurrence of HSC. This diagram itemizes the different transformation processes between the different forms of HSCs. In particular, it describes how HSCs are activated by TGFB1 and differentiate into myofibroblasts that undergo different processes. An important point is the inactivation pathway leading to the generation of iHSCs that differ from the quiescent phenotype. Inactivated cells are more quickly reactivated by TGFB1 than the quiescent ones, thereby amplifying cell response in chronic diseases. The apoptotic/senescence pathway allows for the feedback control (to keep the explanation simple) of the activation pathway.

4.3.2   Some Elements of Biochemistry

Before describing the interaction rules, we give some reminders of basic elements of biochemistry. Our goal is to explain how rate constants are defined and computed. As often as possible, we try to define them with respect to

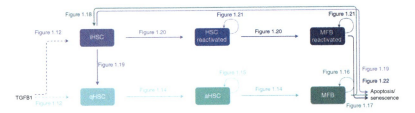

Figure 4.11 The behavior of hepatic stellate cells. TGFB1 protein induces the activation of quiescent hepatic stellate cells (qHSC) (green dashed arrow). The aHSCs differentiate to myobroblasts (MFB) (light green arrow). MFB have two potential behaviors (green arrow): either they enter in the apoptosis/senescence pathway or are inactivated to inactivated hepatic stellate cells (iHSC). iHSC have three potential behaviors: either they reverse into qHSC (light purple arrow), or they enter in the apoptosis/senescence pathway (light purple arrow), or they are activated by the protein TGFB1 (purple dashed arrow), leading to reactivated hepatic stellate cells (HSC reactivated) (light purple) that may differentiate to reactivated myobroblasts (MFB reactivated) (purple). The reactivated MFB have only one behavior, that is the apoptosis/senescence pathway (purple).

the reaction half-time, that is the time after which on average, half of the reactants of a reaction have been consumed.

4.3.2.1 Reaction Half-Time

Let us consider a first-order reaction of the form $A \to B$. We assume that the time when each instance of the reaction is applied is drawn randomly according to an exponential law with parameter k. That is to say that given an occurrence of the component A, the probability that this occurrence has turned into an occurrence of the component B after time t is equal to $(1 - e^{-k \cdot t})$.

Assuming that A is in large quantity in the system, the expectation $E[A](t)$ of the quantity of A remaining in the system at time t is defined by the following equation:

$$E[A](t) = E[A](0) \cdot e^{-k \cdot t}$$

where $E[A](0)$ denotes the initial quantity of A (we write it as an expectation for the sake of homogeneousness).

The half-time of the reaction, which is written as $t_{1/2}$, is then the time so that the expectation of the quantity of the component A that has been consumed at time $t_{1/2}$ is equal to half of the initial quantity of this component. That is to say that:

$$\frac{E[A](0)}{2} = E[A](0) \cdot e^{-k \cdot t_{1/2}}$$

It follows that:
$$k = \frac{\ln 2}{t_{1/2}}$$

In the case of a degradation reaction, that is to say, when the reaction has no product, the half-time of a reaction is also called its half-life time.

4.3.2.2 Conversion

In practice, reaction time is documented in the literature in various forms. Sometimes, it is documented as the time taken to transform all the occurrences of the component A into occurrences of the component B. Since there always remains a residual quantity, we interpret this as the time $t_{99\%}$ so that the expectation of the quantity of the component A that has been consumed by the reaction at time $t_{99\%}$ is equal to 99% of the initial quantity of this component.

The conversion from $t_{99\%}$ to $t_{1/2}$ can be made thanks to the following reasoning. Given q between 0 and 1, we can define t_q the time so that the expectation of the component A that has been consumed is equal to the fraction q of the initial quantity of this component. The duration t_q is defined by the following equation:

$$(1-q) \cdot E[A](0) = E[A](0) \cdot e^{-k \cdot t_q}$$

It follows that:
$$\frac{\ln 2}{t_{1/2}} = \frac{-\ln(1-q)}{t_q}$$

Thus,
$$t_{1/2} = \frac{\ln 2}{\ln \frac{1}{1-q}} \cdot t_q$$

We can conclude that:
$$t_{1/2} = \frac{\ln 2}{\ln 100} \cdot t_{99\%}$$

4.3.2.3 Production Equilibrium

It often happens that a degradation rule is counter-balanced by a synthesis rule in order to maintain an expected average amount of components in stationary regime.

Let us consider two reactions, a degradation reaction $A \to .$ with a half-life time $t_{1/2}$ and a synthesis reaction $. \to A$ at a rate k. The goal is to set the rate constant k so that the expected average amount of the component A is equal to $E[A]_{eq}$ in a stationary regime. This means that, when the quantity of A is equal to $E[A]_{eq}$, the overall propensity of the degradation rule shall be equal to the one of the synthesis rule. That is to say that:

$$k = \frac{\ln 2}{t_{1/2}} \cdot E[A]_{eq}$$

It is worth noting that this parameterization does not enforce a rigid equilibrium. This ensures only the eventual behavior of the system in the absence of other mechanisms that could modify the quantity of the component A.

4.3.2.4 Erlang Distributions

The exponential law is defined by only one parameter. As a consequence, the standard deviation of the time a given event may take is fully defined by its average time. This is not always satisfying from a modeling point of view as some processes may require time distributions with different standard deviations.

A solution to this issue consists in decomposing a given interaction into several intermediary steps, each of these being executed according to an exponential distribution. The resulting composite process satisfies a so-called Erlang distribution Erlang (1909) that is defined by two parameters (the average time of each intermediary step and the number of these steps). For instance, we may consider the sequential composition of two steps, the duration of each of which being defined by an exponential law with a same parameter, and compare it with a single process the duration of which is defined by an exponential law, twice as slow as each step of the composite process. Then, the standard deviation for the time so that half of the quantity of the initial component has completed the two intermediary steps is less than the standard deviation of the time so that half of the quantity of the initial component has completed the single-step process. Moreover, the time so that the expectation of the quantity of the initial component that has completed the two-step process is equal to 99% is shorter than the time that is defined the same way for the single-step process.

We do not know how to define analytically the reaction half-time of the intermediary processes with respect to the overall completion time of the process. Instead, we fit these values empirically by simulating the behavior of the intermediary interactions (without considering the rest of the model).

4.3.3 Interaction Rules

We now itemize the interaction rules of our model. The rate constants are parameterized with some values essentially found in the literature. The value of these parameters is given after the description of the rules in Figure 4.25 on page 52.

4.3.3.1 Behavior of TGFB1 Proteins

Figure 4.12 specifies the behavior of TGFB1 proteins.

The degradation of the protein TGFB1 is described in Figure 4.12(a). In general, the occurrences of the protein TGFB1 are spontaneously degraded only in their active form. Yet, in this model, only the active form of TGFB1

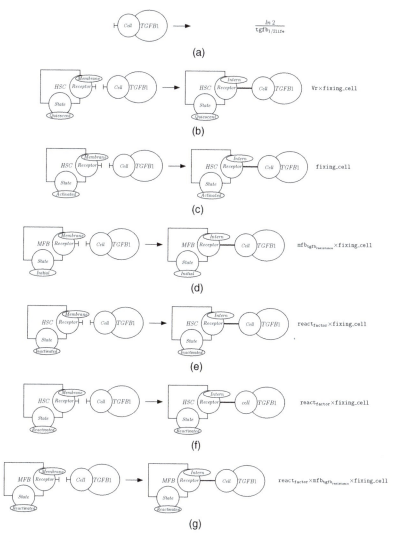

Figure 4.12 The behavior of the protein *TGFB1*. (a) Degradation. (b) Fixing quiescent *HSC*s. (c) Fixing activated *HSC*s. (d) Fixing *MFB*s fixing. (e) Fixing inactivated *HSC*s. (f) Fixing reactivated *HSC*s. (g) Fixing reactivated *MFB*s.

is involved; this is why the activation state of the occurrences of the protein TGFB1 is omitted. Inactive form of the protein TGFB1 plays an important role in the extracellular matrix. Additionally, this rule specifies that only the occurrences of the agent *TGFB1* that are not linked to any other agent may be degraded. The degradation rate is set according to the half-time of the protein.

The rules in Figure 4.12(b)–(g) describe the interactions between TGFB1 and cells, the latter being in different states. We use several rules in order to give them different rates. Interactions with *TGFB1* agents are possible only when cell agents have a free *receptor* at their *membrane* and *TGFB1* agents are in their *active state* (which is omitted in our model) and not bound to any other agents. The receptors that bind TGFB1 are internalized, leading to the state called *intern*. In order to preserve a pool of qHSC for cell renewal, we modulate the rule rate from (Figure 4.12(b)) with a variable *Vr* that stands for the degree of renewal. This variable ranges between 0 and 1 according to the amount of cells that are currently bound to TGFB1 proteins. The more cells are bound to some TGFB1, the lower *Vr* will be, till reaching the value 0. Compared to HSCs, myofibroblasts are less sensitive to the protein TGFB1 Schnabl et al. (2001); therefore, we reduced the rate of transformation induced by TGFB1 binding by 41.6% (41.6% = HSC_proliferation/MFB_proliferation), regardless of the states of the *MFB* (*initial* or *reactivated*). Compared to the cells in the state qHSC and MFB, the cells in the state iHSC, HSC reactivated, and MFB reactivated are more sensitive to TGFB1 Schnabl et al. (2001), that is why we increased the corresponding rate constant (4 fold time).

4.3.3.2 Renewal of Quiescent HSCs

The quantity of quiescent HSC in the liver results from an equilibrium between the production and the degradation of these cells. While qHSCs have a mesenchymal stem cell origin Pittenger et al. (2019), Häussinger and Kordes (2019), there is no information about the qHSC production and their renewal in normal and pathological livers. Upon TGFB1 stimulation, qHSC are transformed toward aHSC and the pool of qHSC is likely consumed. We introduced a variable *Vr* to preserve a residual pool of qHSC (see page 36). The degradation and production rules for HSCs in their quiescent form are described in Figure 4.13. The degradation of qHSC occurs when they have a *null* activation and no links with any other agents (see Figure 4.13(b)). In Figure 4.13(a), the state of the qHSCs that are

4.3 Model of Activation of Stellate Cells

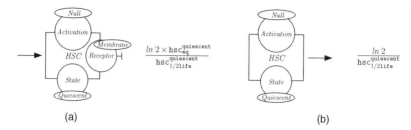

Figure 4.13 The renewal of quiescent hepatic stellate cells. (a) Production. (b) Degradation.

synthetized has to be fully specified. It is written that their activation state is *null* and their receptors are free and on the membrane.

The rates of both rules are set according to the explanations that were given in Section 4.3.2 on page 33 so as to ensure the renewal rate of the HSCs in their quiescent form and also the overall amount of them in a regular regime.

4.3.3.3 Activation and Differentiation

The activation of an occurrence of a HSC may happen when the three following conditions are satisfied: it shall be in a quiescent form (which is written as *quiescent*), its activation level shall be null (which is written as *null*), and it shall be activated by some TGFB1 proteins (Figure 4.14(a)). This activation modifies the conformation of the HSC, which is now an activated one (which is written as *activated*) in the first stage of activation (which is written as *stage0*).

Activation is a gradual process (Figure 4.14b,c), the HSCs enter in their second activation stage (which is written as *stage1*), and their third and last one (which is written as *complete*). During this process, they remain in their activated form.

Once in the last stage of activation, HSCs may differentiate into myofibroblasts. This process is done in three stages. The first step consists in replacing a fully aHSC into a myofibroblast (*MFB* agent) in initial state (which is written as *initial*) and in the first differentiation stage (which is written as *stage0*). The description of this first step requires four rules (Figure 4.14(d)). The main reason is that the state of the cell receptors shall be maintained. Yet, since *HSC* and *MFB* are two different agents, it is not possible to inherit information from the *HSC* agents that are consumed to the *MFB* agents that are created. Thus, the solution is to write one rule for each potential state of the interaction site *receptor*, and there are four of them.

98 | *4 The Rule-Based Model Approach*

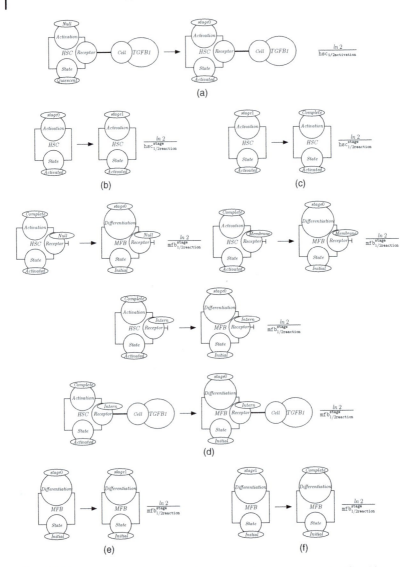

Figure 4.14 Activation of hepatic stellate cells and formation of myofibroblasts. (a) Initial step of activation. (b) Second step of activation. (c) Third step of activation. (d) First step of differentiation. (e) Second step of differentiation. (f) Final step of differentiation.

After differentiation of HSC toward MFB, the latter undergoes two other steps of differentiation. The rule in Figure 4.14(e) describes the passage from the initial stage into the second one (which is written as *stage1*). Then, the rule in Figure 4.14(f) describes the passage from the second stage into the last one (which is written as *complete*).

The rates of activation and differentiation steps are defined by means of a reaction half-time, following the guidelines that were given in Section 4.3.2. Each activation step shares the same rate, while each differentiation step shares another one.

4.3.3.4 Proliferation of Activated Hepatic Stellate Cells

The activation of HSCs by some occurrences of the *TGFB1* protein may induce their proliferation (Figure 4.15).

There are three different rules according to the activation stage of the cells. As a result of proliferation, the occurrences of the *TGFB1* agent are consumed, and the occurrences of the *HSC* agent are duplicated. It is worth

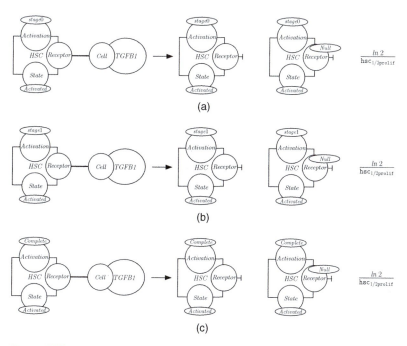

Figure 4.15 Proliferation of activated hepatic stellate cells. (a) When in the initial stage. (b) When in the second stage. (c) When in the final stage.

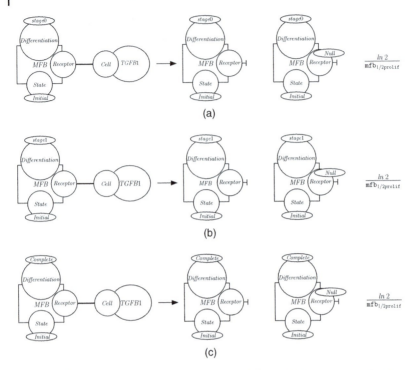

Figure 4.16 Proliferation of myofibroblasts. (a) When in the initial stage. (b) When in the second stage. (c) When in the final stage.

noting that we have assumed that the new occurrences are in the same state and in the same activation stage as the occurrences that have given them birth. However, their receptors are not operational yet (which is written as *null*) (Figure 4.16).

The rate of the rules are computed from the half-time of the proliferation reaction.

4.3.3.5 Proliferation of Myofibroblasts

The proliferation of myofibroblasts works exactly as the proliferation of aHSCs. Myofibroblasts must be in the *initial* state and activated by some occurrences of the TGFB1 protein. Occurrences of myofibroblasts may then be duplicated, conserving the state and the differentiation stage. The receptors of the newly created myofibroblasts are not operational yet (which is written as *null*). Proliferation speed is set by means of the proliferation half-time of myofibroblasts.

4.3.3.6 Apoptosis and Senescence of Myofibroblasts

Upon the action of the TGFB1 protein, HSCs are activated and differentiated into myofibroblasts, which are in charge of tissue repair. Moreover, when TGFB1 is consumed, some of these MFB are eliminated (around 50%) through the apoptosis/senescence pathway, which is modeled in Figure 4.17, the rest of them becomes inactivated (for 50%) through an inactivation pathway, which is depicted in Figure 4.18.

Because of the lack of information, the processes of apoptosis and senescence have been merged. The resulting pathway is made of two steps. The first step (Figure 4.17(a)) consists in marking the myofibroblast for apoptosis/senescence. It requires the myofibroblast to be in its initial form (which is written as *initial*), in the final stage of differentiation (which is written as *complete*), and its receptors to be free. As a result, the state is

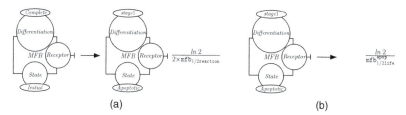

Figure 4.17 Apoptosis/senescence pathway. (a) Apoptosis/senescence selection. (b) Apoptosis/senescence completion.

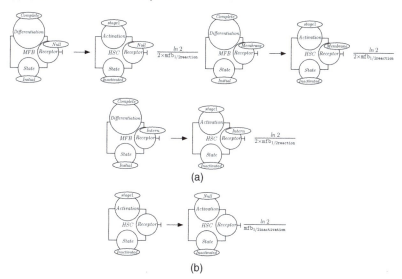

Figure 4.18 Inactivation pathway. (a) First step of inactivation. (b) Second step of inactivation.

changed into *apoptosis*. We use a different scale for the differentiation level of the myofibroblasts that are marked for apoptosis; hence, the differentiation level is set to *stage1*. The rate constant of the interaction accounts for the half-time reaction of myofibroblasts and the fact that only half of them follow this pathway.

The second step consists in the degradation of the myofibroblast (Figure 4.17(b)). The degradation rate is computed from the half-life time of the myofibroblasts once they have entered the apoptosis/senescence pathway.

4.3.3.7 Inactivation of Myofibroblasts

Inactivation of myofibroblasts is a two-step process that is depicted in Figure 4.18. This process turns them back to HSCs. Yet, as the cells that come from the inactivation of myofibroblasts have a different behavior, we call them inactivated cells.

The first step (Figure 4.18(a)) consists in turning occurrences of myofibroblasts into HSCs in an inactivated form (which is written as *inactivated*). This requires the occurrences of myofibroblasts to be in their initial state (hence, they have not entered the apoptosis pathway, and they cannot come from the reactivation of an inactivated cell) and in the final stage of differentiation. Moreover, binding with occurrences of the TGFB1 protein blocks this process; thus, their receptors are assumed to be free. No further assumption is required on the state of the receptors of the occurrences of the myofibroblasts. As a result, the occurrences of myofibroblast are replaced with the occurrences of HSC in the initial state and in the first stage of inactivation (which is written as *stage1*). The state of the receptors is maintained. As the agents *MFB* and *HSC* are different, we have to use three rules to model this (as we did already for describing the first step of differentiation of HSCs (Figure 4.14(d))). The rate of these rules are the same than the rules for apoptosis/senescence (Figure 4.17(a)), since it stands for the behavior of the other half of the myofibroblasts.

The second step of the inactivation of HSCs is described in Figure 4.18(b). It still requires the cell not to be activated by the TGFB1 proteins. As a result, the cell enters the final stage of inactivation (which is encoded by the activation state *null*). The rate of the rule for this second step of inactivation is defined by the half-time of inactivation of HSCs.

4.3.3.8 Behavior of Inactivated Hepatic Stellate Cells

iHSCs are either slowly eliminated through the degradation process or returned to the quiescent form or reactivated by a new *TGFB1* stimulation into myofibroblasts.

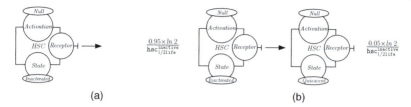

Figure 4.19 Degradation and recycling of inactivated hepatic stellate cells. (a) Degradation. (b) Return to quiescent form.

Inactivated HSC can be eliminated through two processes, apoptosis/senescence or reversing. To undergo those pathways, *HSC* agents must be inactivated (which is written as *inactivated*) and not under the process of activation (which is written as *null*). Moreover, these *HSC* agents must not be bound to *TGFB1* agents. We choose that 95% of iHSC will enter apoptosis/senescence (Figure 4.19(a)), the remaining part will reverse into a quiescent state (Figure 4.19(b)).

The 5% of iHSC undergoing reversion are similar to the one undergoing apoptosis/senescence and reverse into qHSC (Figure 4.19(b)). The process of reversion forms similar qHSC than those previously described (Figure 4.13). Nevertheless, the state of the receptor is conserved during reversing. The rates of degradation and return to quiescence rules depend on iHSC half-life time.

Upon a new TGFB1 stimulation, iHSCs may be reactivated. This is a two-step process. In the first step, which is depicted in Figure 4.20(a), the occurrences of the *TGFB1* agent are consumed and the conformation of the cell changes. They are now in the first stage of reactivation, which is formalized by the state *reactivated* and the activation level *stage0*. The second step, which is written in Figure 4.20(b), puts the cell in the second stage of reactivation, its activation level is set to *complete*. The rate of the reactivation of iHSC is 4 times the rate of a *HSC* activation; that is why, the parameter $ihsc^{stage}_{1/2reaction}$ is divided by the parameter $react_{factor}$ (that is set to the value 4).

Reactivated hepatic stellate cells follow the same process of differentiation than the activated ones, leading to the formation of reactivated myofibroblasts (which is written as *reactivated*) as described in Figure 4.20c,d. Similar to the differentiation of initial myofibroblasts (Figure 4.14), the description of the first step of redifferentiation requires four rules (Figure 4.20(c)). Those four rules aim at conserving the state of the receptors during the change from occurrences of the agent *HSC* into occurrences of the agent *MFB*. The rate of the rules for *MFB reactivated* stage depends on the half-time of the reaction needed to their formation.

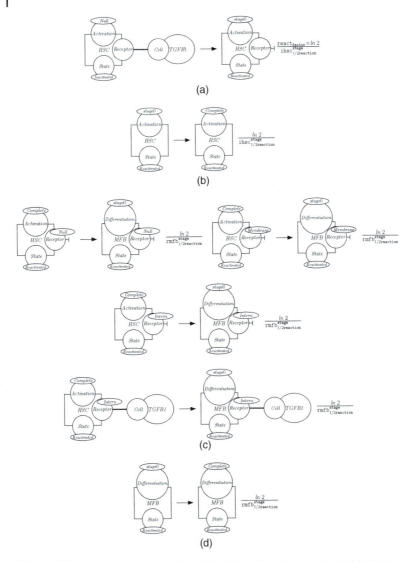

Figure 4.20 Redifferentiation of inactivated hepatic stellate cells. (a) Initial step of reactivation. (b) Second step of reactivation. (c) First step of redifferentiation. (d) Second step of redifferentiation.

After differentiation of HSC toward MFB, the latter undergoes another step of differentiation. The rule in Figure 4.20(d) describes the passage into the last stage (written *complete*). The rate of the complete differentiation rule depends on the half-time of the complete differentiation of the reactivated MFB.

4.3.3.9 Proliferation of Reactivated Cells

The proliferation of reactivated cells is formalized in Figure 4.21. It works exactly as the proliferation of initial cells. Myofibroblasts and HSCs must be in the state *reactivated* and bound to some occurrences of the protein TGFB1. Occurrences of *HSC* and *MFB* may then be duplicated, conserving the state of their sites *state* and *activation* for the formers and the state of their sites *state* and *differentiation* for the latters. The receptors of the newly created cells are not operational yet (which is written as *null*). Proliferation speed is set by means of the proliferation half-time of each stage of activation and differentiation. In particular, the factor

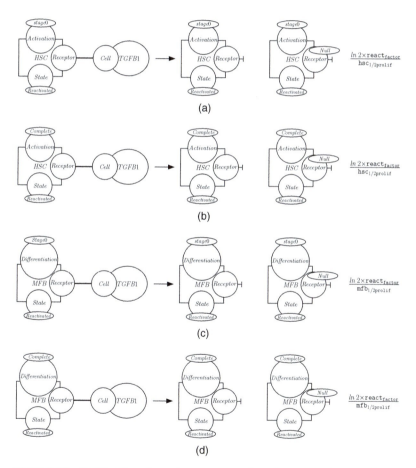

Figure 4.21 Proliferation of reactivated cells. (a) Reactivated *HSC* in stage 0. (b) Reactivated *HSC*. (c) Redifferentiate *MFB* in stage 0. (d) Redifferentiate *MFB*.

Figure 4.22 Apoptosis/senescence of reactivated myofibroblasts.

$react_{factor}$ accounts for the fact that reactivated cells proliferate more than firstly activated ones.

4.3.3.10 Degradation of Reactivated *MFB*

Unlike myofibroblasts, no information shows a potential inactivation for reactivated MFB. Starting from that, the only way for *MFB* in a *reactivated* to disappear is through apoptosis/senescence (Figure 4.22). It requires the myofibroblasts to be in their reactivated form (which is written *reactivated*), in the final stage of differentiation (which is written as *complete*), and their receptors to be free. The reaction rate is computed from the half-life time of the myofibroblasts.

4.3.3.11 Behavior of Receptors

The receptors of cells have also their own behavior that regulates the capability of cells to interact. This holds for the receptors of the protein TGFB1 as well. The behaviors of these receptors in HSCs and in myofibroblasts are sketched in Figure 4.23 by means of transition systems, which show the possible changes from one state to another, labeled with some rate constants. This behavior can be modeled by means of four pairs of rules (Figure 4.24) Vilar et al. (2006). Those rules regulate the capability of the occurrences of the *HSC* and *MFB* agents to interact with the occurrences of the *TGFB1* agent.

The first pair of rules describes the production of the receptors for *HSC* and *MFB* agents (Figure 4.24a,b). HSCs and myofibroblasts may produce their receptors when they are not present (which is written as *null*).

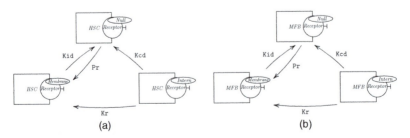

Figure 4.23 Receptors cycle. (a) The receptors of HSC. (b) The receptors of MFB.

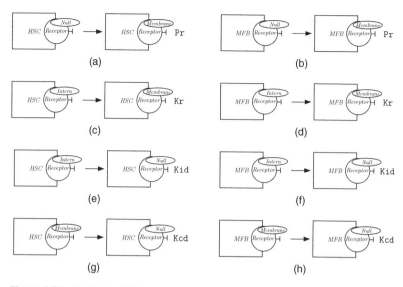

Figure 4.24 Behavior of the receptors. (a) HSC receptors expression. (b) MFB receptors expression. (c) HSC receptors recycling. (d) MFB receptors recycling. (e) HSC internalized receptors degradation. (f) MFB internalized receptors degradation. (g) HSC externalized receptors degradation. (h) MFB externalized receptors degradation.

Newly produced receptors are expressed at the membrane of the cells (which is written as *membrane*). The rate of production of receptors is given by the parameter *Pr*.

The second pair of rules describes the recycling of the receptors from the pool of internalized receptors to the membrane (Figure 4.24c,d). Interacting with their ligands may induce the internalization of the receptors (which is written as *intern*). This internalization may lead to the recovery of the receptor permitting its expression at the membrane (which is written as *membrane*). The rate of this process is given by the parameter *Kr*.

The third pair of rules describes the degradation of the receptors from the pool of internalized receptors (Figure 4.24e,f). This internalization may also lead to the degradation of the receptors (which is written as *null*). The rate of this degradation process is given by the parameter *Kid*.

The fourth pair of rules describes the degradation of the receptors lying on the membrane (Figure 4.24g,h). Receptors have a turnover, leading to their degradation even without interaction with the protein *TGFB1*. In that case, they may change from the state *membrane* into the state *null*. The rate of this degradation process is given by the parameter *Kcd*.

4.3.4 Parameters

In the previous section, we described the rules present in our model, the table in Table 4.1 contains the parameters used to define the rate of these rules. Most of these parameters were found in the literature; for some of them, some calculations and estimations were done. Sometimes, information was lacking; that is why, for few parameters, we estimated their values using the literature and our knowledge. As explained previously in Section 4.3.2, defining the reaction half-time of intermediary stages with respect to the overall completion time of the process is complicated. The computation of those reaction half-times has been made empirically.

Table 4.1 Parameterization of the model.

Definition	Symbol	Values	Acquisition	Reference
Environment				
Number of Disse's Space	Number_of_Disse_study	10	Calculated	[39]
Number of cells by Disse's Space	HSC_number_by_Disses	4	Calculated	[39]
Total number of cells	$hsc_{eq}^{quiescent}$	40	Calculated	
qHSC				
Half-life	$hsc_{1/2life}^{quiescent}$	90 days	Estimated	[62]
Half-activation time	$hsc_{1/2activation}$	0.17 h	Calculated	[49]
Max of cell activated	Vr	0, 0.166 , 1	bibliography	[5]
aHSC				
Doubling time	$hsc_{1/2prolif}$	12.64 h	Estimated	[80]
MFB				
Doubling time	$mfb_{1/2prolif}$	8.43 h	Estimated	[80]
Half-life	$mfb_{1/2reaction}$	30 h	Estimated	[62]
Apoptosis proportion		50 of total MFB	Bibliography	[40]
Inactivation proportion		50 of total MFB	Bibliography	[40]
MFB react				
Half-life	$mfb_{1/2reaction}$	25.29 h	Estimated	
Doubling time	$mfb_{1/2reaction}^{stage} react_{factor}$	1.58 h	Estimated	[40]
iHSC				
Half-life	$hsc_{1/2life}^{inactive}$	90 days	Bibliography	[62]
Apoptosis proportion		95 of total iHSC	Estimated	[36], [66]
Quiescent return proportion		5 of total iHSC	Estimated	[36], [66]
Half-activation time	$ihsc_{1/2reaction}^{stage} react_{factor}$	0.0425 h	Estimated	[40]
TGFB1				
Half-life	$tgfb_{1/2life}$	5 min	Bibliography	[85]
Inflammatory input	TGFB_per_wave	100 * number of qHSC	Calculated	
Fixing time	fixing_cell	3 min	Estimated	
Number of receptor by cell	TGFB_factor	7730	Bilbiography	[69]
Receptor				
Recycling rate	Kr	0.5 h	Bibliography	[84]
Production rate	Pr	0.066 h	Bibliography	[84]
Degradation rate	Kcd	0.6 h	Bibliography	[84]
Ligand induced degradation rate	Kid	0.066 h	Bibliography	[84]
Time of transformation				[15]
aHSC transition stage	$hsc_{1/2reaction}^{stage}$	36.64 h	Calculated	[5]
MFB differentiation stage	$mfb_{1/2reaction}^{stage}$	46.22 h	Calculated	[5]
MFB inactivation stage	$mfb_{1/2inactivation}$	18.06 h	Calculated	[40]
MFB apoptosis stage	$mfb_{1/2life}^{apop}$	3.62 h	Estimated	
IHSC reactivation stage	$ihsc_{1/2reaction}^{stage}$	6.31 h	Calculated	[40]
MFB second differentiation stage	$rmfb_{1/2reaction}^{stage}$	6.31 h	Calculated	[40]
Factors				
IHSC reactivation stage	$react_{factor}$	4	Bibliography	[40]
MFB TGFB1 resistance	$mfb_{tgfb_{resistance}}$	0.416	Bibliography	[80]

4.4 Results

4.4.1 Static Analysis

Before sampling the trajectories of the model, we use the static analyzer KaSa Boutillier et al. (2018a) so as to check the structural invariants. The objective is two-fold. Firstly, the analyzer may detect some rules that will never be applied in the model. If so, this would mean that some parts are missing and that the model should be completed accordingly. It may also be due to some typos that should be corrected. Secondly, we want to check whether the intended relationships among the state of interactions sites in the different cell conformations hold effectively.

The analysis takes about 0.08 seconds on a 2.3 GHz Intel Core i9 8 cores MacBook Pro. The analysis detects no unapplicable rule and infers the structural invariants that are shown in Figure 4.25. These invariants take the form of some refinement lemmas. They are written as logical implications. The left-hand side is made of a pattern that specifies some conditions about the conformation of a cell. The right-hand side completes this pattern with some additional constraints. These constraints are necessarily satisfied in every occurrence of the left-hand side pattern in a state that the system may take during a potential execution. They take the form of sites that are decorated with an exhaustive list of the states that they may take.

In particular, the analysis detects and proves the following properties about the different stages of the different forms of cells.

- qHSCs may be only in the stage *null* (Figure 4.25(a));
- aHSCs may be only in the stages *stage0*, *stage1*, and *complete* (Figure 4.25(b));
- iHSCs may be only in the stages *null* and *stage1* (Figure 4.25(c));
- Reactivated hepatic stellate cells may be only in the stages *stage0* and *complete* (Figure 4.25(d));
- Myofibroblasts in their initial form may be only in the stages *stage0*, *stage1*, and *complete* (Figure 4.25(e));
- Myofibroblasts on the way to apoptosis may be only in the stage *stage1* and their receptors may not be bound (Figure 4.25(f));
- Reactivated myofibrolasts may be only in the stages *stage0* and *complete* (Figure 4.25(g)).

The analysis also discovers that the receptors that are either missing (*null*), or on the membrane (*membrane*) of the cells, are necessarily free both in the case of HSCs (Figure 4.25(h)) and myofibroblats (Figure 4.25(i)).

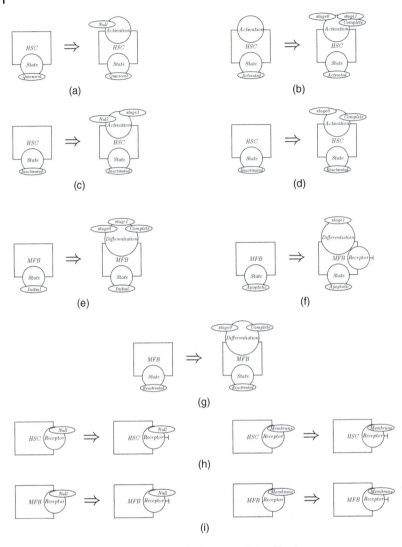

Figure 4.25 The result of static analysis. For each implication, every occurrence of the pattern on the left hand side of an implication in a reachable state necessarily satisfies the conditions described in the right hand side. (a) Quiescent hepatic stellate cells. (b) Activated hepatic stellate cells. (c) Inactivated hepatic stellate cells. (d) Reactivated hepatic stellate. (e) Initial myofibroblasts. (f) Apoptosic myofibroblasts. (g) Reactivated myofibroblasts. (h) Receptors of hepatic stellate cells. (i) Receptors of myofibroblasts.

4.4.2 Underlying Reaction Network

The set of rules may be compiled into a reaction network or equivalently into a system of ordinary differential equations thanks to the tool KaDe Camporesi et al. (2017). This computation takes about 0.04 seconds on a 2.3 GHz Intel Core i9 8 cores MacBook Pro. The resulting system involves 56 kinds of complexes.

4.4.3 Simulations

Simulating our model provides the results given in Figure 4.26. In Figure 4.26(a), we mimic an acute inflammatory aggression by one

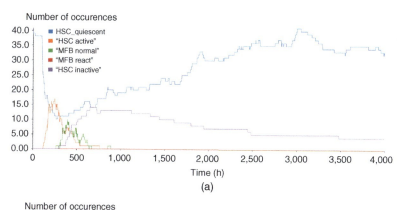

(a)

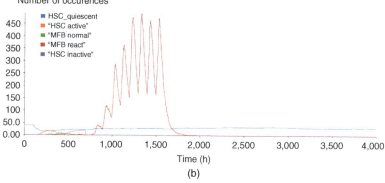

(b)

Figure 4.26 Simulations of the model. Curves represent the time evolution of the number of occurrences of each form of cells in response to TGFB1 inputs. Time is expressed in hours. (a) One input of TGFB1 at time 100. (b) There are 11 inputs of TGFB1, first at time 100 then starting from time 700, each 100 hours there will be an input of TGFB till getting 10 inputs. (a) Response to an acute inflammatory aggression. (b) Response to chronic inflammation.

input of TGFB1. This is indeed the regime under which we have calibrated our model thanks to the information available in the literature. In Figure 4.26(b), chronic inflammation was created by multiple inputs of TGFB1 (10 times here), and the expected behavior was successfully reproduced by our model. Each simulation takes around 14 seconds of CPU to simulate 4000 hours of biological time on a 2.3 GHz Intel Core i9 8 cores MacBook Pro. These curves show the evolution of the number of occurrences of each form of cells with respect to time (in hours). To parameterize the model, we choose to work in a volume corresponding to 10 Disse Spaces, containing among 40 quiescent HSC (calculated from Ehrlich et al. (2019)). Adding TGFB1 to the model at time 100 initializes the activation process, transforming quiescent HSC (qHSC, blue) toward activated HSC (aHSC, orange). After that, the newly activated HSC enter in differentiation process leading to the formation of MFB (green). When TGFB1 is completely consumed, MFB are either eliminated by apoptosis/senescence or inactivated (Figure 4.26(a)). The inactivation of MFB leads to the formation of inactivated HSC (iHSC, purple). The iHSC are either eliminated by apoptosis/senescence or reverse toward quiescent HSC. Upon a new stimulation by TGFB1, iHSC can also reverse toward reactivated HSC first, then in reactivated MFB (red). Iterative inputs of TGFB1 in the model quickly favor the accumulation of reactivated myofibroblasts, the inactivation and apoptosis/senescence pathways being surpassed by MFB proliferation and reactivation. Moreover, TGFB1 iterative inputs induce a population switch with diminution of initial cells and augmentation of reactivated cells. After few inputs of TGFB1, some inactivated HSC remain, but after more inputs, they completely disappear, leading to the saturation of the environment by reactivated MFB Figure 4.26(b). The timing for iterative inputs of TGFB1 is crucial. When too close, only MFP and HSC proliferation is observed. When too separate, much inactivated HSC are eliminated, leading to decrease the inflammatory answer.

The primary results are encouraging, showing the dynamic of HSCs in function of the TGFB1 inputs. Our model successfully describes the behavior of HSCs shown in Section 4.3.1 and respects the time link to this dynamic; activation process takes around 168 hours, differentiation around 336 hours, and inactivation around 664 hours Bachem et al. (1993), El Taghdouini et al. (2015). However, some parts of the model need to be reviewed. Firstly, the proliferation process is one of our main problems. Cell proliferation is lower than expected, for example, the number of activated HSC should be 3-fold higher than the number of cells undergoing the activation process. It could come from a part that is missing to explain the dynamics of the model (as discussed in conclusion Section 4.5). It could also be solved by considering more intermediary stages in Erlang time distributions (which would lower the standard deviation of their overall process durations, hence favoring cell synchronization and amplifying the amplitude of their abundance peaks). Also, reactivated cell elimination should be homogenized with the initial MFB apoptosis/senescence. Last but not least, the model is quite rigid. Some encoding artifacts have been used. For instance, controlling HSC activation by a factor should be removed and the rules firing should control it without any help. Globally, this model needs to be perfected, but the primary results are promising.

4.5 Conclusion

This chapter was devoted to the rule-based language Kappa. As a realistic case study, we developed a model for the activation of HSCs by the protein TGFB1. We have used knowledge from the literature to calibrate the model according to its expected response to a single acute inflammatory aggression, and we were able to reproduce the expected behavior in the case of chronic inflammation (up to minor differences in response amplitudes). Beyond the benefit of formalizing executable models, Kappa offers a convenient syntax, close to biochemistry, which eases the modeling process, the potential updates of the model, and the documentation of it. In Section 4.3, the description of the Kappa interactions provides a practical road map to navigate among the different elements of the model.

Yet modeling requires a constant search for the most adequate abstraction trade-off. Models may be made arbitrarily precise when detailed information about biochemical mechanisms is available in the literature and when this amount of details does not overcome the *in silico* computational resources. In this model, we favored simplicity while sticking to the experimental observations. We aim to investigate further different parts of the model and include more details gradually. During this, we will check that the overall behavior of the model is preserved and explore the impact of these updates in specific regimes. Rule-centric approaches make this empiric modeling approach easier.

More specifically, we plan to investigate further on two parts of the model. The first one is about the receptors of cells. With a view to simplifying, we have abstracted all the receptors of each occurrence of cells, by a single Kappa site, called *receptor*. Moreover, each agent *TGFB1* indeed stands for a pack of occurrences of the protein TGFB1. The main motivation is to spare computation time. The impact on the model is that all the receptors of an occurrence of a cell are considered to be in the same state. This abstraction could be refined by modeling the receptors of each cell and each occurrence of the TGFB1 protein individually. Yet, the cost may become prohibitive both with respect to memory (each occurrence of the TGFB1 protein would be described explicitly) and to computation time (binding between occurrences of the TGFB1 protein and their receptors would occur at a very fast scale and the simulation would have to account for very frequent instances of this event). The issue with memory could be solved easily by means of counters. The total number of occurrences of free TGFB1 could be modeled as a numerical value (called "token" in Kappa), and the number of receptors in each of the four different states (*null*, *membrane*, *intern* and free, and *intern* and bound) should be described by means of some counters, as enabled by a recent extension of Kappa Boutillier et al. (2019).

The second part concerns the memory of cells. As we have seen, activated and reactivated hepatic stellate cells exhibit a different behavior. This means that these cells have a memory and that the history of each cell impacts its further behavior. In the current version, this is modeled by introducing different forms of cells and providing each form with different capabilities of interaction and rates. This modeling directly operates at the level of the phenotypes of the cells. It would be interesting to better understand where these different behaviors come from by modeling an abstraction of the protein content of the cells. This way, the behavior of each cell would directly emerge from its protein content. Because of the difference of time scales and the lack of details in the literature, the protein content cannot be modeled precisely. Instead, it could be described by equipping each cell with some abstract counters Boutillier et al. (2019). The value of these counters could increase upon activation and decrease during cool-down phases, while influencing the capabilities of interaction of the cell.

In Kappa, basic elements are interactions. Their rates provide information about their time distribution independently from the rest of the system. In the literature, information about reaction durations takes different forms. It can be specified as the time to complete a given ratio of a given process or defined more phenotypically by the time period between peaks of concentrations. In Kappa, only exponential time distributions can be assigned to an interaction. This means that the average time of a reaction and its standard deviation are fully entangled. More diverse time distributions may emerge as a result of sequential composition of reactions. For instance, Erlang time distributions Erlang (1909) may be obtained by modeling a process as the sequential composition of k intermediary steps. Intermediary steps reduce the time variability of the full process and eventually lead to a fixed duration. Yet, this comes with a computational overhead. From a parameterization point of view, rates of intermediary steps are more difficult to guess and must be data-fitted.

One critical point of modeling TGFB1-dependent activation of HSC is the identification of the parameters because of the lack of quantitative values. To overcome this issue, we merged information about HSC activation from both *in vivo* and *in vitro* experiments; however, the difference in dynamics of activation/regulation between these two approaches has been already widely documented such as for gene expression De Minicis et al. (2007). Indeed, seeding quiescent HSC on plastic dishes induces HSC activation through molecular mechanisms that differ in part from mechanisms within liver tissue because of the presence of other cell types and microenvironments. An evolution of the model might be to integrate other cellular components and the biomechanical constraints that play a critical role in activation of HSC and in regulating myofibroblasts phenotype Wells (2005), Olsen et al. (2011). Importantly, the phenotype of activated HSC is more complex than initially reported and the recent development of single cells RNAseq analyses allows now to demonstrate the heterogeneity of activated HSC Krenkel et al. (2019). We need to get more information on phenotyping the different HSC species, including the MFB states in order to better characterize the different reversibility pathways and to understand what contributes to the disequilibrium toward the disease progression.

While the biomechanical aspect is generally discarded in modeling cell activation differentiation and cell signaling, modeling biological processes needs to take into account the physical constraints occurring *in situ*. Obviously, extracellular matrix is the paradigm of such constraints and is the major regulator of cell responses Muncie and Weaver (2018), Chaudhuri et al. (2020). Extracellular matrix is not an inert material supporting cells within the tissue but a plastic and complex network associating insoluble molecules such as collagen mostly arranged as supramolecular assemblies, glycoproteins such as fibronectin, and proetoglycans that consist in polypeptide backbone decorated by glycoaminoglycans that confer viscoelasticity and hydrophilic properties. This core matrisome characterized by Hynes and Naba (2011) comprises among 300 proteins and is associated with ECM-affiliated proteins, ECM regulators, and secreted factors Shao et al. (2019). ECM composition and mechanical properties are specific of tissue and change during physiopathological processes such as development, inflammation, wound-healing, fibrosis, and cancer. In the liver, the matrisome analyses were recently reviewed in Arteel and Naba (2020) and showed the incomplete and heterogeneous description of this network. Because of the impossibility to catch a spatial and evolutionary view of this network, models based on differential equations searched for driving molecules of the network behavior such as the core matrisome proteins, fibrin, and collagen Malandrino et al. (2017), and the ECM regulators Friedman and Hao (2017), Jin et al. (2011) in order to reduce the complexity.

Using Kappa language allows for the creation of an agent ECM that could interact with cellular and molecular agents. Different states could be attributed to ECM agent such as low, intermediate, and high stiffness that control in turn the activation of HSC. While the relationship between ECM stiffness and HSC activation is known since a long time mainly by using 2D cell culture Wells (2005), Olsen et al. (2011), the quantitative evaluation of each molecule implication in stiffness remains to be clarified. Of course, the role of several molecules in liver stiffness during fibrosis has been characterized and lysyl oxidases (LOX) that catalyze cross-linking of collagen and elastin are some of these major actors Saneyasu et al. (2016). The development of 3D-multicellular hepatic models and microfluidic organ-on-a-chip liver models might be useful to get quantitative data about the contribution of molecular agents in stiffness and activation of HSC Moradi et al. (2020), Cuvellier et al. (2021). In line with this, our future challenge aims to integrate the Kappa model for extracellular matrix-dependent TGFB1 activation that we recently developed Boutillier et al. (2018a). TGFB1 is synthesized as a latent form (LAP-TGFB1) associated with the latent TGFB1 binding protein (LTBP1) that sequesters it within the extracellular matrix networks. The release of active TGFB1 depends on enzymatic activities but above all on mechanical strengths involving matrix components and membrane receptors Hinz (2015). As a regulatory loop, the activated HSC synthesize extracellular matrix components including TGFB1 and are involved in regulation of matrix plasticity, thereby affecting TGFB1 activation. Integrating matrix components implicated in TGFB1-dependent HSC activation might improve the present model and allow us to identify new regulators of the equilibrium between repair and fibrosis.

References

Jakob L. Andersen, Christoph Flamm, Daniel Merkle, and Peter F. Stadler. A software package for chemically inspired graph transformation. In Rachid Echahed and Mark Minas, editors, *Graph Transformation - 9th International Conference, ICGT 2016, in Memory of Hartmut Ehrig, Held as Part of STAF 2016, Vienna, Austria, July 5–6, 2016, Proceedings,* volume 9761 of *Lecture Notes in Computer Science*, pages 73–88. Springer, 2016. URL https://doi.org/10.1007/978-3-319-40530-8_5.

Oana Andrei and Hélène Kirchner. A rewriting calculus for multigraphs with ports. *Electronic Notes in Theoretical Computer Science*, 219: 67–82, 2008. doi: https://doi.org/10.1016/j.entcs.2008.10.035.

Gavin Arteel and Alexandra Naba. The liver matrisome, looking beyond collagens. *JHEP Reports*, 2: 100115, 2020. doi: https://doi.org/10.1016/j.jhepr.2020.100115.

M. G. Bachem, D. Meyer, W. Schäfer, U. Riess, R. Melchior, K. M. Sell, and A. M. Gressner. The response of rat liver perisinusoidal lipocytes to polypeptide growth regulator changes with their transdifferentiation into myofibroblast-like cells in culture. *Journal of Hepatology*, 18: 40–52, 1993.

Nicolas Behr and Jean Krivine. Compositionality of rewriting rules with conditions. *CoRR*, abs/1904.09322, 2019. URL http://arxiv.org/abs/1904.09322.

Andreea Beica, Jérôme Feret, and Tatjana Petrov. Tropical abstraction of biochemical reaction networks with guarantees. In *Proceedings of SASB 2018, the 9th International Workshop on Static Analysis and Systems Biology, Freiburg, Germany - August 28th, 2018*, volume 350 of *Electronic Notes in Theoretical Computer Science*, pages 3–32. Elsevier, 2020. doi: https://doi.org/10.1016/j.entcs.2020.06.002.

Michael Blinov, James R. Faeder, Byron Goldstein, and William S. Hlavacek. BioNetGen: software for rule-based modeling of signal transduction based on the interactions of molecular domains. *Bioinformatics (Oxford, England)*, 20 (17): 3289–3291, November 2004.

Pierre Boutillier. The Kappa simulator made interactive. In Luca Bortolussi and Guido Sanguinetti, editors, *Computational Methods in Systems Biology - 17th International Conference, CMSB 2019, Trieste, Italy, September 18–20, 2019, Proceedings*, volume 11773 of *Lecture Notes in Computer Science*, pages 296–301. Springer, 2019. doi: https://doi.org/10.1007/978-3-030-31304-3_16.

Pierre Boutillier, Thomas Ehrhard, and Jean Krivine. Incremental update for graph rewriting. In Hongseok Yang, editor, *Programming Languages and Systems - 26th European Symposium on Programming, ESOP 2017, Held as Part of the European Joint Conferences on Theory and Practice of Software, ETAPS 2017, Uppsala, Sweden, April 22–29, 2017, Proceedings*, volume 10201 of *Lecture Notes in Computer Science*, pages 201–228. Springer, 2017. URL https://doi.org/10.1007/978-3-662-54434-1_8.

Pierre Boutillier, Ferdinanda Camporesi, Jean Coquet, Jérôme Feret, Kim Quyên Lý, Nathalie Théret, and Pierre Vignet. KaSa: A static analyzer for Kappa. In Milan Ceska and David Safránek, editors, *Computational Methods in Systems Biology - 16th International Conference, CMSB 2018, Brno, Czech Republic, September 12–14, 2018, Proceedings*, volume 11095 of *Lecture Notes* in *Computer Science*, pages 285–291. Springer, 2018a. URL https://doi.org/10.1007/978-3-319-99429-1_17.

Pierre Boutillier, Mutaamba Maasha, Xing Li, Héctor F. Medina-Abarca, Jean Krivine, Jérôme Feret, Ioana Cristescu, Angus G. Forbes, and Walter Fontana. The Kappa platform for rule-based modeling. *Bioinformatics*, 34 (13): i583–i592, 2018b. URL https://doi.org/10.1093/bioinformatics/bty272.

Pierre Boutillier, Ioana Cristescu, and Jérôme Feret. Counters in Kappa: semantics, simulation, and static analysis. In Luís Caires, editor, *Programming Languages and Systems - 28th European Symposium on Programming, ESOP 2019, Held as Part of the European Joint Conferences on Theory and Practice of Software, ETAPS 2019, Prague, Czech Republic, April 6–11, 2019, Proceedings*, volume 11423 of *Lecture Notes in Computer Science*, pages 176–204. Springer, 2019. doi: https://doi.org/10.1007/978-3-030-17184-1_7.

Peter Byass. The global burden of liver disease: a challenge for methods and for public health. *BMC Medicine*, 12: 159, 2014. doi: https://doi.org/10.1186/s12916-014-0159-5.

calculator.net, 2008–2021. URL https://www.calculator.net/half-life-calculator.htm.

Ferdinanda Camporesi, Jérôme Feret, and Jonathan Hayman. Context-sensitive flow analyses: a hierarchy of model reductions. In Ashutosh Gupta and Thomas A. Henzinger, editors, *Computational Methods in Systems Biology - 11th International Conference, CMSB 2013, Klosterneuburg, Austria, September 22-24, 2013. Proceedings*, volume 8130 of *Lecture Notes in Computer Science*, pages 220–233. Springer, 2013. doi: https://doi.org/10.1007/978-3-642-40708-6_17.

Ferdinanda Camporesi, Jérôme Feret, and Kim Quyên Lý. KaDe: A tool to compile Kappa rules into (reduced) ODE models. In Jérôme Feret and Heinz Koeppl, editors, *Computational Methods in Systems Biology - 15th International Conference, CMSB 2017, Darmstadt, Germany, September 27–29, 2017, Proceedings*, volume 10545 of *Lecture Notes in Computer Science*, pages 291–299. Springer, 2017. doi: https://doi.org/10.1007/978-3-319-67471-1_18.

Luca Cardelli. Brane calculi. In Vincent Danos and Vincent Schächter, editors, *Computational Methods in Systems Biology, International Conference, CMSB 2004, Paris, France, May 26–28, 2004, Revised Selected Papers*, volume 3082 of *Lecture Notes in Computer Science*, pages 257–278. Springer, 2004. URL https://doi.org/10.1007/978-3-540-25974-9_24.

Luca Cardelli and Andrew D. Gordon. Mobile ambients. In Maurice Nivat, editor, *Foundations of Software Science and Computation Structure, 1st*

International Conference, FoSSaCS'98, Held as Part of the European Joint Conferences on the Theory and Practice of Software, ETAPS'98, Lisbon, Portugal, March 28 - April 4, 1998, Proceedings, volume 1378 of *Lecture Notes in Computer Science*, pages 140–155. Springer, 1998. URL https://doi.org/10.1007/BFb0053547.

Luca Cardelli and Andrew D. Gordon. Mobile ambients. *Theoretical Computer Science*, 240 (1): 177–213, 2000. doi: https://doi.org/10.1016/S0304-3975(99)00231-5.

Ovijit Chaudhuri, J. Cooper-White, P. Janmey, D. Mooney, and V. Shenoy. Effects of extracellular matrix viscoelasticity on cellular behaviour. *Nature*, 584: 535–546, 2020.

Federica Ciocchetta and Jane Hillston. Bio-PEPA: A framework for the modelling and analysis of biological systems. *Theoretical Computer Science*, 410 (33–34): 3065–3084, 2009. Concurrent Systems Biology: To Nadia Busi (1968–2007).

Andrea Corradini, Ugo Montanari, Francesca Rossi, Hartmut Ehrig, Reiko Heckel, and Michael Löwe. Algebraic approaches to graph transformation - Part I: basic concepts and double pushout approach. In Grzegorz Rozenberg, editor, *Handbook of Graph Grammars and Computing by Graph Transformations, Volume 1: Foundations*, pages 163–246. World Scientific, 1997.

Andrea Corradini, Tobias Heindel, Frank Hermann, and Barbara König. Sesqui-pushout rewriting. In Andrea Corradini, Hartmut Ehrig, Ugo Montanari, Leila Ribeiro, and Grzegorz Rozenberg, editors, *Graph Transformations, 3rd International Conference, ICGT 2006, Natal, Rio Grande do Norte, Brazil, September 17–23, 2006, Proceedings*, volume 4178 of *Lecture Notes in Computer Science*, pages 30–45. Springer, 2006. URL https://doi.org/10.1007/11841883_4.

Patrick Cousot and Radhia Cousot. Abstract interpretation: a unified lattice model for static analysis of programs by construction or approximation of fixpoints. In Robert M. Graham, Michael A. Harrison, and Ravi Sethi, editors, *Conference Record of the 4th ACM Symposium on Principles of Programming Languages, Los Angeles, California, USA, January 1977*, pages 238–252. ACM, 1977. URL https://doi.org/10.1145/512950.512973.

Marie Cuvellier, Frédéric Ezan, Hugo Oliveira, Sophie Rose, Jean-Christophe Fricain, Sophie Langouët, Vincent Legagneux, and Georges Baffet. 3D culture of HepaRG cells in GelMa and its application to bioprinting of a multicellular hepatic model. *Biomaterials*, 269: 120611, 2021. ISSN 0142-9612. doi: https://doi.org/10.1016/j.biomaterials.2020.120611. URL https://www.sciencedirect.com/science/article/pii/S0142961220308577.

Troels Christoffer Damgaard, Espen Højsgaard, and Jean Krivine. Formal cellular machinery. *Electronic Notes in Theoretical Computer Science*, 284: 55–74, 2012. doi: https://doi.org/10.1016/j.entcs.2012.05.015.

Vincent Danos and Cosimo Laneve. Formal molecular biology. *Theoretical Computer Science*, 325 (1): 69–110, 2004. ISSN 0304-3975. doi: https://doi.org/http://dx.doi.org/10.1016/j.tcs.2004.03.065. URL http://www.sciencedirect.com/science/article/pii/S0304397504002336. Computational Systems Biology.

Vincent Danos and Sylvain Pradalier. Projective brane calculus. In Vincent Danos and Vincent Schächter, editors, *Computational Methods in Systems Biology, International Conference, CMSB 2004, Paris, France, May 26–28, 2004, Revised Selected Papers*, volume 3082 of *Lecture Notes in Computer Science*, pages 134–148. Springer, 2004. URL https://doi.org/10.1007/978-3-540-25974-9_11.

Vincent Danos, Jérôme Feret, Walter Fontana, and Jean Krivine. Scalable simulation of cellular signaling networks. In Zhong Shao, editor, *Programming Languages and Systems, 5th Asian Symposium, APLAS 2007, Singapore, November 29-December 1, 2007, Proceedings*, volume 4807 of *Lecture Notes in Computer Science*, pages 139–157. Springer, 2007. URL https://doi.org/10.1007/978-3-540-76637-7_10.

Vincent Danos, Jérôme Feret, Walter Fontana, and Jean Krivine. Abstract interpretation of cellular signaling networks. In Francesco Logozzo, Doron A. Peled, and Lenore D. Zuck, editors, *Verification, Model Checking, and Abstract Interpretation, 9th International Conference, VMCAI 2008, San Francisco, USA, January 7–9, 2008, Proceedings*, volume 4905 of *Lecture Notes in Computer Science*, pages 83–97. Springer, 2008. URL https://doi.org/10.1007/978-3-540-78163-9.

Vincent Danos, Jérôme Feret, Walter Fontana, Russell Harmer, and Jean Krivine. Abstracting the differential semantics of rule-based models: exact and automated model reduction. In *Proceedings of the 25th Annual IEEE Symposium on Logic in Computer Science, LICS 2010, 11–14 July 2010, Edinburgh, United Kingdom*, pages 362–381. IEEE Computer Society, 2010. ISBN 978-0-7695-4114-3. URL https://doi.org/10.1109/LICS.2010.44.

Vincent Danos, Jérôme Feret, Walter Fontana, Russell Harmer, Jonathan Hayman, Jean Krivine, Christopher D. Thompson-Walsh, and Glynn Winskel. Graphs, rewriting and pathway reconstruction for rule-based models. In Deepak D'Souza, Telikepalli Kavitha, and Jaikumar Radhakrishnan, editors, *IARCS Annual Conference on Foundations of Software Technology and Theoretical Computer Science, FSTTCS 2012, December 15–17, 2012, Hyderabad, India*, volume 18 of *LIPIcs*, pages 276–288. Schloss Dagstuhl - Leibniz-Zentrum fuer Informatik, 2012. URL https://doi.org/10.4230/LIPIcs.FSTTCS.2012.276.

Vincent Danos, Ricardo Honorato-Zimmer, Sebastian Jaramillo-Riveri, and Sandro Stucki. Rigid geometric constraints for Kappa models. In *SASB'12: Post Proceedings of the 3rd International Workshop on Static Analysis and Systems Biology*, volume 313 of *ENTCS*, pages 23–46. Elsevier, 2015.

Samuele De Minicis, Ekihiro Seki, Hiroshi Uchinami, Johannes Kluwe, Yonghui Zhang, David Brenner, and Robert Schwabe. Gene expression profiles during hepatic stellate cell activation in culture and in vivo. *Gastroenterology*, 132: 1937–1946, 06 2007. doi: https://doi.org/10.1053/j.gastro.2007.02.033.

Brenda de Oliveira da Silva, Letícia Ferrreira Ramos, and Karen C.M. Moraes. Molecular interplays in hepatic stellate cells: apoptosis, senescence, and phenotype reversion as cellular connections that modulate liver fibrosis. *Cell Biology International*, 41 (9): 946–959, 2017.

Lorenzo Dematté, Corrado Priami, and Alessandro Romanel. The BlenX language: a tutorial. In Marco Bernardo, Pierpaolo Degano, and Gianluigi Zavattaro, editors, *Formal Methods for Computational Systems Biology: 8th International School on Formal Methods for the Design of Computer, Communication, and Software Systems, SFM 2008 Bertinoro, Italy, June 2–7, 2008 Advanced Lectures*, pages 313–365, Berlin, Heidelberg, 2008. Springer. URL http://dx.doi.org/10.1007/978-3-540-68894-5_9.

John W. Eaton, David Bateman, Soren Hauberg, and Rik Wehbring. *GNU Octave Version 4.0.0 Manual: A High-level Interactive Language for Numerical Computations*. Free Software Foundation, 2015. URL http://www.gnu.org/software/octave.

Avner Ehrlich, Daniel Duche, Gladys Ouedraogo, and Yaakov Nahmias. Challenges and opportunities in the design of liver-on-chip microdevices. *Annual Review of Biomedical Engineering*, 21 (1): 219–239, 2019. doi: https://doi.org/10.1146/annurev-bioeng-060418-052305. PMID: 31167098.

Adil El Taghdouini, Mustapha Najimi, Pau Sancho-Bru, Etienne Sokal, and Leo van Grunsven. In vitro reversion of activated primary human hepatic stellate cells. *Fibrogenesis and Tissue Repair*, 8: 2015. doi: https://doi.org/10.1186/s13069-015-0031-z.

Agner Krarup Erlang. *Sandsynlighedsregning og Telefonsamtaler*. 1909. URL https://books.google.fr/books?id=xBJZQwAACAAJ.

James R. Faeder, Michael L. Blinov, Byron Goldstein, and William S. Hlavacek. Rule-based modeling of biochemical networks. *Complexity*, 10 (4): 22–41, 2005. ISSN 1099-0526. doi: https://doi.org/10.1002/cplx.20074.

James R. Faeder, Michael L. Blinov, and William S. Hlavacek. Rule-based modeling of biochemical systems with BioNetGen. *Methods Mol Biol*, 500: 113–167, 2009.

Jérôme Feret. Gkappa. URL https://github.com/Kappa-Dev/GKappa.

Jérôme Feret. An algebraic approach for inferring and using symmetries in rule-based models. *Electronic Notes in Theoretical Computer Science*, 316: 45–65, 2015. doi: https://doi.org/10.1016/j.entcs.2015.06.010.

Jérôme Feret and Kim Quyên Lý. Reachability analysis via orthogonal sets of patterns. *Electronic Notes in Theoretical Computer Science*, 335: 27–48, 2018. doi: https://doi.org/10.1016/j.entcs.2018.03.007.

Jérôme Feret, Vincent Danos, Jean Krivine, Russ Harmer, and Walter Fontana. Internal coarse-graining of molecular systems. *Proceedings of the National Academy of Sciences of the United States of America*, 106 (16): 6453–6458. doi: https://doi.org/10.1073/pnas.0809908106 2009.

Jérôme Feret, Heinz Koeppl, and Tatjana Petrov. Stochastic fragments: a framework for the exact reduction of the stochastic semantics of rule-based models. *International Journal of Software and Informatics*, 7 (4): 527–604, 2013. URL http://www.ijsi.org/ch/reader/view_abstract.aspx?file_no=i173.

Avner Friedman and Wenrui Hao. Mathematical modeling of liver fibrosis. *Mathematical Biosciences & Engineering*, 14 (1): 143–164, 2017.

Scott L. Friedman, Glenn Yamasaki, and Linda Wong. Modulation of transforming growth factor beta receptors of rat lipocytes during the hepatic wound healing response. Enhanced binding and reduced gene expression accompany cellular activation in culture and in vivo. *Journal of Biological Chemistry*, 269 (14): 10551–10558, 1994.

Daniel T. Gillespie. Exact stochastic simulation of coupled chemical reactions. *The Journal of Physical Chemistry*, 81 (25): 2340–2361, 1977.

Russ Harmer, Vincent Danos, Jérôme Feret, Jean Krivine, and Walter Fontana. Intrinsic information carriers in combinatorial dynamical systems. *Chaos*, 20, September 2010. URL http://link.aip.org/link/CHAOEH/v20/i3/p037108/s1.

Dieter Häussinger and Claus Kordes. Space of Disse: a stem cell niche in the liver. *Biological Chemistry*, 401 (1): 81–95, 2019.

Monika Heiner and Ina Koch. Petri net based model validation in systems biology. In Jordi Cortadella and Wolfgang Reisig, editors, *Applications and Theory of Petri Nets 2004: 25th International Conference, ICATPN 2004, Bologna, Italy, June 21–25, 2004. Proceedings*, pages 216–237. Berlin Heidelberg, Springer, 2004. URL http://dx.doi.org/10.1007/978-3-540-27793-4_13.

Tobias Helms, Tom Warnke, Carsten Maus, and Adelinde M. Uhrmacher. Semantics and efficient simulation algorithms of an expressive multilevel modeling language. *ACM Transactions on Modeling and Computer Simulation*, 27 (2): 8:1–8:25, 2017. doi: https://doi.org/10.1145/2998499.

Boris Hinz. The extracellular matrix and transforming growth factor-β1: tale of a strained relationship. *Matrix Biology*, 47: 54–65, 2015. ISSN 0945-053X. doi:

https://doi.org/10.1016/j.matbio.2015.05.006. URL https://www.sciencedirect.com/science/article/pii/S0945053X15001055.

Michael Hucka, Frank T. Bergmann, Stefan Hoops, Sarah M. Keating, Sven Sahle, James C. Schaff, Lucian P. Smith, and Darren J. Wilkinson. The systems biology markup language (SBML): language specification for level 3 version 1 core, 2010.

Richard Hynes and Alexandra Naba. Overview of the matrisome–an inventory of extracellular matrix constituents and functions. *Cold Spring Harbor Perspectives in Biology*, 4: a004903, 2011. doi: https://doi.org/10.1101/cshperspect.a004903.

Yu-Fang Jin, Hai-Chao Han, Jamie Berger, Qiuxia Dai, and Merry Lindsey. Combining experimental and mathematical modeling to reveal mechanisms of macrophage-dependent left ventricle remodeling. *BMC Systems Biology*, 5: 60, 2011. doi: https://doi.org/10.1186/1752-0509-5-60.

Mathias John, Cédric Lhoussaine, Joachim Niehren, and Cristian Versari. Biochemical reaction rules with constraints. In *Programming Languages and Systems - 20th European Symposium on Programming, ESOP 2011, Held as Part of the Joint European Conferences on Theory and Practice of Software, ETAPS 2011, Saarbrücken, Germany, March 26-April 3, 2011. Proceedings*, volume 6602 of *Lecture Notes in Computer Science*, pages 338–357. Springer, 2011. URL http://dx.doi.org/10.1007/978-3-642-19718-5.

Ozan Kahramanogullari and Luca Cardelli. An intuitive modelling interface for systems biology. *International Journal of Software and Informatics*, 7 (4): 655–674, 2013. URL http://www.ijsi.org/ch/reader/view_abstract.aspx?file_no=i177.

Tatiana Kisseleva and David Brenner. Molecular and cellular mechanisms of liver fibrosis and its regression. *Nature Reviews Gastroenterology & Hepatology*, 18 (3): 151–166, March 2021. ISSN 1759-5045. doi: https://doi.org/10.1038/s41575-020-00372-7.

Tatiana Kisseleva, Min Cong, YongHan Paik, David Scholten, Chunyan Jiang, Chris Benner, Keiko Iwaisako, Thomas Moore-Morris, Brian Scott, Hidekazu Tsukamoto, Sylvia M. Evans, Wolfgang Dillmann, Christopher K. Glass, and David A. Brenner. Myofibroblasts revert to an inactive phenotype during regression of liver fibrosis. *Proceedings of the National Academy of Sciences of the United States of America*, 109 (24): 9448–9453, 2012. ISSN 0027-8424. doi: https://doi.org/10.1073/pnas.1201840109. URL https://www.pnas.org/content/109/24/9448.

Agnes Köhler, Jean Krivine, and Jakob Vidmar. A rule-based model of base excision repair. In Pedro Mendes, Joseph O. Dada, and Kieran Smallbone, editors, *Computational Methods in Systems Biology - 12th International Conference, CMSB 2014, Manchester, UK, November 17–19, 2014, Proceedings,* volume 8859 of *Lecture Notes in Computer Science*, pages 173–195. Springer, 2014. URL https://doi.org/10.1007/978-3-319-12982-2_13.

Oliver Krenkel, Jana Hundertmark, Thomas Ritz, Ralf Weiskirchen, and Frank Tacke. Single cell RNA sequencing identifies subsets of hepatic stellate cells and myofibroblastsin liver fibrosis. *Cells*, 8: 503, 2019. doi: https://doi.org/10.3390/cells8050503.

Marta Z. Kwiatkowska, Gethin Norman, and David Parker. PRISM 4.0: Verification of probabilistic real-time systems. In Ganesh Gopalakrishnan and Shaz Qadeer, editors, *Computer Aided Verification - 23rd International Conference, CAV 2011, Snowbird, UT, USA, July 14–20, 2011. Proceedings*, volume 6806 of *Lecture Notes in Computer Science*, pages 585–591. Springer, 2011. URL https://doi.org/10.1007/978-3-642-22110-1_47.

Xiao Liu, Jun Xu, David Brenner, and Tatiana Kisseleva. Reversibility of liver fibrosis and inactivation of fibrogenic myofibroblasts. *Current Pathobiology Reports*, 1: 209–214, 2013. doi: https://doi.org/10.1007/s40139-013-0018-7.

Michael Löwe. Algebraic approach to single-pushout graph transformation. *Theoretical Computer Science*, 109 (1&2): 181–224, 1993. doi: >https://doi.org/10.1016/0304-3975(93)90068-5.

Andrea Malandrino, Michael Mak, Xavier Trepat, and Roger D. Kamm. Non-elastic remodeling of the 3D extracellular matrix by cell-generated forces. *bioRxiv*, 2017. doi: https://doi.org/10.1101/193458. URL https://www.biorxiv.org/content/early/2017/09/27/193458.

Joan Massague and Betsy Like. Cellular receptors for type beta transforming growth factor. Ligand binding and affinity labeling in human and rodent cell lines. *Journal of Biological Chemistry*, 260 (5): 2636–2645, 1985.

MATLAB. *version 9.2*. The MathWorks Inc., Natick, Massachusetts, 2017.

Michael B. Monagan, Keith O. Geddes, K. Michael Heal, George Labahn, Stefan M. Vorkoetter, James McCarron, and Paul DeMarco. *Maple 10 Programming Guide*. Maplesoft, 2005.

Ehsanollah Moradi, Sasan Jalili-Firoozinezhad, and Mehran Solati-Hashjin. Microfluidic organ-on-a-chip models of human liver tissue. *Acta Biomaterialia*, 116: 67–83, 2020. ISSN 1742-7061. doi: https://doi.org/10.1016/j.actbio.2020.08.041. URL https://www.sciencedirect.com/science/article/pii/S1742706120305110.

Jonathon M. Muncie and Valerie M. Weaver. Chapter one - the physical and biochemical properties of the extracellular matrix regulate cell fate. In Eveline S. Litscher and Paul M. Wassarman, editors, *Extracellular Matrix and Egg Coats*, volume 130 of *Current Topics in Developmental Biology*, pages 1–37. Academic Press, 2018. doi: https://doi.org/10.1016/bs.ctdb.2018.02.002. URL https://www.sciencedirect.com/science/article/pii/S0070215318300346.

Abby Olsen, Steven Bloomer, Erick Chan, Marianna Gaca, Penelope Georges, Bridget Sackey, Masayuki Uemura, Paul Janmey, and Rebecca Wells. Hepatic stellate cells require a stiff environment for myofibroblastic differentiation. *American Journal of Physiology-Gastrointestinal and Liver Physiology*, 301: G110–G118, 2011. doi: https://doi.org/10.1152/ajpgi.00412.2010.

Tatjana Petrov, Jérôme Feret, and Heinz Koeppl. Reconstructing species-based dynamics from reduced stochastic rule-based models. In Oliver Rose and Adelinde M. Uhrmacher, editors, *Winter Simulation Conference, WSC '12, Berlin, Germany, December 9–12, 2012*, pages 225:1–225:15. WSC, 2012. ISBN 978-1-4673-4779-2. URL https://doi.org/10.1109/WSC.2012.6465241.

Mark F. Pittenger, Dennis E. Discher, Bruno M. Péaultand Donald G. Phinney, Joshua M. Hare, and Arnold I. Caplan. Mesenchymal stem cell perspective: cell biology to clinical progress. *npj Regenerative Medicine*, 4: 22. URL https://www.nature.com/articles/s41536-019-0083-6. 2019.

Aviv Regev, William Silverman, and Ehud Shapiro. Representation and simulation of biochemical processes using the pi-calculus process algebra. In R. B. Altman, A. K. Dunker, L. Hunter, and T. E. Klein, editors, *Pacific Symposium on Biocomputing, Volume 6*, pages 459–470, Singapore, 2001.

Aviv Regev, E. M. Panina, William Silverman, Luca Cardelli, and E. Y. Shapiro. Bioambients: an abstraction for biological compartments. *TCS*, 325 (1): 141–167, 2004.

Takaoki Saneyasu, Riaz Akhtar, and Takao Sakai. Molecular cues guiding matrix stiffness in liver fibrosis. *BioMed Research International*, 2016: 1–11, 2016. doi: https://doi.org/10.1155/2016/2646212.

Bernd Schnabl, Cynthia A. Bradham, Brydon L. Bennett, Anthony M. Manning, Branko Stefanovic, and David A. Brenner. TAK1/JNK and p38 have opposite effects on rat hepatic stellate cells. *Hepatology*, 34 (5): 953–963, 2001. ISSN 0270-9139. doi: https://doi.org/10.1053/jhep.2001.28790. URL https://www.sciencedirect.com/science/article/pii/S0270913901529845.

Xinhao Shao, Isra N. Taha, Karl R. Clauser, Yu (Tom) Gao, and Alexandra Naba. MatrisomeDB: the ECM-protein knowledge database. *Nucleic Acids Research*, 48 (D1): D1136–D1144, 2019. ISSN 0305-1048. doi: https://doi.org/10.1093/nar/gkz849.

Donald Stewart. *Spatial biomodelling*. Master thesis, School of Informatics, University of Edinburgh, 2010.

Takuma Tsuchida and Scott L. Friedman. Mechanisms of hepatic stellate cell activation. *Nature Reviews Gastroenterology & Hepatology*, 14 (7): 397–411, July 2017. ISSN 1759-5045. doi: https://doi.org/10.1038/nrgastro.2017.38.

Jose Vilar, Ronald Jansen, and Chris Sander. Signal processing in the TGF-β superfamily ligand-receptor network. *PLoS Computational Biology*, 2: e3, 2006. doi: https://doi.org/10.1371/journal.pcbi.0020003.

Laiage M. Wakefield, Thomas S. Winokur, Robin S. Hollands, Karen Christopherson, Arthur D. Levinson, Michael B. Sporn, et al. Recombinant latent transforming growth factor beta 1 has a longer plasma half-life in rats than active transforming growth factor beta 1, and a different tissue distribution. *The Journal of Clinical Investigation*, 86 (6): 1976–1984, 1990.

Rebecca Wells. The role of matrix stiffness in hepatic stellate cell activation and liver fibrosis. *Journal of Clinical Gastroenterology*, 39: S158–61, 2005. doi: https://doi.org/10.1097/01.mcg.0000155516.02468.0f.

Inc. Wolfram Research. *Mathematica*. Wolfram Research, Inc., 2017.

5

Pathway Logic: Curation and Analysis of Experiment-Based Signaling Response Networks

Merrill Knapp[1], Keith Laderoute[2], and Carolyn Talcott[1]

[1] *Information and Computing Sciences, SRI International, Menlo Park, CA, USA*
[2] *Numentus Technologies, Inc., Menlo Park, CA, USA*

5.1 Introduction

An increasing trend in biology focuses on describing or defining cells and collections of cells (e.g. biofilms and tissues) as complex systems (Keating and et al. (2020)): assemblies of diverse molecular and higher order components that interact with internal and external environments in a highly dynamic way over time (Wikipedia (2021)). The importance of applying this systems-level approach to understanding biology can only grow as "omics" and other comprehensive datasets accumulate for all domains of life in the biosphere.

Computational modelling is essential to organize large experimental datasets and enable researchers to reason about biological processes considered as complex networks of components. Modelling of biological networks as graphs is a common method used to display and understand interactions of network components, thus enabling testable predictions (hypotheses) concerning a biological system (e.g. Koutrouli et al. (2020)). The approach discussed in this chapter will focus on a *rule-based* computational modelling of biological networks (e.g. Hlavacek et al. (2006), Keating and et al. (2020)). In rule-based modelling, the rules constrain the number of validated or potential biochemical or physiological interactions that need to be computed, which is a powerful advantage compared with models that compute all possible combinations or correlations of the components being represented (the problem of combinatorial complexity Hlavacek et al. (2006)).

Pathway Logic (PL, Mason and Talcott (2021)) is a unique, rule-based modelling system based on the formal methods' approach of rewriting

Systems Biology Modelling and Analysis: Formal Bioinformatics Methods and Tools, First Edition. Edited by Elisabetta De Maria.
© 2023 John Wiley & Sons, Inc. Published 2023 by John Wiley & Sons, Inc.

logic (Meseguer (1992)). In addition to rule-based modelling, PL has the following key features: (Talcott (2016), Talcott and Knapp (2017)): (i) PL rules are evidence based, with each rule being linked to experimental results supporting the biochemical or physiological mechanism formalized by the rule; (ii) PL models are executable, so that the application of rules describing a process (e.g. a signaling pathway) propagates information from a defined initial state of the system to a selected system state of interest (e.g. a phenotype); (iii) potential pathways can be discovered by requiring one or more executions that converge on a state of interest; (iv) in silico experiments can be performed by altering the composition of an initial state or modifying one or more rules to observe the effect of such perturbations on reaching a state of interest (similar to a biological knockout/knock-in or equivalent transfection experiment); (v) novel pathways found by in silico experiments represent hypotheses, which could be tested experimentally for a model under consideration (hypothesis generation). PL has been used to model some metabolic networks, protease signaling in bacteria, protein glycosylation, and certain signaling processes within the human immune system. The signal transduction model (STM), which is he most developed PL model, will be used to exemplify the features and applications of PL modelling in this chapter.

Other formal methods' approaches being used to develop executable models of biological systems include Petri nets, BioCham, Kappa, Bionet-Gen, BioPepa, the Cell Net Optimizer, and Indra. Petri nets (Petri (1980)), a general model of concurrent processes with a number of variants, offer different levels of expressiveness and analysis complexity (see Chapter 2) which is entirely devoted to Petri nets). Overviews of different Petri net formalisms and their application to modelling biological processes can be found in Gilbert et al. (2007) and Chaouiya (2007). PL uses Petri nets to display the results of executing models and to efficiently find pathways in a network (Talcott and Dill (2006)). BioCham (Fages and Soliman (2008)), a rule-based language founded on constraint logic programming for modelling biochemical systems, uses simulators for both qualitative and quantitative semantics and a temporal logic language for specifying system properties. Kappa (Boutillier et al. (2018)) is a rule-based language for modelling systems of interacting agents. Kappa models represent details of protein states (e.g. post-translational modifications), binding interactions, and complex formation. The semantics of Kappa is given by stochastic graph rewriting. BioNetGen (Harris and et al. (2016)) is a system for the specification and stochastic simulation of rule-based models of biochemical systems including signal transduction and metabolic and genetic regulatory networks. Bio-PEPA (Ciocchetta and Hillston (2009)) is a process algebra language for the modelling and analysis of biochemical networks. Bio-PEPA supports different kinds of analyses, including stochastic simulation, analysis based on ordinary differential equations (ODEs), and model checking in PRISM (Kwiatkowska et al. (2011)). The Cell Net Optimizer (Saez-Rodriguez et al. (2009)) is a software system for assembling Boolean logic models from signaling interaction networks and calibrating the models against experimental data. INDRA (Integrated Network and Dynamical Reasoning Assembler Gyori et al. (2017)) is an automated model assembly system that uses natural language processing and structured databases to collect mechanistic and causal assertions, represent them in a standardized form including associated evidence, and assemble them into causal graphs or rule-based dynamical models. These systems all use executable models based on rules describing local changes in the states of proteins and other cellular components. They all provide tools for analysis based on execution/simulation. BioCham and Bio-PEPA also provide support for model-checking-based analysis. PL shares with Kappa and BioNetGen the use of a rich and controlled language for describing protein states.

This chapter emphasizes the important distinguishing features of PL, which include focus on the inference of rules from experimental data; the formal representation of such experimental findings; linking of model elements (e.g. proteins) to external reference sources and linking of rules to the supporting evidence (Datums Knapp (2021a)); and the symbolic representation of states and rules using place holders whose values are filled in according to the context of use. Although network analysis tools typically look for holes in the network, the PL approach highlights holes in the experimental data and helps to design experiments that potentially address such problems. PL response networks are living documents; the user can see the evidence used to make the rules, and a modeler can change the rules as more evidence is collected. The remainder of the chapter is organized as follows:

- Section 5.2 gives an overview of PL components and the modelling process.
- Section 5.3 describes the formal representation of a cellular state; the use of rewrite rules to model signaling mechanisms; and the algorithms used to assemble and analyze executable models.
- Section 5.4 presents the Pathway Logic Assistant (PLA), which enables an interactive graphical representation of PL models.
- Section 5.5 discusses the datum data structure for representing experimental evidence; the process of collecting experimental evidence and the

resulting searchable knowledge base (DKB); and how datums are used to derive rules.
- Section 5.6 describes the STM8 collection of signaling network models and demonstrates the use of two selected models: *Lps response* and *Death Map*.
- Section 5.7 provides a conclusion.
- Appendix 5.A provides a brief summary of each of the STM8 models.

5.2 Pathway Logic Overview

The PL project's objective is to gain a better understanding about how cells function as complex systems or networks of interacting components. In practice, PL is used to build models consisting of steps involving cellular components (e.g. proteins, genes, and small molecules) that combined constitute processes (e.g. signal transduction) initiated in response to specific extracellular stimuli, such as drugs, pathogens, or stress conditions. The process steps are assembled into executable models that can be interrogated to enable a user to better understand the roles or functions of each component, their interdependencies, and how they work together as a process.

PL is a formal system for representing and understanding cellular pathways (signaling pathways in this chapter) based on rewriting logic (Meseguer (1992, 2012)). Signaling pathway models in PL are networks of rules curated from experimental findings; each rule describes a local or specific biochemical change in a cell (e.g. receptor binding, protein dynamics such as post-translational modification, translocation, complex formation, or secretion). PL networks are not hardwired; they are assembled (discovered) by querying a network about how a given final state might be reached from a defined initial state. PL networks are not kinetic models; they represent possible changes over time as a given network process unfolds (what must or might happen before, what is the state after), but not how long the process takes or the concentrations of its components. The PLA is a tool for visualizing and querying PL networks.

The PL framework has three main parts: (i) the datum evidence language and knowledge base; (ii) the PL language and the rules knowledge base for representing cellular states and changes; and (iii) algorithms and tools for visualizing and querying the resulting executable models.

Figure 5.1 illustrates the process of developing PL models. The first step is to select the topic to model, such as, a specific response to a stimulus (the Lps model, Section 5.6.1), the paths to a phenotype (the Death Map, Section 5.6.3), or how a specific process might occur (upregulation of Ifnb1). The

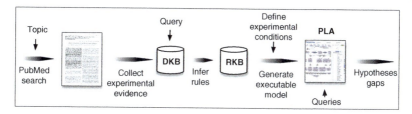

Figure 5.1 Pathway Logic components and processes.

next step is to find publications with relevant experiments, perhaps by first selecting reviews to obtain an overview and an initial set of necessary components (e.g. proteins and genes). Relevant experimental evidence is then curated from the detailed descriptions of experiments (methods and figures) found in the collected research publications. This evidence is represented using the datum language and collected in the datum knowledge base (DKB) (Knapp (2021a)).

When sufficient evidence has been collected, the modeler uses the evidence to derive one or more rules. A rule describes a local change of state, such as a specific biochemical alteration of a signaling protein (e.g. phosphorylation/dephosphorylation, translocation, complex formation, and cleavage). A rule also specifies the conditions under which a given change occurs, such as whether a particular kinase or protease is required. The query interface to the DKB helps a modeler find and organize the diverse evidence required for model building. A collection of derived rules forms a rule knowledge base (RKB) (Knapp (2021b)).

To build a specific PL model, a user defines an initial condition, called a *dish*. A dish models an experimental set up, specifying the overall state of a cell type or compartment of interest by explicitly indicating the states of relevant components (proteins, genes, and small molecules). Given a dish and an RKB, rules can be applied to show how an initial state of the dish might evolve over time in response to a stimulus. The PL language for describing dishes and rules is formalized in the Maude rewriting logic system (Clavel et al. (2007), Durán et al. (2020), Maude-Team (2016)), which can be used to apply rules and to search for states of interest reachable by one or more rule applications (Riesco et al. (2017), Santos-Buitrago et al. (2019), Riesco et al. (2020), Jo et al. (2020)).

When rule sets and dishes are large, using Maude to investigate modelling outcomes can be cumbersome. The PLA solves this problem by enabling a curator and other users of PL to display, browse, and query a model. For this purpose, rules relevant to a dish (see Section 5.3.4) are assembled into a response network, and then, PLA can construct subnets representing all pathways leading from an initial state of a dish to a state of interest, or to find a specific pathway (set of rules) that could lead to a state that satisfies some simple conditions (goal). The set of rules applied to reach a given goal is called an execution path. The paths in a subnet can be enumerated and subnets or paths can be compared. Moreover, in silico experiments can be performed by selecting knockouts or inhibitors, and knock-ins to evaluate perturbations of a response network. For example, a user investigating the Egf (epidermal growth factor) response network can ask for the subnet of paths leading to phosphorylation of `Erk1` (extracellular signal-regulated kinase), followed by its translocation to the nucleus (Knapp et al. (2021)). The user could then delete (knockout) one of more of the Erk1-targeting kinase (Mek1) to determine if Erk1 could still be phosphorylated and, thus, translocated.

PL has been used to develop a variety of cell signaling models, some of which have been used to understand certain drug responses (Talcott and Knapp (2017)) and to identify viral attack surfaces. PL models have also been used as a gold standard for learning natural language processing models (Freitag and Niekrasz (2016), Freitag et al. (2017)). Two collections of models are available on the PL website (pl.csl.sri.com). The STM8 collection (see Section 5.6) that consists of over 40 pathway response networks and Death Net, a model of cell death mechanisms. The PL Sampler collection includes

- SmallKB, a very small model of epidermal growth factor receptor (EgfR) signaling through the classic EgfR system in response to its ligands Egf or Tgfa.
- SKMEL133, a model of an exponentially growing cell based on the SKMEL133 cell line, which was developed to study effects of certain drugs on signal transduction (Talcott and Knapp (2017)).
- STM7, an early version of STM8 with 33 response networks and a base model of a proliferating cell type.
- Protease, a model of protease activity in various bacteria (Panikkar et al. (2011)).
- Mycolate, a model of the mycolate synthesis pathway in *Mycobacterium tuberculosis* (Mtb).
- GlycoSTM, models of protein glycosylation-based pathways from the KEGG database (www.kegg.jp/kegg/pathway.html).
- VMac, a model of macrophage signaling based on the Macrophage Map (vcells.net/macrophage/).

The Pathway Logic website (pl.csl.sri.com) also offers tutorial guided tours of the PL models, the datum query interface (DQI), and related publications and presentations. Users can download PL models represented in Maude,

PLAOnline clients for interacting with the published models, and the full PL/PLA system to develop and analyze their own models.

5.3 PL Representation System

PL models are curated using the Maude rewriting logic system (Clavel et al. (2007), Durán et al. (2020), Maude-Team (2016)). This framework provides an expressive and extensible domain-specific language for representing cellular states and their changes. Maude also provides execution and search capabilities to explore possible behaviors of a model.

This section begins with an introduction to rewriting logic and Maude (Section 5.3.1), followed by a description of the formal representations of PL concepts in Maude (Section 5.3.2), and the basic algorithms used to construct models from rule sets (Sections 5.3.3 and 5.3.4).

PL names of proteins are used in this chapter. The DKB query interface (Knapp (2021a)) can be used to access metadata linked to PL protein names by searching for the PL name and clicking on the name in one of the resulting datums. The same metadata can be accessed from the STM8 PLA Online viewer (see Section 5.4).

5.3.1 Rewriting Logic and Maude

Rewriting logic (Meseguer (1992, 2012)) is a logical formalism that is based on two simple ideas: (i) states of a system are represented as elements of an algebraic data type, specified in an equational theory, and (ii) the behavior of a system is specified by local transitions between states described by *rewrite rules*. An equational theory specifies data types by declaring constants and constructor operations that build complex structured data from simpler parts. Mathematical structures such as sets and maps can be represented directly by declaring, using axioms, that the constructors are associative and commutative, and naming the identity element. Functions on the specified data types are defined by *equations* that allow the result of applying the function to be computed. A *term* is a variable, a constant, or the application of a constructor or function symbol to a list of terms.

A rewrite rule has the form $t \Rightarrow t'$ *if* c, where t and t' are terms possibly containing variables and c is a condition (a Boolean term). For a system in state s, such a rule is enabled if t can be matched to a part of s by supplying the right values for the variables (using a matching substitution) and if the condition c holds when supplied with those values. In this case, the rule can be applied by replacing the part of s matching t by t' using the matching values for the place holders in t'. The process of application of rewrite rules generates computations (also considered deductions). In PL, these computations are called *execution pathways*.

As a simple example, consider a theory with two constants a and b, of sort *Atom*, a function $p : Atom, Atom \rightarrow Atom$, and a rule $p(x,x) \Rightarrow x$ in which x is a variable of sort *Atom*. Then, $p(p(a,a),b)$ can be rewritten to $p(a,b)$ by matching the subterm $p(a,a)$ to the rule left-hand side using a substitution mapping x to a.

Maude is both a language and tool based on rewriting logic (Clavel et al. (2007), Maude-Team (2016)). Maude provides a high-performance rewriting engine featuring matching modulo associativity, commutativity, identity axioms, and search and model-checking capabilities. Thus, given a specification S of a concurrent system, a user can execute S, using one of Maude's built-in rewriting strategies, to find one possible behavior; use search to see if a state meeting a given condition can be reached; or model-check S to see if a temporal property is satisfied, and if not to see a counter-example computation. Maude also supports reflection with a simple representation of modules and their components and access to key functions of the core Maude system that allow the user to specify execution and search strategies and module transformation.

5.3.2 Pathway Logic Language

PL provides two formal representations of rules: location centric (preferred by curators) and occurrence centric that simplifies visualization and analysis.[1] We start with the location centric representation and later show how to transform that to the occurrence centric representation for visualization and analysis. We will use fragments from the Lps response network to introduce the formal notation and concepts. This network will be discussed in more detail in Section 5.6.1.

The state of an experiment is represented by a data structure called a *Dish* (as in a Petri dish) whose contents are a set of *Locations*. The locations model actual locations in a cell such as position with respect to a membrane (attached to the inside or outside or passing through), or the space enclosed by a membrane, such as the cytoplasm. A Location has a name and associated contents. The contents consist of a *Soup* of terms representing biological components such as proteins (possibly modified), genes, protein complexes, or chemicals. This description is formalized in Maude using the following declarations of sorts (types of things) and constructors (of things).

---

[1] An occurrence represents a biological component, its modifications, and its location.

5.3 PL Representation System

```
sort Dish. sort Soup.
op PD : Soup -> Dish [ctor].

sorts Location Locations LocName.
subsort Location < Locations < Soup.
op {_|_}: LocName Soup -> Location
          [ctor format (n d d d d d)].
op __ : Locations Locations -> Locations
          [ctor assoc comm id: empty].
```

The keyword sort (or sorts) introduces names for sorts; the Maude name for type. Sorts denote sets of elements and are partially ordered by the subsort (subset) relation. Thus, Location is a subsort of Locations, meaning that every element of sort Location is also of sort Locations and of sort Soup. The sort Soup is used to represent a (multi)set of cellular components when we want to mix finer grained sorts.

The keyword op introduces a function declaration. This declaration is followed by the list of *argument* sorts, an arrow, and the *result* sort (functions with an empty list of argument sorts are called *constants*.) Thus, PD is a function that encapsulates a Soup into a Dish (PD for Petri Dish). The annotation [ctor] says that PD is a constructor; that is, it canonically represents an element. Functions that are not constructors generally have associated equations defining their meaning in terms of constructor terms. The constructor {_|_} illustrates the use of mix-fix notation. The underscores are argument place holders. The constructor __ illustrates the use of axioms (assoc, comm, id: empty) to define structures such as multisets. The empty syntax of the constructor (__) means that forming a multiset union is just writing two multisets side by side. The axioms for associativity (assoc) and commutativity (comm) say that the elements of a multiset can be written in any order, and without any parentheses indicating the pairwise union formation.

Here is a small *dish* representing part of an Lps treatment experiment. Lps (lipopolysaccharide) is representative of a component of the outer membrane of Gram-negative bacteria and used to study cellular response to detecting this common endotoxin (Mazgaeen and Gurung (2020)).

```
LpsDishX =  PD(
    {XOut    | Lps Lbp Md2         }
    {CLo     | Cd14                }
    {TLR4C   | TLR4                }
    {CLc     | Trif Rnf41 Ifnb1    }
).
```

XOut is the location outside the cell (the supernatant of a culture dish). Here, the Dish contains Lps and the helper proteins Lbp (LPS binding protein) and Md2 (a cell surface protein that binds LPS together with Toll-like receptor 4/TLR4). The CLo contains proteins bound to the outside of the cell (plasma) membrane; in this case, just the protein Cd14 (a co-receptor with Md2). CLc represents the cytoplasm. In this PL experiment, we are only interested in the cytoplasmic proteins Trif (a TLR adaptor), Rnf41 (an E3 ubiquitin ligase), and Ifnb1 (interferon β 1; a secreted cytokine). TLR4C is an ephemeral location, where a group of proteins collect and interact in response to a signal, and later disperse. The four locations make up the soup argument to the dish constructor PD.

Each protein and other symbol used in a term must be declared so that Maude's type checking mechanism will recognize it. Declarations may have associated metadata, including links to well-known databases. PL requires that the metadata for a protein contains at least the UniProt accession number, the HUGO name, and known synonyms. Chemical metadata contains a PubChem accession number. Here is the declaration for the Toll-like receptor 4, TLR4.

```
op TLR4: -> BProtein [ctor metadata "(\
    (category Receptor)\
    (spnumber 000206)\
    (hugosym TLR4)\
    (synonyms \"Toll-like receptor 4\"\
              \"hToll\"\
              \"TLR4_HUMAN\"))"].
```

The metadata item labeled spnumber is the Uniprot (SwissProt) accession number of the canonical sequence of the protein. One can go to Uniprot and type in this number to find out many properties of the protein. The sort BProtein consists of wild-type proteins, without modifications or mutations.

The first rule of the Lps model is as follows:

```
rl[612.TLR4.irt.Lps]:
    {XOut   | xout  Lps Lbp Md2              }
    {CLo    | clo   Cd14                     }
    {TLR4C  | tlr4c TLR4                     }
    =>
    {XOut   | xout                           }
    {CLo    | clo                            }
```

{TLR4C | tlr4c (Lps: Lbp: Md2: Cd14: TLR4)}.

***../evidence/Lps-Evidence/612.TLR4.irt.Lps.txt

This rule shows the formation of the complex Lps: Lbp: Md2: Cd14: TLR4 in response to Lps. Rules typically have informative labels (in []s following the key rl). In STM8 models, rule labels begin with a number that by itself uniquely identifies the rule. The three location terms above the => form the rule input; the local state before the rule is applied. The three location terms below the => form the rule output; the local state after the rule is applied. The symbols xout, clo, and tlr4c are variables denoting the remainder of the state of the three locations, which do not change. Three (or more) stars (***) or dashes (---) indicate that the rest of the line is a comment. The line of dashes is for visual convenience. The comment following *** gives the name of the file that contains the evidence for this rule. An evidence file for response network rules contains structures called *datums* that summarize the experimental findings used for constructing rules. (Datums will be explained in Section 5.5.1).

The term (Lps: Lbp: Md2: Cd14: TLR4) denotes a complex formed from Lps with four proteins. The complex constructor is an infix binary operator declared by

op _:_ : Thing Thing -> Complex [ctor assoc comm].

Similar to union for Locations, the axioms assoc and comm say that the operator (_:_) is associative and commutative. Thus, one can form complexes of complexes, remove the parentheses, and reorder the elements without changing the meaning. The sort Thing collects together different biological subsorts such as Protein, Gene, RNA, or Chemical.

Ephemeral locations such as TLR4C are an alternative representation of complexes. They are used in PL to model less well-defined complexes, where not all the elements are known. It is often the case that either not all of the components are required to be in place for a rule to be enabled, or it may not be known which components are required.

Applying rule 612 to the dish LpsDishX results in the following dish by binding the three variables to the empty soup, empty.

```
    PD (
       {XOut  | empty}
       {CLo   | empty}
       {TLR4C | (Lps: Lbp: Md2: Cd14: TLR4)}
       {CLc   | Trif Rnf41 Ifnb1}
).
```

5 Pathway Logic: Curation and Analysis

The rule 642 illustrates how translocation to a receptor complex is modeled. Here, the protein `Trif` moves from the `CLc` to the `TLR4C`.

```
rl[642.Trif.to.TLR4.irt.Lps]:
  {TLR4C | tlr4c (Lps: Lbp: Md2: Cd14: TLR4)      }
  {CLc   | clc   Trif                             }
  =>
  {TLR4C | tlr4c (Lps: Lbp: Md2: Cd14: TLR4) Trif }
  {CLc   | clc                                    }.
```

Figure 5.2 shows a network with a three-step execution path to secrete `Ifnb1` in response to `Lps`. The figure shows a visualization where each oval represents a protein occurrence (a specific protein with its modifications and its location), round rectangles with layers represent complexes, and boxes with numbers represent rules. More details concerning such representations will be presented in Section 5.4.

The `LpsDishX` experiment was designed with `Ifnb1` already in the cytoplasm in order to generate a small network. Normally, the `Ifnb1` gene must be "turned on" in response to `Lps` to generate the `Ifnb1` protein. Rule 1009 states that the `Ifnb1` gene will be turned on once the required proteins are correctly located in a cell. In particular, several proteins must be associated

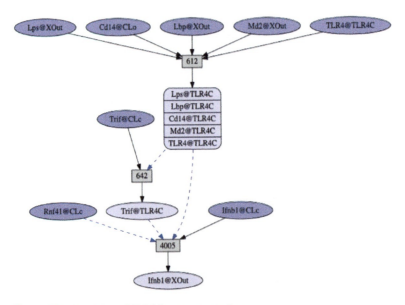

Figure 5.2 Secretion of Ifnb1 in response to Lps.

as a receptor complex, and transcription factors Irf1 and Irf3 must be in the nucleus, as shown in the following:

```
rl[1009.Ifnb1-gene.irt.Lps]:
  {TLR4CI | tlr4ci  (Lps  :  Lbp  :  Md2  :  Cd14  :  TLR4)
                    Pik3cd Ticam2 Trif}
  {CLc    | clc     [Irak4 - ms1] Peli1 Rnf41 Stub1
                    Tbk1 Tirap [Traf3 - ms2] Trim6}
  {NUc    | nuc     [Irf3 - ms3] Ifnb1-gene}
  =>
  {TLR4CI | tlr4ci  (Lps  :  Lbp  :  Md2  :  Cd14  :  TLR4)
                    Pik3cd Ticam2 Trif}
  {CLc    | clc     [Irak4 - ms1] Peli1 Rnf41 Stub1
                    Tbk1 Tirap [Traf3 - ms2] Trim6}
  {NUc    | nuc     [Irf3 - ms3]  [Ifnb1-gene - on]}.
```

The term [Traf3 - ms2 K63ubiq not(K48ubiq)] states that the protein Traf3 bound with ubiquitin polymerized via K63 linkages (and not bound to ubiquitin polymerized via K48 linkages) must be present in the CLc.

PL provides language to represent the modification of a protein at the level of detail that information is known.

The term [Akt1 - phos(S 473)] represents phosphorylation of Akt1 on serine 473. If the available evidence only confirms phosphorylation on a serine residue, the term [Akt1 - Sphos] would be used, and if the assay simply detects phosphorylation, then the term [Akt1 - phos] would be used. Rule 030 shows how phosphorylation is represented and also introduces the concept of a protein *family*.

```
rl[030.Akts.irt.Lps]:
  {TLR4C | tlr4c  (Lps  :  Lbp  :  Md2  :  Cd14  :  TLR4)    }
  {CLc   | clc    Ampk [Akts - ms1]                          }
  =>
  {TLR4C | tlr4c  (Lps  :  Lbp  :  Md2  :  Cd14  :  TLR4)    }
  {CLc   | clc Ampk [Akts - ms1 phos(FSY)phos(KTF)]}.
```

The symbol Akts represents a member of the Akt protein kinase family: Akt1, Akt2, and Akt3. The concept of *family* is used in the case that, e.g. the antibody used to identify a protein may not be specific but may bind any member of a set of genetically related proteins, which are usually functionally similar. In such cases, a sort is introduced to name the family (AktS), and individual family member constants are declared to be of that sort. A constant is introduced to refer to a generic family member (Akts). The symbols FSY and KTF represent serine/S and threonine/T residues within proteins at sites corresponding to a particular family member. This site information is declared using metadata annotations as follows:

```
op FSY:    -> Site [ctor metadata "(\
   (FamilyName Akts)\
   (epitopes \"Akt1 phos(S473)\"\
             \"Akt2 phos(S474)\"\
             \"Akt3 phos(S472)\"))"].
op KTF:    -> Site [ctor metadata "(\
   (FamilyName Akts)\
   (epitopes \"Akt1 phos(T308)\"\
             \"Akt2 phos(T309)\"\
             \"Akt3 phos(T307)\"))"].
```

Rule 030 can be satisfied by Akt1 phosphorylated on S473 and T308, Akt2 phosphorylated on S474 and T309, or Akt3 phosphorylated on S472 and T308.

The rule 030 requires Ampk (*AMP-activated protein kinase*) in the CLc location. Ampk consists of three proteins (subunits) that together make a functional enzyme. In PL, these types of named protein complexes are modeled by the sort Composite; here is the PL declaration for Ampk.

```
op Ampk: -> Composite [ctor metadata "(\
  (subunits Ampkas Ampkbs Ampkgs)\
  (comment \"AMK activated protein kinase consists of:\"\
    \"one catalytic alpha chain (Ampka1 or Ampka2)\"\
    \"one beta regulatory subunit (Ampkb1 or Ampkb2)\"\
    \"one gamma regulatory subunit (Ampkg1,Ampkg2,Ampkg3)\")\
  (synonyms \"Ack3\"\
            \"Acetyl CoA Carboxylase Kinase \"))"].
```

The subunits, Ampkas, Ampkbs, and Ampkgs, are themselves families. Thus, the composite Ampk has 12 instances!

5.3.3 Petri Net Representation

PL rules capture mechanisms of cellular change and information/state propagation at a level of detail based on known (published) experimental evidence. Because change may occur in many contexts, there are variables in PL rules; for example, there could be variables for unspecified modifications or for unspecified elements in a location. The Maude inference engine can be used to execute rules starting with a specific dish, instantiating variables by a matching process. It can also be used to search for paths to states of interest. This capability is especially helpful for developing small submodels. When the set of rules and the number of entities in a dish becomes large (and difficult for a user to evaluate), it is important have a browsable visual representation that also supports efficient query evaluation.

To address this need, PL uses a concurrent computational model called Petri nets (Petri (1980), Peterson (1981), Gilbert et al. (2007)). Petri nets have a natural graphical representation as a bipartite graph with nodes called places, representing local system state, connected to nodes called transitions, representing change rules. In PL, the places in Petri nets are called occurrences. A transition has associated with it two sets of occurrences: inputs and outputs. A transition together with its inputs and outputs correspond to an instance of a rewrite rule. Some of the occurrences are marked, indicating that they form the current state. When all the inputs of a transition are marked, the transition can fire (the rewrite rule can be applied), removing marks from the input occurrences and placing marks on the output occurrences. As in rewriting logic, multiple transitions may fire at the same time as long as no two transitions change the same components. PL rules are constrained to obey laws related to conservation of matter laws. This results in Petri nets of a special form (1 safe), which have very efficient analysis algorithms, and thus, large networks can be handled (Schmidt (2000b)).

To connect with the PL model presented in Section 5.3.2, it is necessary to explain occurrences. An occurrence is a protein (possibly modified), a complex, a chemical, or a gene, paired with its location name. Thus, the locations in a dish or a rule are turned into occurrences by distributing the location name across the contents and dropping content variables (they become implicit). Rule 612 (Section 5.3.2) as a Petri net rule becomes the following:

```
rl[612p.TLR4.irt.Lps] :
    < Lps,XOut > < Lbp,XOut > < Md2,XOut >
    < Cd14,CLo >
    < TLR4,TLR4C >
    =>
    < (Lps:  Lbp: Md2: Cd14: TLR4),TLR4C >.
```

Returning to Figure 5.2, ovals represent protein or gene occurrences, and divided rounded rectangles represent complex occurrences. The labels are the printed name of the occurrence where the entity state is separated from the location name by an @ symbol. Boxes labeled with (truncated) rule identifiers represent transitions. A box with its directly connected occurrences

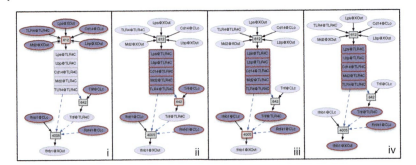

Figure 5.3 Example Petri Net execution starting with the dish, LpsDishX.

represents a rewrite rule. Solid arrows from occurrences to transitions represent inputs. Solid arrows from transitions to occurrences represent outputs. A dashed arrow from an occurrence to a transition denotes an occurrence that is *both* an input and an output. Such occurrences are required for the represented change to occur but are themselves unchanged; for example, in rule 642, the TLR4C complex is required to attract Trif but is not changed by the rule. The top part of Figure 5.2 is the visualization of rule 612 in Petri net form.

Figure 5.3 shows how the process depicted in Figure 5.2 changes state as transitions fire, showing a possible evolution of the cellular state as an Lps receptor-based signal propagates. In (i), the dark ovals (with red borders for emphasis) represent the experimental setup, LpsDishX. All the inputs to rule/transition 612 are marked; thus, the rule is enabled (indicated by the red border). The result of firing the transition is shown in (ii). The marks have been removed from the individual protein occurrences of Lps, Lbp, Md2, Cd14, and TLR4 (lighter color), and the newly formed complex is marked. This enables rule 642 to fire, marking Trif@TLR4C as shown in (iii). Now 4005 is enabled, and firing this rule leads to a state with Ifnb1 secreted (XOut) as shown in (iv).

5.3.4 Computing with Petri Nets

PL uses several algorithms to move between the location centric Maude representation and the Petri net representation and to generate and analyze Petri nets. Section 5.3.3 shows how dishes and rules can be transformed to Petri net form. This mapping, its correctness, and properties of the basic algorithms are given in Talcott and Dill (2006). This section summarizes *forward* and *backward* collection and their use in PL. These algorithms are inspired by well-known static analysis techniques.

Forward collection starts with a concrete set of occurrences (no variables), termed O, and a collection of rules (that may have variables), termed R. The result of forward collection is the set of all instances of rules in the collection that can be "reached" from the initial set of occurrences. As a side effect, the algorithm also computes all the occurrences that might appear in an execution pathway starting from O. The idea is simple: at stage 0, we have $O_0 = O$ and R_0 is empty. At stage $i + 1$, all the rule instances r from R whose inputs are in O_i and whose outputs are not all in O_i are found. These rule instances are added to R_i to obtain R_{i+1}, and their outputs are added to O_i to obtain O_{i+1}. Because there are finitely many rules with finitely many instances, at some n, there are no new rule instances to add: $R_n = R_{n+1}$ and $O_n = O_{n+1}$. R_n is the output of the forward collection from O.

Backward collection starts with a goal occurrence (or set thereof), termed G, and a set of concrete rules (rule instances) termed R. The result of backward collection is the set of rules that might contribute to reach the specified goal(s). Starting with $G_0 = G$, at each stage i, R_i is the set of rules from R whose outputs have non-empty intersection with G_i, and G_{i+1} is obtained by adding the inputs of R_i to G_i. At some point, n we will have $G_n = G_{n+1}$ because there are only finitely many occurrences. The result of the backward collection from G is the union of the R_i.

A *response net* is a Petri net generated from an initial set of occurrences (corresponding to the dish modelling the experimental set) and a collection of Petri net form rules transformed from the location centric RKB. The transitions are the output of forward collection from the initial occurrences and the rule set. The marked occurrences are the initial set of occurrences.

A *relevant subnet* of a PL Petri net, for a given set of goal occurrences, is the subnet that contains all execution pathways from the initial state to the goals. It is obtained by using backward collection from the goals using the Petri net rules, followed by forward collection from the initial marking using the rules produced by the backward collection (to eliminate unreachable rules). This process is a useful way to reduce the complexity of a large model such as the Lps model and to focus attention on endpoints of interest.

PL Petri nets generally have the property known as 1-safe (safe, 1-bounded) which means that in any state reachable from the initial marking, a given occurrence appears at most once. This enables efficient reachability analysis (Schmidt (2000b)) used in the PLA (Section 5.4) to find single execution paths. It also makes possible an algorithm to find all the execution paths within a relevant subnet. The main ideas of this algorithm are outlined below.

The *allpaths algorithm* (Donaldson et al. (2010), Donaldson (2012)) computes paths by exploring the state space to find *reaction minimal paths*

(RMP). These are sets of rules that lead from the initial state to the goal set such that no subset of the rules will do so. Thus, any loops or side effects are eliminated.

The basic RMP algorithm outline is the following:

- Explore the space of *(state,path)* pairs using breadth first search. Here, *state* is a set of occurrences, and *path* is the set of rules fired to reach *state* from the initial marking.
- Throw out *(state,path)* pairs, where path is not reaction minimal. (A path is reaction minimal if it is not a superset of a path to the same state.)
- Do not explore past goal states.

RMP generates all RMPs to a goal set G. This simple RMP algorithm, however, suffers from the well-known state space explosion problem. In Donaldson (2012), partial order reduction algorithms are used to improve the efficiency.

The set of RMPs in a network allows us to easily compute interesting properties:

- *Essential transitions*: the set of rules that are in all RMP pathways to a goal set.
- *Used occurrences*: occurrences that are in at least one RMP pathway.
- *Inhibition targets*: occurrences that are in all pathways to an output; thus, inhibition of one will make the goal set unreachable.
- *Biological knockouts*: proteins or genes (in the initial state) that are in all pathways to the goal set, in some form. Removing these entities from the dish (cell) will make the goal set unreachable.
- *Multisignal cellular responses*: at least one RMP pathway to a goal set has more than one stimulus.
- *Pathway enumeration*: The set of RMP pathways containing one or more specified occurrences.

5.4 Pathway Logic Assistant

The PLA provides an interactive user interface to visualize, explore, and query PL networks. For those interested in developing PL models, the PL framework and the PLA can be downloaded from the PL webpage (pl.csl.sri.com/download.html). To only access the published PL models (Sampler, Section 5.2, STM8, Section 5.6), PLAOnline launchers for either of the collections can be downloaded from (pl.csl.sri.com/online.html).

On opening a launcher, a list of models is presented; selecting a model displays a brief description and a button to launch the PLA loaded with the

5.4 Pathway Logic Assistant

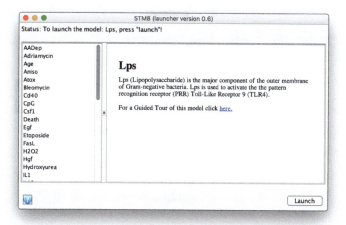

Figure 5.4 The PLA Online STM8 launcher.

selected model. Figure 5.4 shows a screenshot from the PLA launcher for the STM8 collection. Clicking on the *guided tour* link leads to a guided tour with links to additional documentation. Some key features of the PLA are described below, using the Lps model, which is discussed in more detail in Section 5.6.

A small manager window (entitled Lps Manager) will appear, probably in the upper left corner of the screen. Selecting Lps in the pre-defined item of the select dish menu in the manager window will launch the Lps network. (A user can also create a starting state by selecting edit and then selecting occurrences from those currently available.)

The Lps net window will initially display a list of occurrences in the network, rather than rendering the network graph. This is because the network is large and takes significant time to render. The network graph can be shown by pressing *Render* in the tool bar.

Figure 5.5 shows a screenshot of the PLA window displaying the Lps subnet for turning on the proTnf gene. This subnet is produced by selecting the occurrence ProTnf-gene-on@NUc, making it a goal, and using the subnet algorithm (more details below). Here, the PLA window has three main regions:

- The upper right panel displays a thumbnail of the network graph and allows the user to choose what is visible in the main panel by dragging the red box around.

5 Pathway Logic: Curation and Analysis

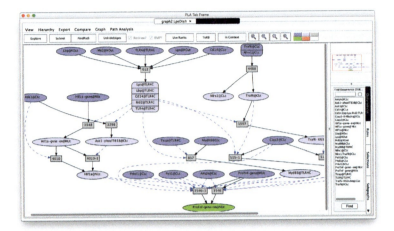

Figure 5.5 Screenshot of the PLA window showing a subnet for turning on transcription of the `ProTnf` gene (green oval).

- The southeast panel (SE, lower right) is an information panel. The tabs on the right edge allow the user to choose what information to display. In Figure 5.5, the Occurrences tab is displayed, showing a list of all the occurrences in the network. Double clicking on an occurrence in the list will make that occurrence visible in the graph panel. The rules tab displays a list of all the rule instances in the network. Double clicking on a rule in the list will make that rule visible in the graph panel.
- The graph panel occupies the left half of the window. It shows a view into the entire network (what is inside the red box). Clicking on an occurrence node displays information about the occurrence in the south east panel. The user can choose to see the associated metadata; explore the network stepwise from the chosen occurrence; make the occurrence a goal; or to knock it out. Clicking on a rule node displays information options for the rule. The user can ask for the evidence page or explore the network stepwise from the rule node.
- A menubar and a toolbar are displayed across the top of the window. These give access to the functions described in Section 5.3.4 above. The Path Analysis menu gives access to the RMP functions; the Subnet button in the toolbar invokes the relevant subnet algorithm; and the FindPath button invokes the LoLa model checker (Schmidt (2000b)). In both cases, the goal and knockout parameters are those that the user has set, working in the southeast panel.

The ovals, boxes, and arrows in the network graph were introduced in Section 5.3.2. Ovals represent the simple (single protein or gene) occurrences given by their label. Rounded rectangles represent complexes with components named by a list of labels. The occurrences in the Lps dish (the initial state) are dark colored. The green oval is the chosen goal of the Petri net. The label, ProTnf-gene-on@NUc, of the green oval names the gene ProTnf-on with transcribing turned on, occurring in the location NUc in the location-centric representation. Boxes with numeric labels represent rules. The rule input consists of occurrences connected by incoming arrows (solid or dashed). The rule output consists of occurrences connected by outgoing arrows plus occurrences connected by dashed incoming arrows. The occurrences connected by dashed incoming arrows model the biological context required for the change modeled by the rule, such as enzymes, scaffolds, or upstream stimuli. These components are themselves unchanged at the end.

The red box and tabs on the SE panel allow the user to browse the network. The spy glasses in the toolbar allow one to control the zoom level. The colored rectangles in the tool bar are reminders about the meaning of the node colors.

Clicking on a rule node and selecting *Rule Evidence* in the context menu (SE panel) provides a link to the evidence page for that rule. This page contains the datums used to support the rule. (See Section 5.5.1 for more information about datums.)

Clicking on an occurrence node and selecting *About Occurrence* displays metadata for the occurrence. The metadata for Ciap2@CLc, one of the requirements for the goal ProTnf-gene-on@NUc, is shown in Figure 5.6.

The Lps network contains several interesting subnets, including subnets for the secretion of Ifnb1, Tnf, IL6, IL10, and other cytokines. These subnets are large and difficult to understand as static figures. Examples will be discussed in more detail in Section 5.6.1, which demonstrates the real value

Figure 5.6 Metadata for the Ciap2 occurrence in the cytoplasm.

Name: Ciap2 *of sort* Ciap12S
UniProt: O75582
Synonyms:

— Baculoviral IAP repeat-containing protein 3
— Inhibitor of apoptosis protein 1
— HIAP1 HIAP-1 C-IAP2 cIAP2 IAP homolog C
— TNFR2-TRAF signaling complex protein 1
— Apoptosis inhibitor 2
— API2 IAP1 MIHC BIRC3_HUMAN

Gene Symbol: BIRC3
Location: CLc Cytoplasm

of models such as the Lps network. The small subnet shown in Figure 5.5 will be used to illustrate the path analysis capability of PLA.

From the Path Analysis menu, selecting *Number of Paths* indicates that there are six paths. A particular path can be found by using FindPath in the toolbar. This choice invokes the Lola model checker and typically finds a shortest path. The Path Analysis menu can be used to ask for paths containing Lps@XOut, which will list all the paths because they all start with TLR4 binding to Lps. The six paths are listed in the SE panel, as lists of rule identifiers, and are shown below.

```
Paths using Lps@XOut
1. [515-1, 4010, 1546, 4469, 857, 612, 1548]
2. [4010, 1546-1, 857, 612, 1548]
3. [3298, 515-1, 1546, 4469, 4010-1, 857, 612, 1548]
4. [3298, 4010-1, 1546-1, 857, 612, 1548]
5. [515, 4010, 1546, 1597, 4469, 857, 612, 1548]
6. [3298, 515, 1546, 1597, 4469, 4010-1, 857, 612, 1548]
```

PLA highlights a path when it is selected. In addition, the selected path can be launched in a separate tab. Figure 5.7a shows path 2 highlighted in red and the result of launching and comparing paths 2 and 3.

Path Analysis shows us that rules 612, 857, and 1548 are the only ruler that appears in all six paths and that Ask1 and Ciap2 are a single knockouts. Lps, ProTnf-gene, and several others are logically knockouts, but they are not meaningful as biological knockouts in the given experiment.

The Explore capability of PLA allows the user to unfold a network one (or more) step(s) at a time, starting with a specific occurrence (set) or rule (set). Exploring can be initiated from the manager window using the *Explore(Occs)* or *Explore(Rules)* buttons, or by clicking on an occurrence or rule in a displayed graph and pressing the *Explore* button in the resulting context menu. When the manager interface is used, a selection window will open, listing the occurrences or rules that are available and one or more items can be selected to start the exploration.

Once exploration is initiated, an exploration step consists of selecting one or more occurrences and choosing to explore up or down stream. The result of upstream exploration is to add all rules (not in the current network) with outputs including at least one selected occurrence. The occurrences connected to the new rules that are not already in the current network are also added. Figure 5.8 shows an exploration in the Lps network initiated by exploring the occurrence Tnfaip3-gene@NUc. This adds the rule that turns on the gene in response to Lps. The green color of Tnfaip3-gene-on@NUc indicates that this occurrence has

5.4 Pathway Logic Assistant | 149

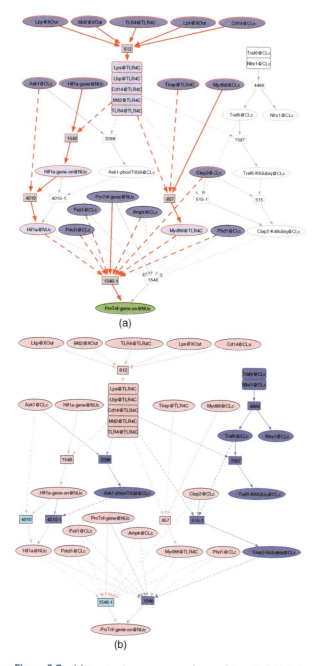

Figure 5.7 (a) ProTnf-gene-on subnet with path 2 highlighted. (b) ProTnf-gene-on subnet comparing paths 2 and 3. Pink nodes belong to both paths, cyan nodes belong only to path 2, and blue/violet nodes belong only to path 3.

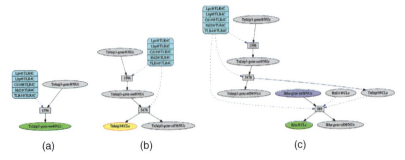

Figure 5.8 Generating a subnet by exploration. (a) Explore(Occs) `Tnfaip3-gene@NUc` up; (b) explore selected `Tnfaip3-gene-on@NUc` down; (c) explore selected `Tnfaip3@CLc` down. Gray nodes have no additional connections to explore, bluish nodes have incoming connections to explore, green nodes have outgoing connections to explore, and cyan nodes have both incoming and outgoing connections to explore. Yellow nodes are selected for exploring next.

downstream connections. Selecting this occurrence and exploring downward adds the protein occurrence, `Tnfaip3@CLc`. Finally, `Tnfaip3@CLc` is selected (yellow color in Figure 5.8b) and exploring downward shows that `Tnfaip3@CLc` regulates translation of the `Ikba` protein.

PL graphs can be exported as png, svg, eps, or xdot (graphviz dot layout program format) for input to other tools. That is how many of the figures in this chapter were produced.

5.5 Datum Curation and Model Development

As mentioned in Section 5.3, a collection of experimental evidence as datums is the starting point for developing PL rules and models. In this section, we explain the elements of a datum, the process of curating datums from publications, and how datums are used to develop PL rules.

5.5.1 Datum Curation

Datums are a human and computer readable summary of an experimental finding representing a single biological assay. Figure 5.9 shows an example datum with elements and their parts annotated. The elements of a datum are expressed using a controlled vocabulary that extends the vocabulary used for PL rules. The information provided by the top-level datum elements is summarized in the following:

- the *Subject*: a protein, gene, or cellular phenotype
- the *Assay*: what was measured and how it was detected

5.5 Datum Curation and Model Development

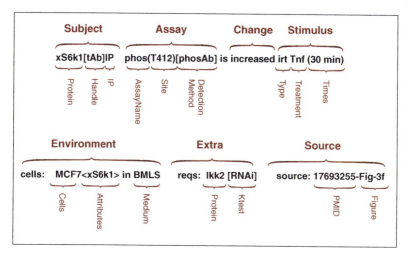

Figure 5.9 Main elements of a datum.

- the *Change*: increased, decreased, unchanged, and detectable but unchanged
- the *Treatment*: what was added to or subtracted from the environment (e.g., a small molecule/drug, peptide, or stress) and for how long
- the *Treatment Type*
  - in response to (*irt*): adding something directly to a cell culture supernatant
  - in the presence of (*itpo*): adding a protein by transfection
  - in the absence of (*itao*): deleting a protein by knockout
  - cell free (*by*): interaction is direct and not due to other proteins in the system
- the *Environment*: cells and culture conditions used at the start of a treatment
- the *Source*: a PubMed ID and the number of the figure or table containing the experimental result

Often, experiments of interest are performed with non-human cells or use non-human proteins. To be able to use evidence from non-human data to model processes in human cells, these proteins and modification sites are converted into the closest human homolog. When a close homolog does not exist, a protein name with the species name is attached to the protein name using "!". Examples are `Naip5!mouse` and `Cpr1!C.violaceum`. This notation allows PL to naturally model host–pathogen interactions.

To support the development of signal transduction models, PL uses cell-based assay experiments. A correct interpretation of experimental results requires knowing the starting state of an experiment (e.g. how the cells were cultured/prepared; the growth medium; and auxiliary treatments). This requirement is the reason for the *environment* element of a datum.

Datums can contain additional information in the form of "extras." Extras report the effects of a system perturbation, stating whether the change increased, decreased, or was not affected. The perturbation can be the removal of a protein by knockout or knockdown, the addition of chemical inhibitors, or the substitution of a protein with a modified version.

The following datum provides evidence that Md2 and Cd14 are required for Lps binding to TLR4. HEK293<xTLR4><xCd14><xMd2> represents HEK293 human embryonic kidney cells transfected with expression vectors for TLR4, Cd14, and Md2. The annotation [omission] says that the corresponding transfection was not done.

```
*** xTLR4[125I] boundby[Xlink] Lps[125I] is detectable
*** cells: HEK293<xTLR4><xCd14><xMd2> in BMS
*** comment: Lbp was added with Lps
*** reqs: xMd2 [omission]
*** reqs: xCd14 [omission]
*** source: 11274165-Fig-1a
```

The PL datum curation methods are focused on experimental data found in the figures and tables of a publication. The required information is found in the caption, the results, and the methods (including methods and figures from online supplementary information). As many datum elements as possible are curated from these sections of publications, including negative data. The datum vocabulary also includes specific terms to be used when a datum element cannot be found in a paper. This convention makes it clear that the information was not provided in the paper, as opposed to the curator not recording the information. For example, if the site(s) detected by an antibody to a phosphorylated protein are not found in a publication, "sitenr" is entered; if the time of a treatment is not supplied, "tnr" is entered.

The remaining datum element to be discussed is the *change* element. This element is the only one that requires an interpretation by the curator. Interpretation is not a problem when the data is numerical and statistics are reported; for example, any change in a readout that is reported with a statistical p-value of less than or equal to 0.05 is acceptable. Determining a change becomes problematic when results are presented in images, as is the case for two commonly used protein detection methods.

One method is immunocytochemistry, which uses images of labeled antibody binding to depict changes in the intensity and location of a protein within a cell. The second method is the Western or immunoblot in which changes in intensity and the location of a protein separated in a gel provide information about, for example, post-translational modifications. In such cases, the curator relies on expertise to determine if a change is significant. However, because a corresponding datum includes the PubMed Id and figure number, users can independently evaluate if they agree with a curatorâs conclusion.

Datums are stored in the DKB where they can be searched using the DQI (datum.csl.sri.com). The DKB can be searched for text or by using tags to limit the search term to a single datum element. A Query Builder helps the user enter well-formed, meaningful queries by listing available tags, qualifiers, and entry values when there are few of options. The results of a query can be downloaded in the JSON format for input to other tools and in the original text form that is readable by humans. A tutorial describing datums and giving example queries of the DKB can be found at pl.csl.sri.com/datumkb.html.

5.5.2 Model Development – Inferring Rules

Once sufficient datums about a particular topic have been curated, they are used to create rules. The first step in creating the rules for a response network is to collect all the experiments in the DKB in which cells have been treated with a single stimulus. This collection can be done using the DQI described in Section 5.5.1. To build the Lps response network, for example, the DKB was asked for all datums from experiments measuring some cellular feature in response to treatment with Lps by using the query

```
treatment:Lps treattype:irt
```

The result consisted of 4190 datums. The datums are then separated into separate groups for each subject, and the subject groups are further separated into subgroups for each assay. Every *subject-assay pair* is a potential rule. If there are sufficient datums to provide evidence that a result was obtained by at least three independent experiments, a rule is written.

Consider the induction of the gene for Tnf (ProTnf-gene) by Lps. The assay for gene induction is *mRNA*, so the query would be

```
treatment:"Lps" treattype:"irt" subject:ProTnf-gene assay:mRNA
```

This query produced 101 datums. All of the datums reported an increase in mRNA expression. Because it is considered unethical to report the results of an experiment more than once in a data-paper from a journal indexed in Pubmed, these results are believed to be from independent experiments. Although 85% of the experiments were performed in mouse cells and only 14% in human cells, the evidence supports a rule in which the `ProTnf-gene` is turned on. The rule should reflect that the change in the gene state is in response to Lps treatment. Section 5.3.2 described the first rule of the `Lps` model, in which a complex is formed between `Lps`, `TLR4`, and required cofactors. This complex is included in the rule about the induction of the `ProTnf-gene` as a requirement for the rule to fire. Thus, the first version of the rule is:

```
rl [1546.ProTnf-gene.irt.Lps]:
  {TLR4C | tlr4c (Lps:  Lbp: Md2: Cd14: TLR4) Myd88}
  {CLc   | clc Ampk [ciap12s:Ciap12S - ms1] Peli1 Phd1 Prkd1}
  {NUc   | nuc Hif1a ProTnf-gene }
  =>
  {TLR4C | tlr4c (Lps:  Lbp: Md2: Cd14: TLR4) Myd88}
  {CLc   | clc Ampk [ciap12s:Ciap12S - ms1] Peli1 Phd1 Prkd1}
  {NUc   | nuc Hif1a [ProTnf-gene - on] }.
```

The extras of the datums contain more information about requirements, particularly extras that contain "reqs:". This term is used when the result seen in the first line is abrogated by removal of a protein by RNA interference, gene knockout at the organism level, or by CRISPR. The proteins required for this reaction to take place are `Myd88`, `Peli1`, `Phd1`, and `Hif1a`. The following datum reports that a double knockout was performed for `Ciap1` and `Ciap2`:

```
*** ProTnf-gene mRNA[RTPCR] is increased irt Lps (tnr)
*** cells: RAW2647 in BMS
*** reqs: Ciap1 and/or Ciap2 [RNAi]
*** source: 19898473-Fig-S9b
```

This datum shows that either or both proteins are required. In this case, the variable `ciap12s:Ciap12S` is used to state that either protein will satisfy the rule. It is possible that both proteins are required, but it is unlikely that a double knockout would be developed unless there were an indication that the proteins were redundant. A knockdown or knockout experiment would indicate whether a protein is needed, but not whether it needs to be modified. A datum element of the form

```
*** inhibited by: x<protein(mutation)> [substitution]
```

indicates that some function of the mutated site is required, so it can be hypothesized that a modification associated with this site is required. No experiments were found using substitution of mutated proteins for this rule, so the proteins were left unmodified. The required proteins are put in the locations where they are expected to be found:

```
rl[1546.ProTnf-gene.irt.Lps]:
   {TLR4C | tlr4c  (Lps:   Lbp: Md2: Cd14: TLR4)      }
   {CLc   | clc    ciap12s:Ciap12S Myd88 Peli1 Phd1 }
   {NUc   | nuc    Hif1a ProTnf-gene                   }
   =>
   {TLR4C | tlr4c  (Lps:   Lbp: Md2: Cd14: TLR4)      }
   {CLc   | clc    ciap12s:Ciap12 Myd88 Peli1 Phd1   }
   {NUc   | nuc    Hif1a [ProTnf-gene - on]           }.
```

The rule 1546 will not fire if any of the required proteins are modified by an upstream event. To solve this problem, variables are introduced to allow the rule to match both modified and unmodified forms of a protein. In the case of rule 1546, the only protein shown to be modified is Ciap2, which is ubiquitinated by Traf6 in response to Lps (rule 515). Thus, ciap12s:Ciap12S, being a variable ranging over members of the Ciap12S family (Ciap1 and Ciap2), is replaced by [ciap12s:Ciap12S - ms1]. The final form of the rule is

```
rl[1546.ProTnf-gene.irt.Lps]:
   {TLR4C | tlr4c  (Lps:   Lbp: Md2: Cd14: TLR4)              }
   {CLc   | clc    [ciap12s:Ciap12S - ms1] Myd88 Peli1 Phd1}
   {NUc   | nuc    Hif1a ProTnf-gene                           }
   =>
   {TLR4C | tlr4c  (Lps:   Lbp: Md2: Cd14: TLR4)              }
   {CLc   | clc    [ciap12s:Ciap12S - ms1] Myd88 Peli1 Phd1}
   {NUc   | nuc    Hif1a [ProTnf-gene - on]                   }.
```

Once rules have been written that capture available experimental evidence, a model is generated by defining the starting state (dish). This state is created by collecting all the occurrences that cannot be created by the rules. The stimulus is added to set off the execution pathways described by the rules, and PLA is used to browse and query the result. In practice, developing a model is an iterative process, using PLA to discover gaps and unreachable rules, gathering more evidence, and refining the rules and initial state.

5.6 STM8

STM8 is a collection of PL models containing information about the changes that occur in the proteins inside a cell in response to exposure to receptor ligands, chemicals, or various stresses. STM8 can be explored using the PLA via the client launcher application available at pl.csl.sri.com/online.html.

The STM8 collection has two types of networks: response networks and endpoint networks. The STM8 *response networks* are formal representations of the changes in the state of the proteins and genes in a cell in response to a single stimulus. *Endpoint networks* use multiple stimuli to reach a single endpoint, which is usually a phenotype. The Death Network discussed in Section 5.6.3 shows how different types of cell death occur in response to certain agents.

Key features of STM8 (and PLA), including the systematic use of evidence and uniform representation, distinguish it in various ways from other signal transduction pathway knowledge bases and viewers (e.g. Reactome Jassal et al. (2020), KEGG Kanehisa et al. (2020)), Cytoscape Shannon et al. (2003), CellDesigner Funahashi et al. (2008)). All rules in the response networks are derived from reported experiments. As discussed in Section 5.5.2, the experimental evidence for each rule is supplied in its evidence page allowing a user to decide whether or not the evidence is convincing. All the networks use the same nomenclature allowing different networks to be compared and combined.

The full list of STM8 networks is summarized in Appendix 5.A. The remainder of this section discusses the Lps and Death Map networks as examples.

5.6.1 LPS Response Network

As mentioned in Section 5.3.2, lipopolysaccharide (Lps) is the major component of the outer membrane of Gram-negative bacteria. Lps is used to activate the pattern recognition receptor (PRR) Toll-like receptor 4 (TLR4). Lps binds to cells that express the members of the TLR4 complex (TLR4, CD14, and MD2), such as monocytes, dendritic cells, macrophages, and B cells, to promote the secretion of pro-inflammatory cytokines (e.g. Tnf, IL1, and type I interferons) and upregulate Nos2 (inducible nitric oxide synthase).

The STM8 Lps response network has 398 rules involving 373 occurrences. To illustrate how this network might be further explored, a user could open the PLAOnline application and look at the subnet relevant to secretion of Tnf by launching the Lps net and generating the subnet for Tnf@XOut. That is, using the selections tab in the SE panel, make Tnf@XOut a goal and then press the Subnet button in the tool bar.

Using the Path Analysis menu will show that this subnet contains 30 different paths to secretion of Tnf. The reasons for these multiple paths can be explained by rule 093 (see also Section 5.5.2).

```
rl[093.Tnf.secretion.irt.Lps]:
  {XOut   | xout   }
  {TLR4C  | tlr4c (Lps:  Lbp: Md2: Cd14: TLR4)
                  Irak1 [Irak4 - ms1] Myd88 [Traf6 - ms2]}
  {CLc    | clc [Ask1 - ms3] [ciap12s:Ciap12S - ms4] Phd1 Prkd1
          Ripk2 [Tpl2 - ms5 phos(S 400) act] Trif [ProTnf - ms6] }
```

```
=>
{XOut   | xout   Tnf}
{TLR4C  | tlr4c (Lps:   Lbp: Md2: Cd14: TLR4)
                Irak1 [Irak4 - ms1] Myd88    [Traf6 - ms2]}
{CLc    | clc [Ask1 - ms3] [ciap12s:Ciap12S - ms4] Phd1 Prkd1
          Ripk2 [Tpl2 - ms5 phos(S 400) act] Trif }.
```

The variable `ciap12s:Ciap12S` means that either `Ciap1` or `Ciap2` will satisfy the rule (Figure 5.10).

The ModSet declarations `ms<n>` state that `Ciaps`, `Irak4`, `Traf6`, and `Ask1` may or may not be modified for the rule to fire. All of these proteins are modified in response to `Lps`, but until there is evidence that the modifications are required, all possibilities are considered in the subnet. Thus, the large number of instances of rule `093` each producing `Tnf@XOut`.

The 30 paths can be listed in the information window by rule numbers using Path Analysis/Paths menu item and selecting the occurrence `Lps@Xout`. As shown in Section 5.4, the different paths can be highlighted in the context of the subnet. A single path can be displayed in a new window by clicking on the Launch button.

Figure 5.11 compares paths 1 and 4 in the `Tnf@XOut` subnet. Path 1 consists of the pink or cyan nodes, while path 4 is the pink or bluish nodes. The picture on the left shows the result of clicking on the cyan rectangle-labeled 4010 in path 1 and choosing "about rule" in the context menu. This shows that unmodified `Ask1@CLc` was used for the production of `Hif1a` in this

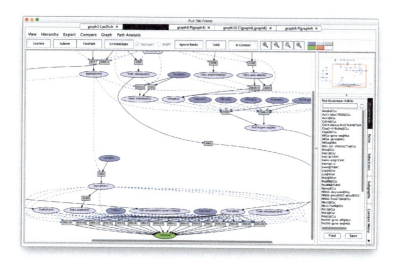

Figure 5.10 Subnet for secretion of Tnf in response to Lps.

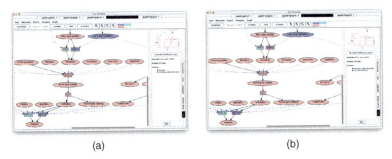

Figure 5.11 Comparison of pathways for secretion of Tnf via rule 4010 instances using `Ask1@CLc` (a, left) vs. `Ask1-phos!T838@CLc` (b, right).

path. Path 4 uses a different instance, `4010-1`, of rule `4010`. The right part of Figure 5.11 displays the "about rule" information for instance `4010-1`, which shows that phosphorylated `Ask1`, `Ask1-phos!T838@CLc`, is used as a requirement for the production of `Hif1a` in this instance. We also see from the figure that both paths use `Ciap2` and not `Ciap1`.

5.6.2 Combining Network Analyses

The STM8 networks all use the same vocabulary, so it is possible to combine the networks to explain experimental results (note that this cannot be done in Online PLA). Consider a result from Mizgalska et al. (2009). Figure 1c in that paper shows a time course of expression of the `Mcpip1-gene` in response to different stimuli. Adding `Tnf` to U937 cells causes an increase in 1 hour with a peak at 2 hours. Addition of `Lps` in the same experiment does not cause a significant increase until 4 hours.

The `Lps` and `Tnf` response networks can be combined by making a new dish containing the components used in both networks. The new dish is loaded into PLA with the RKBs for both networks. If we ask PLA to find all the paths to `Mcpip1-gene-on@NUc`, we get two paths of two rules, shown in Figure 5.12.

These paths say that the gene for `Mcpip1` is turned on in response to either `Lps` or `Tnf` with no evidence about other required components. We know that `Tnf` is secreted in response to `Lps` (Figure 5.10), but PLA will not pick up those paths because `Tnf@XOut` is already in the combined dish. By removing `Tnf@XOut` from the combined dish using the edit function in the STM Manager, redoing the path analysis returns 73 paths to `Mcpip1-gene-on@NUc`. The first and shortest path says that the gene is induced in response to `Lps`. The other 72 paths all involve the secretion of `Tnf`.

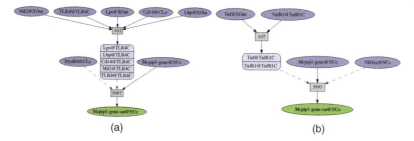

Figure 5.12 Pathways leading to induction of Mcpip1 (a) in response to Lps and (b) in response to Tnf.

The result is a hypothesis that activation of the Mcpip1-gene by Lps requires Tnf, which would explain why Tnf works so much faster than Lps. When Lps is added to cells, sufficient ProTnf protein has to be made to turn on the gene, which occurs over hours. If Tnf is added to the cells instead, there is no need for new protein synthesis. The only time constraint is how long it takes to generate detectable amounts of mRNA, which occurs in less than 30 minutes. This hypothesis could be tested or proven by repeating the experiment in cells with TnfR1 knocked out. This example illustrates how PL could be used to assist in designing a laboratory experiment to test a plausible hypothesis.

5.6.3 Death Map: A Review Model

PL can be used for more than creating response networks from experimental data. Models can also be constructed from hypotheses and assertions made in review articles. Recently, the PL project was asked to make a model of the current perspectives about how mammalian cells in culture die in response to different toxic stimuli. Although there are many terms used to describe modes of cell death, the most common are *apoptosis*, *necroptosis*, and *pyroptosis*. Rules were made from assertions from three current review articles (Ashkenazi and Salvesen (2014), Green and Llambi (2015), Mathur et al. (2018)) to generate models that lead to each of these cell death phenotypes. These rules were supplemented with existing rules from response networks of stimuli mentioned in the reviews. The resulting network has 161 rules and 214 occurrences.

Review networks are built by collecting statements made in review articles, translating them into rules, and using PLA to display them. Once the network is assembled, the next step is to try to verify it by collecting the experimental evidence from the supporting references. If experimental evidence is found that supports the statement, then the rule is considered valid.

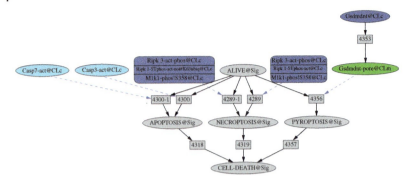

Figure 5.13 Exploration network showing final steps in three major cell death pathways. Gray nodes have no additional connections to explore, blue/violet nodes have incoming connections to explore, green nodes have outgoing connections to explore, and cyan nodes have both incoming and outgoing connections to explore.

If the experimental evidence does not support the statement or if the reference is to a different review article, the rule is considered to be unsupported and is colored yellow in the PLA display. In either case, the rules will have links to evidence pages containing the statements from the review article, the PMIDs of the supporting references, and any datums that are relevant.

To use phenotype rules, each phenotype must be translated into the molecular changes that are unique to it. Activation of `Casp1` or `Casp7` was used for apoptosis, and the creation of a Gasdermin pore was used for pyroptosis. Necroptosis was defined by a state containing a complex of `Ripk1`, `Ripk3`, and `Mlk1`, which allows `Mlk1` to permeabilize the cell membrane. These molecular changes are shown in Figure 5.13, which was produced using the PLA explore function, starting with the cell death signature occurrence (`CELL-DEATH@Sig`) and exploring backward adding rules whose output includes this occurrence, along with all the other occurrences connected to these rules. This exploration process was repeated for each of the three death phenotype occurrences.

The Death Map itself can be thought of as an interactive review document. The similarities between the paths to these modes of cell death can be examined using the PLA compare function (Figure 5.14).[2] The subnets leading to necroptosis and apoptosis are disjoint; they share only the cell death signature occurrence. In this model, the pyroptosis and apoptosis subnets have paths stimulated by the formation of an inflammasome by fragments of pyrogenic bacteria (represented by `Cpr1!C.violaceum`) binding to `Naip`

---

[2] Such networks are best viewed in PLA. The figures are included to give a high-level impression of the relations among the Death Map subnets.

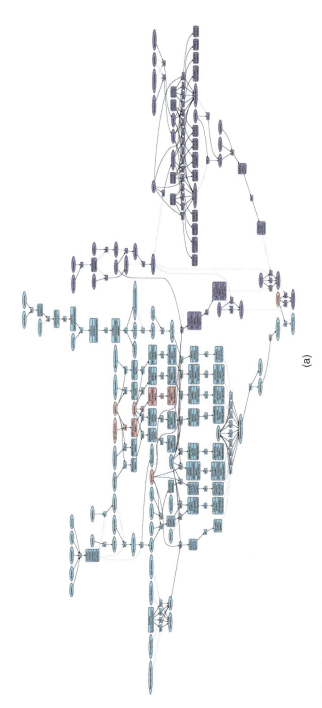

Figure 5.14 Comparing subnets of the Death Map. (a) apoptosis (blue) vs. pyroptosis (cyan). (b) Pyroptosis (cyan) vs. necroptosis (blue). Pink nodes belong to both compared subnets.

(a)

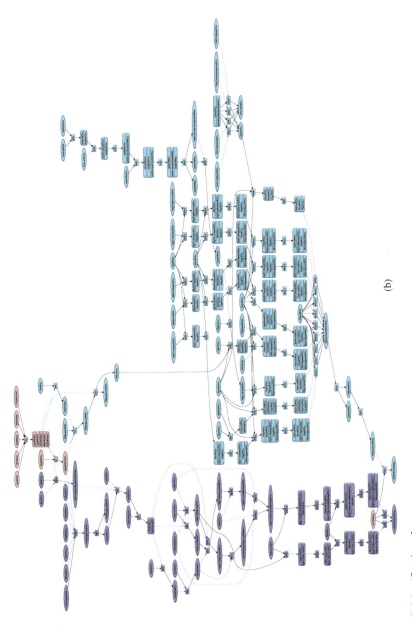

Figure 5.14 (*Continued*)

and `Nlrc4`. If the complex binds and activates `Casp8`, it leads to activation of `Casp7`, which is a hallmark of apoptosis. If it binds and activates `Casp1`, it leads to the cleavage of `Gsdmd` and the formation of the `Gsdmdnt-pore`, which is the hallmark for pyroptosis. The pyroptosis and necroptosis subnets have paths stimulated by `Lps` using `Myd88`. In the first path, `Lps` upregulates the expression of `Nlrp3` forming an inflammasome that binds and activates `Casp1` and thus induces pyroptosis. The path to necroptosis starts with the upregulation of `Tnf`, leading to necroptosis.

The apoptosis subnet shows pathways resulting from three main stimuli: `FasL-XL`, `DNA-DAMAGE`, and `Cprl!C.violaceum` (in the cytoplasm). `DNA-DAMAGE` and some of the `FasL-XL` pathways work by activating the BH3-only proteins (`Bid`, `Noxa`, and `Puma`), which cause the oligomerization of `Bax` or `Bak1` into pores that allow the release of `Cytc` from the mitochondria. This mechanism is called the intrinsic or mitochondrial pathway.

Pathways in the necroptosis subnet are stimulated in the Death Map model by `etoposide` alone or in combination with `Lps` and `Tnf`. The pyroptosis subnet has multiple pathways, each of which is stimulated by different pathogenic proteins. There are also pathways stimulated by `dsDNA` in the cytoplasm and by `Lps` together with `ATP` or `nigericin`.

5.6.3.1 Review Map as a Summary of the State of the Art

A review network can be used to learn about what we know and/or believe about cellular processes. Many assertions are repeated in the three reviews used to develop the Death Map. The process of collecting datums from the references as support for a statement confirmed many of the assertions. Classifying a rule as unsupported means only that the rule is unconfirmed and calls for further experimentation.

5.7 Conclusion

PL is a unique, rule-based computational system for modelling any biological process that can be represented by networks of molecular events (reactions, interactions) that unfold from defined starting states. As detailed above, PL models use rules that are derived from the best available experimental evidence according to strict criteria of curation. PL models are executable and enable users to explore how a specified starting state could unfold to one or more chosen end states (goals), by selecting and modifying potential pathways in a network or subnetwork. Users are provided these capabilities through a sophisticated graphical interface. PL experiments, which simulate bench-top procedures, such as knockouts, knock-ins, or small-molecule perturbations, offer users the opportunity to generate testable hypotheses in silico and to design practical experiments to challenge them. Users can perform their own curation tasks, tailored to biological systems or subjects of interest. The PL website (pl.csl.sri.com) provides tutorials for understanding and navigating the available tools and knowledge bases and downloadable files for installing the full PL system.

Acknowledgments

This work was supported by the National Institutes of Health [CA112970 and GM068146], the Defense Advanced Projects Agency [W911NF-14-C-0108 and W911NF-14-2-0020], Intelligence Advanced Research Projects Activity [W911NF-17-C-0051], and the National Science Foundation [2034772 and IIS-0513857].

Appendix 5.A: Summary of STM8 Networks

- AADep: Amino Acid Deprivation (usually histidine and/or leucine deprivation) is used as a mechanism to deactivate Mtor kinase activity and cause a decrease in protein synthesis. (21 rules, 35 occurrences)
- Adriamycin: Adriamycin (aka doxorubicin) is a chemotherapy medication used to treat cancer. It works by intercalation with DNA and inhibiting transcription. It is used to elicit the DNA damage response in cells and usually causes cell death. (16 rules, 35 occurrences)
- Age: Advanced glycosylation end-products are proteins or lipids that become glycated as a result of exposure to sugars. These bind to the RAGE receptor and lead to activation of inflammatory genes. (18 rules, 33 occurrences)
- Aniso: Anisomycin is an antibiotic used by cell biologists to activate the stress-activated protein kinases, P38s and Jnks. It interferes with protein and DNA synthesis by inhibiting peptidyl transferase or the 80S ribosome system. (33 rules, 61 occurrences)
- Atox: Anthrax toxin is a three-protein exotoxin secreted by virulent strains of the bacterium, *Bacillus anthracis*, the causative agent of anthrax. Anthrax toxin is composed of a cell-binding protein, known as protective antigen (PA), and two enzyme components, called edema factor (EF) and lethal factor (LF). This network contains some rules without evidence. These describe the steps in which PA translocates EF

and LF into the cell. The steps were worked out in the 1980s, so finding the experiments is challenging. (19 rules, 35 occurrences)
- Bleomycin: Bleomycin is a chemotherapy medication used to treat cancer. It works by induction of DNA strand breaks. It is used to elicit the DNA damage response in cells and usually causes cell death. (8 rules, 16 occurrences)
- Cd40: Cd40Lg is a protein that is expressed on the surface of activated T-cells and binds to Cd40 on the surface of antigen-presenting B-cells. To avoid confounding factors that arise from using one cell to activate another, the response to Cd40Lg is commonly studied by treating cells with an antibody to Cd40. (55 rules, 51 occurrences)
- CpG: CpG oligodeoxynucleotides (or CpG ODN) are short single-stranded synthetic DNA molecules that contain a cytosine triphosphate deoxynucleotide ("C"), followed by a guanine triphosphate deoxynucleotide ("G"). The "p" refers to the phosphodiester link between consecutive nucleotides, although some ODNs have a modified phosphorothioate (PS) backbone instead. CpG is used to activate the PRR Toll-like receptor 9 (TLR9). (25 rules, 54 occurrences)
- Csf1: (Colony stimulating factor 1), also known as macrophage colony-stimulating factor (M-CSF), is a secreted cytokine that causes hematopoietic stem cells to differentiate into macrophages or other related cell types. Interaction of Csf1 with its membrane receptor (Csf1R) causes cells to proliferate. (28 rules, 57 occurrences)
- Egf: Egf stimulates cell growth and differentiation by binding to its receptor, EgfR. (255 rules, 325 occurrences)
- Etoposide: Etoposide is a chemotherapy medication used to treat cancer. It forms a ternary complex with DNA and the topoisomerase II enzyme (which aids in DNA unwinding), prevents re-ligation of the DNA strands, and by doing so causes DNA strands to break. It is used to elicit the DNA damage response in cells and usually causes cell death. (15 rules, 19 occurrences)
- FasL: Fas ligand or FasL is a homotrimeric type II transmembrane protein expressed on cytotoxic T lymphocytes. It signals through trimerization of FasR, which spans the membrane of the "target" cell. This trimerization usually leads to apoptosis and cell death. (9 rules, 16 occurrences)
- H_2O_2: Hydrogen peroxide is used to induce the DNA damage response in cells. (45 rules, 68 occurrences)
- Hgf: Hgf (hepatocyte growth factor or scatter factor) regulates cell growth, cell motility, and morphogenesis by activating a tyrosine kinase signaling cascade after binding to the proto-oncogenic c-Met receptor (HgfR). (48 rules, 95 occurrences)

- Hydroxyurea: Hydroxyurea is an antimetabolite that selectively inhibits ribonucleoside diphosphate reductase, an enzyme required to convert ribonucleoside diphosphates into deoxyribonucleoside diphosphates, thereby preventing cells from leaving the G1/S phase of the cell cycle. (9 rules, 19 occurrences)
- Ifnab: Ifnab is the PL family name for type I interferons that bind to the Ifna receptor that consists of IfnaR1 and IfnaR2. Binding of the type I interferons to the receptor causes the upregulation of many genes associated with an antiviral immune response. (119 rules, 123 occurrences)
- Ifng: Interferon gamma is a dimerized soluble cytokine that is the only member of the type II class of interferons. Binding Ifng to its receptor causes the upregulation of many genes producing a variety of cellular responses. (64 rules, 79 occurrences)
- Igf1: Insulin-like growth factor 1 is a hormone similar in molecular structure to insulin. Igf1 binds to at least two cell surface receptor tyrosine kinases: the IGF-1 receptor (IGF1R) and the insulin receptor. Its primary action is mediated by binding to its specific receptor, Igf1R, which is present on the surface of many cell types in many tissues. Binding to the IGF1R stimulates proliferation. (24 rules, 53 occurrences)
- IL1: Interleukin 1 is the PL family name for IL1a and IL1b. Binding these ligands to IL1R1 and its coreceptor IL1Rap causes the upregulation of proinflammatory cytokines. (285 rules, 139 occurrences)
- IL2: Interleukin 2 mediates its effects by binding to IL-2 receptors, which are expressed by lymphocytes. It causes T-cells to proliferate. (49 rules, 102 occurrences)
- IL4: Interleukin 4 is a cytokine that stimulates the proliferation of activated B-cells (31 rules, 66 occurrences)
- IL6: Interleukin 6 signals through a cell surface type I cytokine receptor complex consisting of the ligand-binding IL6R chain and the signal-transducing component Gp130. (9 rules, 16 occurrences)
- IL12: Interleukin 12 is a heterodimer made of IL12a and IL12b that binds to the IL12 heterodimeric receptor made of IL12Rb1 and IL12Rb2. (18 rules, 38 occurrences)
- IL17: Interleukin 17 is a proinflammatory cytokine that upregulates other proinflammatory cytokines. (23 rules, 43 occurrences)
- IL22: Interleukin 22 is a member of the cytokine IL10 superfamily, a class of potent mediators of cellular inflammatory responses. (15 rules, 32 occurrences)
- Ins: Insulin is a peptide hormone that binds to the insulin receptor and causes an increase in proliferation and metabolic processes. (43 rules, 94 occurrences)

- IRad: (Ionizing Radiation) Treating cells with IRad causes double-stranded DNA breaks. It is used to elicit the DNA damage response in cells and usually causes cell death. (134 rules, 75 occurrences)
- Lps: Lipopolysaccharide is the major component of the outer membrane of Gram-negative bacteria. Lps is used to activate the PRR TLR4. (339 rules, 320 occurrences)
- NCS: Neocarzinostatin is secreted by *Streptomyces macromomyceticus*. When added to cells, it induces sequence-specific single and double-stranded breaks in DNA. (15 rules, 27 occurrences)
- NDV: (Newcastle disease virus). Infection of human cells with NDV causes the upregulation and secretion of Ifnb1. It is used to determine the ability and mechanism that other viruses use to prevent the Ifnb1 response. (23 rules, 42 occurrences)
- Ngf: Nerve growth factor binds to two very different receptors. The first receptor (TrkA) is a tyrosine kinase receptor that leads to cell differentiation when activated. NgfR can mediate cell survival and cell death of neural cells. (50 rules, 97 occurrences)
- Pdgf: Platelet-derived growth factor is a dimeric glycoprotein that can be composed of two A subunits (Pdgfa), two B subunits (Pdgfb), or one of each. Binding Pdgf to the PdgfR (a hetero or homodimer of PdgfRa and PdgfRb) leads to cell proliferation. (33 rules, 69 occurrences)
- PMA: Phorbol 12-myristate 13-Acetate is a diester of phorbol and a potent tumor promoter often used to activate the signal transduction enzyme protein kinase C (PKC). (47 rules, 92 occurrences)
- PolyIC: Polyinosinic:polycytidylic acid is a synthetic analog of double-stranded RNA, which is present in some viruses. It can bind to TLR3, Rig1, Mda5, and Pkr causing the upregulation of inflammatory cytokines. (54 rules, 15 occurrences)
- Rig1: The cellular responses to activation of Rig1 are measured using the specific agonist 5' triphosphate double-stranded RNA or expression of a constitutively active Rig1 mutant. This system is commonly used to determine how viral proteins highjack the host antiviral response. (8 rules, 16 occurrences)
- SEV: Infection of human cells with SEV (Sendai Virus) causes the upregulation and secretion of Ifnb1. It is used to determine the ability and mechanism that other viruses use to prevent the Ifnb1 response. (45 rules, 80 occurrences)
- SFTSV: Infection of human cells with SFTSV (Severe Fever with Thrombocytopenia Syndrome Virus) causes the upregulation and secretion of Ifnb1. It is used to determine the ability and mechanism that other viruses use to prevent the Ifnb1 response. (37 rules, 71 occurrences)

- Sorbitol: Addition of high concentrations of sorbitol to the cell medium leads to high osmolarity and the activation of P38s and Jnks. (19 rules, 38 occurrences)
- Tgfb1: Transforming growth factor beta 1 is a secreted protein that performs many cellular functions, including the control of cell growth, cell proliferation, cell differentiation, and apoptosis. (90 rules, 117 occurrences)
- UV: Treating cells with UV (ultraviolet light) causes single-stranded DNA breaks. It is used to elicit the DNA damage response in cells and usually causes cell death. (82 rules, 135 occurrences)
- Tnf: Binding of Tnf (tumor necrosis factor) to TnfR1 and/or TnfR1 causes the translocation of transcription factors such as Rela, Jun, and Atf2 to the nucleus where they participate in gene regulation. (251 rules, 234 occurrences)

References

A. Ashkenazi and G. Salvesen. Regulated cell death: signaling and mechanisms. *Annual Review of Cell and Developmental Biology*, 30: 337–356, 2014.

Pierre Boutillier, Mutaamba Maasha, Xing Li, Héctor F. Medina-Abarca, Jean Krivine, Jérôme Feret, Ioana Cristescu, Angus G. Forbes, and Walter Fontana. The Kappa platform for rule-based modeling. *Bioinformatics*, 34 (13): i583–i592, 2018. ISSN 1367-4803.

Claudine Chaouiya. Petri net modelling of biological networks. *Briefings in Bioinformatics*, 8 (4): 210–219, 2007.

F. Ciocchetta and J. Hillston. Bio-PEPA: A framework for the modelling and analysis of biochemical networks. *Theoretical Computer Science*, 410 (33–34): 3065–3084, 2009.

Manuel Clavel, Francisco Durán, Steven Eker, Patrick Lincoln, Narciso Martí-Oliet, José Meseguer, and Carolyn Talcott. *All About Maude: A High-Performance Logical Framework*. Springer, 2007.

Robin Donaldson. *Modelling and Analysis of Structure in Cellular Signalling Systems*. PhD thesis, University of Glasgow, 2012.

Robin Donaldson, Carolyn Talcott, Merrill Knapp, and Muffy Calder. Understanding signalling networks as collections of signal transduction pathways. In *Proceedings of the 8th International Conference on Computational Methods in Systems Biology*, pages 86–95. 2010.

Francisco Durán, Steven Eker, Santiago Escobar, Narciso Martí-Oliet, José Meseguer, Rubén Rubio, and Carolyn L. Talcott. Programming and symbolic computation in Maude. *Journal of Logical and Algebraic Methods in Programming*, 110: 100497, 2020.

Francois Fages and Sylvain Soliman. Formal cell biology in BIOCHAM. In M. Bernardo, P. Degano, and G. Zavattaro, editors, *8th International School on Formal Methods for the Design of Computer, Communication and Software Systems: Computational Systems Biology SFM'08*, volume 5016 of *LNCS*, pages 54–80. Springer, 2008.

D. Freitag and J. Niekrasz. Feature derivation for exploitation of distant annotation via pattern induction against dependency parses. In *15th Workshop on Biomedical Natural Language Processing*, pages 36–45, 2016.

D. Freitag, P. Kalmar, and E. Yeh. Discourse-wide extraction of assay frames from the biological literature. In *Biomedical NLP Workshop associated with RANLP*, pages 15–23, 2017.

A. Funahashi, Y. Matsuoka, A. Jouraku, M. Morohashi, N. Kikuchi, and H. Kitano. CellDesigner 3.5: A versatile modeling tool for biochemical networks. *Proceedings of the IEEE*, 96 (8): 1254–1265, 2008.

D. Gilbert, M. Heiner, and S. Lehrack. A unifying framework for modelling and analysing biochemical pathways using Petri nets. In *Proceedings of the CMSB 2007*, pages 200–216. LNCS/LNBI 4695, Springer, 2007.

D. R. Green and F. Llambi. Cell death signaling. *Cold Spring Harbor Perspectives in Biology*, 12 (7), 2015.

B. M. Gyori, J. A. Bachman, K. Subramanian, J. L. Muhlich, L. Galescu, and P. K. Sorger. From word models to executable models of signaling networks using automated assembly. *Molecular Systems Biology*, 13 (11): 954, 2017.

L. A. Harris et al. BioNetGen 2.2: Advances in rule-based modeling. *Bioinformatics*, 32: 3366–3368, 2016.

W. S. Hlavacek, J. R. Faeder, M. L. Blinov, R. G. Posner, M. Hucka, and W. Fontana. Rules for modeling signal-transduction systems. *Science STKE*, 344 (2006): re6 2006.

B. Jassal, P. D'Eustachio, et al. The Reactome pathway knowledgebase. *Nucleic Acids Research*, 48 (D1): 233–257, 2020.

Kyuri Jo, Beatriz Santos Buitrago, Minsu Kim, Sungmin Rhee, Carolyn Talcott, and Sun Kim. Logic-based analysis of gene expression data predicts pathway crosstalk between TNF, TGFB1 and EGF in basal-like breast cancer. *Methods*, 179: 89–100, 2020.

Minoru Kanehisa, Miho Furumichi, Yoko Sato, Mari Ishiguro-Watanabe, and Mao Tanabe. KEGG: Integrating viruses and cellular organisms. *Nucleic Acids Research*, 49 (D1): D545–D551, 2020. ISSN 0305-1048. doi: https://doi.org/10.1093/nar/gkaa970.

Sarah M. Keating et al. SBML level 3: an extensible format for the exchange and reuse of biological models. *Molecular Systems Biology*, 16: e9110, 2020.

Merrill Knapp. The Datum Knowledge Base, 2021a. URL datum.csl.sri.com. Accessed: 2021 March 27.

Merrill Knapp. Downloadable Pathway Logic models, 2021b. URL pl.csl.sri.com/download.html. Accessed: 2021 April 08.

Merrill Knapp, Ian Mason, and Carolyn Talcott. STM7: A Pathway Logic model of intracellular signal transduction, 2021. URL pl.csl.sri.com/stm7-guide.html. Accessed: 2021 April 08.

M. Koutrouli, E. Karatzas, D. Paez-Espino, and Pavlopoulos G. A. A guide to conquer the biological network era using graph theory. *Frontiers in Bioengineering and Biotechnology*, 8 (34), 2020.

M. Kwiatkowska, G. Norman, and D. Parker. PRISM 4.0: Verification of probabilistic real-time systems. In G. Gopalakrishnan and S. Qadeer, editors, *Proceedings of the 23rd International Conference on Computer Aided Verification (CAV'11)*, volume 6806 of *LNCS*, pages 585–591. Springer, 2011.

Ian Mason and Carolyn Talcott. Pathway Logic, 2021. URL pl.csl.sri.com. Accessed: 2021 April 08.

A. Mathur, J. A. Hayward, and S. M. Man. Molecular mechanisms of inflammasome signaling. *Journal of Leukocyte Biology*, 103 (2): 233–257, 2018.

Maude-Team. The Maude System, 2016. URL http://maude.cs.uiuc.edu. Accessed: 2016-11-10.

Lalita Mazgaeen and Prajwal Gurung. Recent advances in lipopolysaccharide recognition systems. *International Journal of Molecular Sciences*, 21 (2), 2020. URL https://www.mdpi.com/1422-0067/21/2/379.

J. Meseguer. Conditional Rewriting Logic as a unified model of concurrency. *Theoretical Computer Science*, 96 (1): 73–155, 1992.

J. Meseguer. Twenty years of rewriting logic. *The Journal of Logic and Algebraic Programming*, 81 (7–8): 721–781, 2012.

D. Mizgalska, P. Wegrzyn, K. Murzyn, A. Kasza, A. Koj, J. Jura, B. Jarzab, and J. Jura. Interleukin-1-inducible MCPIP protein has structural and functional properties of RNase and participates in degradation of IL-1beta mRNA. *FEBS Journal*, 276 (24): 7386–7399, 2009.

Anupama Panikkar, Merrill Knapp, Huaiyu Mi, Dave Anderson, Krishna Kodukula, Amit K. Galande, and Carolyn Talcott. Applications of pathway logic modeling to target identification. In G. Agha, O. Danvy, and J. Meseguer, editors, *Formal Modeling: Actors, Open Systems, Biological Systems*, volume 7000 of *LNCS*, pages 434–445, 2011.

J. L. Peterson. *Petri Nets: Properties, Analysis, and Applications*. Prentice-Hall, 1981.

C. A. Petri. Introduction to general net theory. In W. Brauer, editor, *Net Theory and Applications, Proceedings of the Advanced Course on General Net Theory of Processes and Systems, Hamburg, 1979*, volume 84 of *LNCS*, pages 1–19, Berlin, Heidelberg, New York, 1980. Springer-Verlag.

Adrian Riesco, Beatriz Santos-Buitrago, Javier De Las Rivas, Merrill Knapp, Gustavo Santos-Garcia, and Carolyn Talcott. Epidermal growth factor signaling towards proliferation: modeling and logic inference using forward and backward search. *BioMed Research International*, 2017. doi: https://doi.org/10.1155/2017/1809513.

Adrián Riesco, Beatriz Santos-Buitrago, Merrill Knapp, Gustavo Santos-García, Emiliano Hernández Galilea, and Carolyn Talcott. Fuzzy matching for cellular signaling networks in a Choroidal melanoma model. In *14th International Conference on Practical Applications of Computational Biology and Bioinformatics*, volume 1240 of *Advances in Intelligent Systems and Computing*, 2020.

Julio Saez-Rodriguez, Leonidas G. Alexopoulos, Jonathan Epperlein, Regina Samaga, Douglas A. Lauffenburger, Steffen Klamt, and Peter K. Sorger. Discrete logic modelling as a means to link protein signalling networks with functional analysis of mammalian signal transduction. *Molecular Systems Biology*, 5 (1): 331, 2009.

Beatriz Santos-Buitrago, Adrián Riesco, Merrill Knapp, José Carlos R. Alcantud, Gustavo Santos-García, and Carolyn Talcott. Soft set theory for decision making in computational biology under incomplete information. *IEEE Access*, 7: 18183–18193, 2019.

Karsten Schmidt. LoLA: A low level analyser. In Mogens Nielsen and Dan Simpson, editors, *Application and Theory of Petri Nets, 21st International Conference (ICATPN 2000)*, volume 1825 of *LNCS*, pages 465–474. Springer-Verlag, 2000.

P. Shannon, A. Markiel, O. Ozier, N. S. Baliga, J. T. Wang, D. Ramage, N. Amin, B. Schwikowski, and T. Ideker. Cytoscape: a software environment for integrated models of biomolecular interaction networks. *Genome Research*, 11 (13): 2498–2504, 2003.

Carolyn Talcott. The pathway logic formal modeling system: diverse views of a formal representation of signal transduction. In Qinsi Wang, editor, *2016 IEEE International Conference on Bioinformatics and Biomedicine (BIBM)*, 1468–1476, IEEE, 2016.

Carolyn Talcott and David L. Dill. Multiple representations of biological processes. *Transactions on Computational Systems Biology VI*, 221–245, Springer, 2006.

Carolyn Talcott and Merrill Knapp. Explaining response to drugs using Pathway Logic. In J. Feret and H. Koeppl, editors, *Computational Methods in Systems Biology*, LNCS. 249–264, Springer, 2017.

Wikipedia. Complex systems, 2021. URL en.wikipedia.org/wiki/Complex_system. Accessed: 2021 April 12.

6

Boolean Networks and Their Dynamics: The Impact of Updates

Loïc Paulevé[1] and Sylvain Sené[2]

[1] Bordeaux INP, CNRS, LaBRI, UMR 5800, University of Bordeaux, Talence, France
[2] Aix Marseille University, CNRS, LIS, Marseille, France

6.1 Introduction

The term *bioinformatics* was originally introduced by Hesper and Hogeweg and defined as "the study of informatic processes in biotic systems" (Hesper and Hogeweg, 1970, Hogeweg and Hesper, 1978, Hogeweg, 1978). Although bioinformatics raised in the 1970s, we did not have to wait for the appearing of the term to witness studies on biological information processes. Let us simply take a look at the beginning of modern computer science and its development in the first half of the twentieth century. Admittedly, the science of computation has its roots in the purely theoretical works in the domain of discrete mathematics by Herbrand and Gödel on recursive functions, by Church (1932, 1936) on λ-calculus, and by Kleene (1936a,b) at the frontiers of both. Nevertheless, early major progresses in this discipline have been made thanks to inspirations from natural phenomenology. The first example is undoubtedly Turing machines. In his seminal papers, Turing built a mechanical comprehension of computation by means of successive deconstructions of the human calculation process (Turing, 1936). Another classical and in no way less important family of examples, in particular in the context of biological modelling, is automata networks. Automata networks were introduced distinctly in the 1940s by McCulloch and Pitts with artificial neural networks and by von Neumann with cellular automata (McCulloch and Pitts, 1943, von Neumann, 1966). In both of these works, the idea was twofold:

- to abstract the logical structure of life and thus obtain more inclined mathematical models to capture natural phenomena and
- to understand the computational power of these abstractions.

Systems Biology Modelling and Analysis: Formal Bioinformatics Methods and Tools,
First Edition. Edited by Elisabetta De Maria.
© 2023 John Wiley & Sons, Inc. Published 2023 by John Wiley & Sons, Inc.

The way to achieve that is rather "simple": it consists in designing networks of entities (called automata) that influence each other over time. As suggested by the information and computability theories, those entities evolve locally inside a finite set, and time is discrete. Just as biology inspired and keeps inspiring computer science (DNA computing and molecular programming), computer science and discrete mathematics, through the automata network model, have become central in qualitative modelling in molecular biology since the end of the 1960s. Indeed, automata networks provide an appropriate setting to model the structural and dynamical features of biological complex systems at the level of the cell such as gene or protein networks. Besides, the understanding of regulation processes is a key problem that biology cannot solve by itself. It is a fundamental observation made in parallel by Kauffman (1969a,b) and Thomas (1973) in the framework of genetic regulation, with the support of Delbrück's works emphasizing analogies between differentiated cellular types and attractors of network theoretical models (Delbrück, 1949). The experimental nature of the common techniques in biology cannot address the whole problem as they are conceived to deal with specific matters. In other terms, experimentations are not adapted for this problem. Their repetition allows us to acquire sharp knowledge on biological elements (subject to human interpretation, observation mistakes, statistical biases, etc.), but for complexity reasons notably, they prevent from putting this knowledge in perspective and have hindsight enough to comprehend the causalities and the effect of combinations of regulations. Kauffman and Thomas early mentioned that biology needs to rest on more theoretical sciences to develop more general and systematic approaches of the living underlying issues. Kauffman made explicit in 1971 "the urgent need for theories about the ways in which integrated genetic control systems might function" (Kauffman, 1971). Thomas wrote in 1973: "the mere description of a situation as it is seen at a given state of research has often become heavy and tedious, requiring long sentences which are easily ambiguous or misleading. I have felt for some years an increasing necessity for a formalization of the concepts in the field" (Thomas, 1973). Following these lines, both of them introduced first models of gene regulation networks by using a peculiar restriction of automata networks, Boolean networks, i.e. automata networks in which elements can be either active or inactive, so that they are in fact a generalization of McCulloch and Pitts' neural networks. This choice comes from the idealization consisting in ignoring the activities of mRNAs and proteins. Informally, a Boolean network can be viewed as a directed graph composed of Boolean automata (taking values in the set $\{0, 1\}$) that influence each other through local Boolean functions that govern their evolution over time. Despite this abstraction, which can be considered simplistic, Boolean networks have very relevant characteristics discussed further in the chapter, which make this mathematical model the most used and impacting in the framework of regulation network modelling to this day. Whilst Kauffman and Thomas worked intrinsically on the same mathematical objects, their approaches were not equivalent. The main differences lie in network structures and the ways automata update their local state over time. When Kauffman chose to formalize gene regulation through an approach inheriting from statistical mechanics with randomly constructed regular graphs whose nodes represent genes that update their state synchronously, Thomas used an approach closer to combinatorics and theoretical computer science with general graphs whose nodes evolve in a totally non-deterministic asynchronous manner, arguing the absence of genetic synchronicity. Of course, further studies based on these seminal works have relaxed the hypotheses mentioned above. On the one hand, works on Kauffman's model called *random Boolean networks* have re-examined the assumptions of regular graph structures and synchronicity. Probabilistic asynchronicity was notably considered (Harvey and Bossomaier, 1997, Gershenson, 2004), as it has been widely made in the context of cellular automata (Fatès, 2013). On the other hand, the asynchronicity of Thomas' model has been questioned. Some studies focused on variations reinforcing it by adding delays (Kaufman et al., 1985, Thomas and Kaufman, 2001, Ahmad et al., 2008), and other ones relaxed it by allowing synchronous events (Noual, 2012b). However, whatever the choices made about synchronicity, a fundamental point is that Boolean networks have been and keep being widely used in the framework of biological modelling, for formal analysis and for prediction purposes. Examples of this are the modelling of the control of cellular differentiation processes of T cells (Mendoza, 2006, Naldi et al., 2010, Abou-Jaoudé et al., 2015) and blood cells (Collombet et al., 2017), the modelling of cancer development (Cohen et al., 2015) and drug resistance (Za nudo et al., 2017), the sharp understanding of the floral morphogenesis of *Arabidopsis thaliana* (Mendoza and Alvarez-Buylla, 1998, Mendoza et al., 1999), and the analysis of the dynamics of the genetic control of budding yeast *Saccharomyces cerevisiae* (Kauffman et al., 2003), fission yeast (Davidich and Bornholdt, 2008), and *Candida albicans* yeast (Wooten et al., 2021), to name but a few. More generally, this issue about (a)synchronicity and the chosen method to compute the new local states of automata in Boolean networks has very strong relations with the concept of *discrete time* (and qualitative modelling), and is thus fundamental, on at least three grounds. First, from the biological standpoint, the passage of time and its consequences are certainly a major issue at the different scales of the living. For instance, if we consider genetic regulations, the question of how gene expressions occur over time is central. Some elements have recently been proposed in the research undertaken on chromatin dynamics (Benecke, 2006, Mekhail and Moazed, 2010), but they are only preliminary and do not allow us to pinpoint a fine spectrum of biologically plausible updates. Second, from the mathematical standpoint, understanding the consequences of choosing this or that specific updating mode, and the sensitivities of a network to this choice, is one of the fundamental current issues in the field, following in the discrete world some lines given by René Thom in his essay on structural stability (Thom, 1975). Third, in a modelling framework, from the theoretical computer science point of view, the non-countable infinite number of possible updating ways and their consequences on the dynamics of Boolean networks, along with the underlying intrinsic exponential complexity between static and dynamical views of the latter, require sampling relevant updating modes and understanding (even predicting) their impact. As we will emphasize in this chapter, the updating mode is a crucial part of the dynamical modelling and consequently of the validation of models. In the scope of biology, the validation of a Boolean model has multiple facets. On the one hand, Boolean networks can serve as conceptual and phenomenological models: they can demonstrate sufficient or necessary mechanisms leading to complex emergent dynamical properties, inspired by the observation of natural systems, e.g. the emergence of patterns, rhythms, and resilience. The validation of such models builds primarily on the justification of the different parameters, including the updating mode, and how they can relate to the current knowledge and hypotheses on the biological system. On the other hand, Boolean networks are also employed as "mechanistic" models of biological systems: the automata model the activity of identified biological entities, which can be directly or indirectly measured in laboratory. Boolean networks then describe the logic of activation and inactivation of biological entities over time. In many such applications, such as to gene regulatory networks, the Boolean modelling of the dynamics of gene activity is viewed as a simplification of a quantitative system, making the assumption of non-linear influences and abstracting away influence thresholds based on copy-number or concentration of the transcription factors. The validation of these models then relies on their capability to reproduce the observed behaviors. Such a modelling approach can be qualified as "top-down," with a direct specification of an abstract model to derive predictions on a concrete quantitative system, in opposition to "bottom-up" abstractions. Bottom-up abstractions typically start from a precise quantitative model and derive mathematically abstractions of it, which formally ensure that some properties will be preserved in the abstract model (Cousot and Cousot, 1977, Fages and Soliman, 2008, Abou-Jaoudé et al., 2016). However, in practice, the knowledge of the system is often insufficient to build such a precise quantitative model, even putting aside its parameterization. It turns out that the Boolean network modelling framework fits with the current granularity of the knowledge of biological processes. Moreover, modelers often take advantage of the abstraction level of Boolean networks to account for automata relating to different biological concepts: for instance, mixing automata modelling activity of genes, proteins, and more abstract concepts such as phenotypes. The fact that they enable transforming static information on known and putative influences into a dynamical *in silico* model, which can be easily simulated, is probably one of the reasons for their current growing adoption in theoretical and experimental biology research groups. In summary, since the seminal works of Kauffman and Thomas, Boolean networks have been widely employed and certainly are the most used model in biological regulation network modelling as a matter of fact. Nevertheless, most of the works that have followed their lines have been conducted principally in the direction of applications. There currently exists a deep gap between the applicative and fundamental aspects of Boolean networks in favor of applications. As a consequence, the intrinsic properties of Boolean networks, such as the way that information is transmitted along the entities, the ability to produce this or that global dynamical behavior depending on local interactions, are still not well understood. Of course, results have already been obtained, but they are far from being sufficient. In some sense, this lack of fundamental knowledge tends to limit the innovating power of applications. It can be explained by the fact that, at present, as soon as a question is addressed from the application standpoint, scientists provide ad hoc solutions to them, which prevents to develop a unified, federative, and general framework. This actually is paradoxical because without more theoretical fundamental knowledge, applications cannot evolve deeply and become more impacting. In light of this, keeping in mind what Kauffman and Thomas said (see above), we made the choice in this chapter to address biological modelling and more precisely gene regulation network modelling with a rather theoretical but illustrated standpoint, through the prism of updating modes and dynamics. To this end, this chapter outline is as follows:

- After a short briefing on general notations, Section 6.2 presents the Boolean network framework and gives the main definitions and notations specific to it.

- Section 6.3 puts the emphasis on some relevant case studies from the literature in biological modelling, which show some important features of Boolean modelling.
- Section 6.4 focuses on deep fundamental known results on Boolean networks coming from discrete mathematics and computer science, which establish very important links with qualitative knowledge on genetic regulations.
- Finally, Section 6.5 concludes this chapter with a discussion about time and current research directions and links to software tools for analyzing Boolean networks with the different updating modes presented here.

6.1.1 General Notations and Definitions

The Boolean domain $\{0,1\}$ is denoted by $\mathbb{B}$. Given a finite set S (resp. a vector V), $|S|$ (resp. $|V|$) denotes its cardinality (resp. its dimension). The set $\{1,\ldots,n\}$ is denoted by $[\![n]\!]$. The empty set $\{\}$ is denoted by $\emptyset$. The empty vector $()$ is denoted by $\vec{\emptyset}$. Given two vectors $V = (1,2)$ and $V' = (2,3)$, their concatenation is denoted by $V \parallel V' = (1,2,2,3)$. Given two Boolean vectors $x, y \in \mathbb{B}^n$, the set of their components having a different value is denoted by $\Delta(x,y) = \{i \in [\![n]\!] \mid x_i \neq y_i\}$. Given a Boolean vector $x \in \mathbb{B}^n$ and $i \in [\![n]\!]$, x^i is the Boolean vector identical to x except on the ith component that is inverted, i.e. $\Delta(x, x^i) = \{i\}$. This naturally extends to the subsets of $[\![n]\!]$. Formally, we also use, for any $x = (x_1, \ldots, x_n) \in \mathbb{B}^n$:

- $\forall W = W' \uplus \{i\} \subseteq [\![n]\!]$, $x^{\overline{W}} = (x^{\vec{i}})^{\overline{W'}} = (x^{\overline{W'}})^{\vec{i}}$, where $\uplus$ represents the union of two disjoint sets, i.e. $A = B \uplus C \iff A = B \cup C$ and $B \cap C = \emptyset$;
- $\bar{x} = x^{[\![n]\!]} = (\neg x_1, \ldots, \neg x_n)$, with $\neg a = 1 - a$.

Notice that a Boolean vector $x = (x_1, \ldots, x_n) \in \mathbb{B}^n$ is sometimes written as the binary word $x = x_1 \ldots x_n$ for the sake of clarity in some contexts (notably in the figures). Given a directed graph $G = (V, E)$ and two vertices i and j of V, the set $V^-(i) = \{j \mid (j,i) \in E\}$ (resp. $V^+(i) = \{j \mid (i,j) \in E\}$) denotes the *in-neighborhood* (resp. the *out-neighborhood*) of i. A *path* is a finite or infinite series of edges that joins a sequence of vertices. Consider the relation of being strongly connected of such a graph as the fact that there is a path between every ordered pair of its vertices. This relation is an equivalence relation, and the induced sub-graphs by its equivalence classes are the *strongly connected components* of G. A strongly connected component is said to be *terminal* if and only if there is no path from it to another strongly connected component. The classical Boolean operators used in this chapter are as follows: $\neg$ denotes the unary negation operator (NOT) such that $\neg a = 0$ if and only if $a = 1$ and *vice versa*; $\vee$ denotes the binary disjunction operator (OR) such that $a \vee b = 1$ if and only if at least one of the operands equals 1; $\wedge$ denotes the binary conjunction operator (AND) such that $a \wedge b = 1$ if and only if both the operands equal 1; and $\underline{\vee}$ denotes the binary exclusive disjunction operator (XOR) such that $a \underline{\vee} b = (a \wedge \neg b) \vee (\neg a \wedge b)$, i.e. either a equals 1 or b equals 1.

6.2 Boolean Network Framework

This section aims to present the mathematical and computational model of Boolean networks. After a discussion on the motivations for choosing such a framework in the context of modelling, we put the emphasis on classical formal definitions related to these networks, their updating modes, and their dynamics.

6.2.1 On the Simplicity of Boolean Networks

From a general standpoint, automata networks can be used to model any real system that satisfies the following three properties:

- it is a real system made up of distinct entities that interact with each other;
- each entity is characterized by a variable quantity, which is precisely intended to be expressed formally in terms of states of automata in the model;
- the events undergone by the real system, just like the mechanisms that are responsible for them, are not directly and fully observable with certainty: only the consequences of these events, namely completely accomplished changes, are.

These three properties impose very few restrictions on the set of systems that can be modeled by automata networks. These theoretical objects are therefore generic models for a very large variety of real systems. Let us come back to the "variable quantity" of entities mentioned in the second point above. To be translated in terms of automaton states, it calls for a formalization. This consists in choosing whether what interests us in the variation of this quantity is of Boolean, discrete, or continuous nature. As an illustration, consider the example of genetic regulations and choose the action of a gene as the so-called "variable quantity":

- If, in the action of this gene, what interests us is its expression or non-expression, then we fall directly into the Boolean case.

- If what interests us are the different ways in which the gene acts on other elements of the system, then we can match a state to each of them. We then fall into the discrete case that can be encoded in Boolean (an automaton state k can simply be encoded by $\log_2(k)$ automata).
- Finally, if we measure the action of the gene through the concentration of proteins it produces, then we fall into the continuous case. This concentration is usually presented as a sigmoidal function. To deal with this case, three common methods exist. The first one is to stay in a continuous formalism (see Chapter 9). The second one consists in approximating the sigmoid by cutting it into intervals to which states are made to correspond so as to fall back into the discrete formalism. The third one consists in considering the extremal concentrations of the sigmoid so as to fall in the Boolean case.

This shows that we can give different statuses to the Boolean framework depending on whether we see it as a direct modelling of reality or as an approximation or an encoding of an intrinsically continuous or discrete modelling. Note that the direct Boolean modelling is consistent with the choice to focus on the state changes of the automata rather than on their states themselves. As an illustration, if we compare automata to internal combustion engines, the interest is more in the fact that an engine, taken separately, goes from the "off" state to the "on" state and vice versa than on the quantity of electricity supplied by the battery to start it or on that released by the spark plugs to cause the explosion and initiate movement. Under this assumption, the Boolean abstraction is necessary and sufficient. In addition, the discourse of biologists is generally imbued with syntactic elements of propositional logic, and it is not uncommon to hear sentences such as: "in the absence of repressor α, gene β is expressed" or further "if the products of genes α and β form a complex, the latter promotes the expression of gene γ, whereas these genes tend to inhibit its expression when they are in monomeric form." This again agrees with a direct modelling of reality in Boolean formalism.

In addition, Boolean networks derive other interesting benefits from their simplicity. In particular, they provide a framework with clearly defined contours, ideal for tackling fundamental problems around the modelling of complex interacting systems. Some of these problems are presented in this chapter. Given the variety of their nature, ranging from structural sensitivity analyses to dynamical behavior characterizations, and the current state of our knowledge, such problems could not currently benefit from significantly more elaborate frameworks. This would inevitably lead to diluting the primary questions and de-structuring the problems posed by drawing attention to ancillary questions induced by the set of parameters to be considered and not intrinsically included in the initial problem. For these questions, on the contrary, Boolean networks offer just what is needed and facilitate the manipulation of a minimal concept of causality, which is rooted in the notion of state change. Their merit therefore lies in the reliability of the information they potentially provide and the simplicity of their setting (see below) associated with their ability to capture most of the heterogeneities and intricacies carried by the modeled systems.

6.2.2 Boolean Network Specification

A *Boolean network* of dimension n is specified by a function $f : \mathbb{B}^n \to \mathbb{B}^n$ mapping Boolean vectors of dimension n to Boolean vectors of dimension n. For every $i \in [\![n]\!], f_i : \mathbb{B}^n \to \mathbb{B}$ is the ith component of this function, which we call the *local function* of automaton i. Classically, the 2^n Boolean vectors of $\mathbb{B}^n$ are called the *configurations* of the Boolean network. A configuration denoted by x can also be viewed as a one-to-one function from $[\![n]\!]$ to $\mathbb{B}$. With this notation, we say that x_i is the local state, abbreviated by *state* in the sequel, of automaton i. The local functions may only depend on a subset of automata of the network, and these dependencies may even be monotone. This information can be summarized by a directed and signed graph, called the *influence graph* (often called interaction graph in the literature), (V, E) with its vertices $V = [\![n]\!]$ and edges $E \subseteq [\![n]\!] \times \{+, -\} \times [\![n]\!]$. Whenever $(i, +, j) \in E$, we say that i is an *activator* of j; whenever $(i, -, j) \in E$, i is an *inhibitor* of j. An influence graph is *simple* if there is at most one edge from one vertex to another, i.e. there is no $(i, j) \in [\![n]\!]^2$ with $\{(i, +, j), (i, -, j)\} \subseteq E$. An influence graph $([\![n]\!], E)$ is *compatible* with an influence graph $([\![n]\!], E')$ whenever $E \subseteq E'$. The influence graph of f has a positive (resp. negative) edge from i to j if and only if one can assign a state to each automaton other than i so that the local function of j becomes equal (resp. different) to the state of i. It is formally denoted by $\mathscr{G}_f = ([\![n]\!], E_f)$ where

- $(i, +, j) \in E_f$ if and only if there exists $x, y \in \mathbb{B}^n$ with $\Delta(x, y) = \{i\}$ and $x_i = 0$ such that $f_j(x) = 0$ and $f_j(y) = 1$;
- $(i, -, j) \in E_f$ if and only if there exists $x, y \in \mathbb{B}^n$ with $\Delta(x, y) = \{i\}$ and $x_i = 0$ such that $f_j(x) = 1$ and $f_j(y) = 0$.

By forgetting the signs on the edges, a weaker way to relate the local functions to the edges of the influence graph is given by the following formula:

$$\forall i \in [\![n]\!], \exists x \in \mathbb{B}^n, f_j(x) \neq f_j(\bar{x}^i) \iff (i, j) \in E_f$$

Such a relation implies notably that $\mathcal{G}_f$ is minimal, i.e. any of its edges is supposed to represent an effective influence based on essential Boolean variables (Crama and Hammer, 2011). In other terms, given f and a configuration $x \in \mathbb{B}^n$, the set of edges of $\mathcal{G}_f$ which operate an influence on x equals $E_f(x) = \{(i, s \in \{+, -\}, j) \mid f_j(x) \neq f_j(\bar{x}^i)\}$ A Boolean network f is *locally monotone* if and only if its influence graph $\mathcal{G}_f$ is simple. In such a case, for each local function f_j, for each i so that $(i, +, j) \in E_f$, the sole change of state of automaton i from 0 to 1 cannot cause f_j to switch from 1 to 0; for each i so that $(i, -, j) \in E_f$, the sole change of state of automaton i from 0 to 1 cannot cause f_j to switch from 0 to 1. In other words, the local functions f_j are monotone according to a component-wise ordering of Boolean vectors that depends on j, where activators are ordered increasingly ($\leq$), and inhibitors decreasingly ($\geq$). Remark that f is monotone only if it is locally monotone. Formally, a Boolean network f is locally monotone if and only if, $\forall j \in V$, $\forall i \in V^-(j), f_j$ either satisfies:

$$\forall x \in \mathbb{B}^n, x_i = 0 \implies f_j(x) \leq f_j(\bar{x}^i), \text{ and thus } (i, +, j) \in E_f$$

or

$$\forall x \in \mathbb{B}^n, x_i = 0 \implies f_j(x) \geq f_j(\bar{x}^i), \text{ and thus } (i, -, j) \in E_f$$

Example 6.1 Let us consider the two following distinct Boolean networks f and g of dimension $n = 3$ defined so that they respect the minimality constraint of their influence graph:

$$f(x) = \begin{pmatrix} f_1(x) = \neg x_3 \\ f_2(x) = x_2 \wedge (x_1 \vee x_3) \\ f_3(x) = \neg x_1 \end{pmatrix} \text{ and } g(x) = \begin{pmatrix} g_1(x) = x_1 \vee \neg x_2 \vee \neg x_3 \\ g_2(x) = x_3 \wedge (x_1 \veebar x_2) \\ g_3(x) = \neg x_1 \vee \neg x_2 \vee \neg x_3 \end{pmatrix}$$

Given these Boolean networks, it is easy to construct their associated influence graphs $\mathcal{G}_f = (\llbracket 3 \rrbracket, E_f)$ and $\mathcal{G}_g = (\llbracket 3 \rrbracket, E_g)$. The general idea is to look at every local function f_i, written in conjunctive or disjunctive normal form. If there is a literal x_j (resp. its negation $\neg x_j$) in its definition, with $j \in \llbracket n \rrbracket$, then $(j, +, i) \in E_f$ (resp. $(j, -, i) \in E_f$). Consider Boolean network f. By local function f_1, we deduce that automaton 1 of f is influenced negatively by automaton 3; definition of f_2 shows that automaton 2 is influenced positively by automata 1 and 3, and itself; for automaton 3, we deduce that it is influenced negatively by automaton 1. Now, for Boolean network g, we have by local function g_1, automaton 1 is influenced positively by itself and negatively by the others; by expanding the XOR operator, g_2 can be rewritten into $g_2(x) = x_3 \wedge ((x_1 \wedge \neg x_2) \vee (\neg x_1 \wedge x_2))$, which shows that automaton 2 is influenced positively by automaton 3 and both positively and negatively by

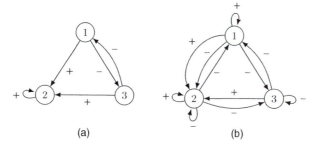

Figure 6.1 The two Boolean networks presented in Example 6.1: (a) Boolean network f and its associated influence graph $\mathscr{G}_f$; (b) Boolean network g and its associated influence graph $\mathscr{G}_g$.

automaton 1 and itself. This leads to obtain $\mathscr{G}_f$ and $\mathscr{G}_g$ depicted in Figure 6.1. From local function definitions as well as influence graph constructions, it appears that f is locally monotone, which is not the case for g. Indeed, function g_2 has a XOR operator that is non-monotone, which leads $\mathscr{G}_g$ not to be simple; indeed, $\{(1, -, 2), (1, +, 2), (2, -, 2), (2, +, 2)\} \subseteq E_g$.

6.2.3 Boolean Network Dynamics

A dynamical system describes the temporal evolution of the set of its *configurations*, being $\mathbb{B}^n$ in the case of Boolean networks. A Boolean network f can lead to several distinct ways of computing the possible evolutions of a configuration $x \in \mathbb{B}^n$ depending on how the latter gets updated according to $f(x)$. Thus, specifying this *updating mode* is a compulsory part of Boolean network analysis and modelling, which asks for introducing several concepts.

6.2.3.1 Updates

In a configuration, events update the state of one or more automata. Suppose that $x \in \mathbb{B}^n$ is the current configuration of a network f of dimension n whose influence graph is $\mathscr{G}_f = (\llbracket n \rrbracket, E_f)$. We say that automaton $i \in \llbracket n \rrbracket$ is *updated* if its state x_i becomes $f_i(x)$. If $f_i(x) = x_i$, then the update of i is not effective in x. Such a local event (even ineffective) leads to what is called an *update* of configuration at the global level, which is described by the *updating function* $\phi_i : \mathbb{B}^n \to \mathbb{B}^n$ of automaton i such that:

$$\forall x \in \mathbb{B}^n, \; \phi_i(x) = (x_1, \ldots, x_{i-1}, f_i(x), x_{i+1}, \ldots, x_n).$$

Do not confuse $f_i(x)$, $\phi_i(x)$, and $f(x)_i$. The first notation refers to the state of automaton i after the execution of its local function on x, the second one refers to the configuration obtained after the execution of the updating function ϕ_i on x, and the third one refers to the state of automaton i after the execution of (global) function f on x. More generally, the updating function extends to subsets of $[\![n]\!]$. Notice that abusing notations here for the sake of clarity, we have $\phi_i(x) = \phi_{\{i\}}(x)$. Formally, given $W \subseteq [\![n]\!]$, the updating function $\phi_W : \mathbb{B}^n \to \mathbb{B}^n$ represents the configuration change caused by the update of all the automata belonging to W and is such that:

$$\forall x \in \mathbb{B}^n, \forall i \in [\![n]\!], \quad \phi_W(x)_i = \begin{cases} f_i(x) & \text{if } i \in W, \\ x_i & \text{otherwise.} \end{cases}$$

As an illustration, consider the network f defined in Figure 6.1 and let us focus on four distinct updates on its configurations. The first update is ineffective and consists in changing nothing. The second update changes the state of automaton 1 by application of ϕ_1, the third one changes the states of both automata 2 and 3 by application of $\phi_{\{2,3\}}$, and the fourth one changes the state of every automaton by application of $\phi_{[\![n]\!]}$. Notice that applying $\phi_{[\![n]\!]}$ to a configuration is equivalent to applying directly f on this configuration. Table 6.1 presents the results of these updates. Until now, we have defined the classical framework of updates, in which updates correspond to executions of one or more local functions simultaneously. We will see later (when we introduce the most permissive updating mode) that we can extend this framework and thus consider more complex ways of updating a Boolean network.

Table 6.1 Configurations, local functions $\left((f_i)_{i \in [\![3]\!]} \right)$, and four updating functions ($\phi_\emptyset, \phi_1, \phi_{\{2,3\}}$, and $\phi_{[\![3]\!]}$) of Boolean network f presented in Example 6.1 and depicted in Figure 6.1.a

$x = (x_1, x_2, x_3)$	$f_1(x)$	$f_2(x)$	$f_3(x)$	$\phi_\emptyset(x)$	$\phi_1(x)$	$\phi_{\{2,3\}}(x)$	$\phi_{[\![3]\!]}(x) \equiv f(x)$
(0, 0, 0)	1	0	1	(0, 0, 0)	(1, 0, 0)	(0, 0, 1)	(1, 0, 1)
(0, 0, 1)	0	0	1	(0, 0, 1)	(0, 0, 1)	(0, 0, 1)	(0, 0, 1)
(0, 1, 0)	1	0	1	(0, 1, 0)	(1, 1, 0)	(0, 0, 1)	(1, 0, 1)
(0, 1, 1)	0	1	1	(0, 1, 1)	(0, 1, 1)	(0, 1, 1)	(0, 1, 1)
(1, 0, 0)	1	0	0	(1, 0, 0)	(1, 0, 0)	(1, 0, 0)	(1, 0, 0)
(1, 0, 1)	0	0	0	(1, 0, 1)	(0, 0, 1)	(1, 0, 0)	(0, 0, 0)
(1, 1, 0)	1	1	0	(1, 1, 0)	(1, 1, 0)	(1, 1, 0)	(1, 1, 0)
(1, 1, 1)	0	1	0	(1, 1, 1)	(0, 1, 1)	(1, 1, 0)	(0, 1, 0)

6.2.3.2 Transitions and Trajectories

Informally speaking, a transition is a couple $(x, y) \in \mathbb{B}^n \times \mathbb{B}^n$ that represents the change of configuration x into configuration y operated by a series of events (possibly composed of only one event). The transitions that come from a unique update are said to be *elementary*. Conversely, those that come from a series of several updates are said to be *non-elementary*. More formally, given a Boolean network f, an *elementary transition* $(x, y) \in \mathbb{B}^n \times \mathbb{B}^n$ of f corresponds to the update in x of any subset $W \neq \emptyset \subseteq [\![n]\!]$ of automata, i.e. such that $y = \phi_W(x)$. By convention, we denote them by $x \to_f y$ or $x \xrightarrow{W}_f y$. There exist two kinds of elementary transitions: the *asynchronous transitions* are such that $W \neq \emptyset \subsetneq [\![n]\!]$; the *synchronous transitions* are such that $W = [\![n]\!]$. The reflexive and transitive closure of $\to_f$ is denoted by $\to_f^*$ and is defined as: given two configurations $x, y \in \mathbb{B}^n$, $x \to_f^* y$ if and only if there exists a series of transitions that changes x into y. Consider now the non-elementary transitions of the form $(x, y) \in \mathbb{B}^n \times \mathbb{B}^n$ of f denoted by $x \to_f^* y$. They correspond to a series of several elementary transitions such that: $x \to_f^* y \iff \exists p \geq 2, \exists x^1, \ldots, x^{p-1} \in \mathbb{B}^n, x \to_f x^1 \to_f \cdots \to_f x^{p-1} \to_f y$. In other terms, any *non-elementary transition* $x \to_f y$ is a vector of sets $(W_k)_{1 \leq k \leq p}$ such that $y = \phi_{W_p} \circ \cdots \circ \phi_{W_1}(x)$ and can be represented by $x \xrightarrow{W_1, \ldots, W_p}_f y$. Trajectories are vectors of (elementary or not) transitions $((x^0, x^1), (x^1, x^2), \ldots, (x^{p-1}, x^p))$ such that $x^0 \to_f x^1 \to_f x^2 \to_f \cdots \to_f x^{p-1} \to_f x^p$. Let us illustrate theoretically these definitions and notations by the following trajectory of an arbitrary Boolean network f of dimension $n \geq 3$, given $i, j \in [\![n]\!]$ and $W_1 = \{1, \ldots, \lfloor \frac{n}{2} \rfloor\}, W_2 = \{\lceil \frac{n}{2} \rceil, \ldots, n\} \subsetneq [\![n]\!]$:

$$x^0 \xrightarrow{[\![n]\!]}_f (x^1 = f(x^0)) \xrightarrow{W_1, W_2}_f (x^2 = \phi_{W_2} \circ \phi_{W_1}(x^1))$$
$$\xrightarrow{\{i,j\}}_f (x^3 = \phi_{\{i,j\}}(x^2))$$

This trajectory is composed of three transitions among which

- two are elementary: $x^0 \xrightarrow{[\![n]\!]}_f x^1$ is a synchronous transition and $x^2 \xrightarrow{\{i,j\}}_f x^3$ is an asynchronous one;
- one is non-elementary, $x^1 \xrightarrow{W_1, W_2}_f x^2$, and could be decomposed into $x^1 \xrightarrow{W_1}_f y \xrightarrow{W_2}_f x^2$, namely, a series of two elementary transitions.

Now, for the sake of clarity, consider configuration $(1, 1, 1)$ of Boolean network f defined in Figure 6.1. We know that $W_1 = \{1, \ldots, \lfloor \frac{n}{2} \rfloor\} = \{1\}$ and $W_2 = \{\lceil \frac{n}{2} \rceil, \ldots, n\} = \{2, 3\}$; suppose that $i = 1$ and $j = 3$. With Table 6.1, it is easy to compute the previous trajectory on it:

$$x^0 = (1,1,1) \xrightarrow{[\![n]\!]}_f x^1 = (0,1,0) \xrightarrow{W_1,W_2}_f x^2 = (1,1,0)$$
$$\xrightarrow{\{1,3\}}_f x^3 = (1,1,0)$$

6.2.3.3 Updating Mode and Transition Graph

A dynamical system associated with a Boolean network f is the combination of f with an updating mode. In other words, the dynamics of f depends on the chosen way to proceed with automata updates, from which of course the trajectories to be considered depend. From a general standpoint, the updating modes define restrictions on the set of imaginable updates (and thus of the set of underlying transitions) allowed in the set of configurations of a network. Thus, given a Boolean network, choosing an updating mode gives formally a framework to the study of its dynamics. Here, we provide the main definitions related to Boolean network dynamics by paying particular attention to updating modes and transition graphs. Observe that, given f of dimension n and an updating mode μ, the dynamical system (f,μ) defines a binary *transition relation* between configurations of $\mathbb{B}^n$ denoted by $\rightarrow_{(f,\mu)} \subseteq \mathbb{B}^n \times \mathbb{B}^n$.

6.2.3.3.1 Main Concepts Related to Dynamical Systems

Let (f,μ) be the dynamical system associated with Boolean network f of dimension n and updating mode μ. This dynamical system can be represented by a directed graph $\mathcal{D}_{(f,\mu)} = (\mathbb{B}^n, \rightarrow_{(f,\mu)})$, where $\rightarrow_{(f,\mu)}$ represents the set of realizable transitions. This graph is usually called the *transition graph* of (f,μ). Now, as for transitions above, we can define the reflexive and transitive closure of relation $\rightarrow_{(f,\mu)}$, denoted by $\rightarrow^*_{(f,\mu)}$ as follows: given two configurations $x,y \in \mathbb{B}^n$, we have $x \rightarrow^*_{(f,\mu)} y$ if and only if there exists a path from x to y in $\mathcal{D}_{(f,\mu)}$. Configuration $x \in \mathbb{B}^n$ is said to be *transient* if there exists a configuration y such that $(x,y) \in \rightarrow^*_{(f,\mu)}$ and $(y,x) \notin \rightarrow^*_{(f,\mu)}$. Configurations that are not transient are called *limit configurations*. Because n is finite, these configurations induce the terminal strongly connected components of $\mathcal{D}$, called the *limit sets* of (f,μ). If there exists at least one trajectory from a transient configuration to a limit set, then this limit set is called an *attractor* of (f,μ) (Cosnard and Demongeot, 1985, Milnor, 1985). The *basin of attraction* of an attractor $\mathcal{A}$ of (f,μ), denoted by $\mathcal{B}(\mathcal{A})$, is the sub-graph of $\mathcal{D}_{(f,\mu)}$ induced by the set of transient configurations x such that, for any limit configuration y belonging to $\mathcal{A}$, $(x,y) \in \rightarrow^*_{(f,\mu)}$. A limit set of cardinal 1, i.e. composed of a unique limit configuration x that only admits $(x,x) \in \rightarrow_{(f,\mu)}$

as possible outgoing transitions, is called a *fixed point* (or stable configuration) of (f, μ). A limit set of cardinal greater than 1, i.e. composed of configurations that can reach each other without reaching the others, is called a *limit cycle* (or sustained oscillation) of (f, μ). Now that the main general concepts on dynamical systems are clarified, some classical updating modes are going to be defined. Let us start by specifying that, considering Boolean networks as mathematical objects evolving over discrete time, it is common to denote by x^t (resp. x_i^t) an image (or the image, depending on the context) of configuration $x = x^0$ (resp. the state of the automaton $i \in [\![n]\!]$) at step (of time) $t \in \mathbb{N}$. It is essential to understand that it is impossible to present all updating modes, simply because of a combinatorial argument. If we consider a Boolean network f of dimension n and only the deterministic updating modes based on updates, such as those that have been formally presented until now, the most general way to define an updating mode is as an infinite vector of subsets of $[\![n]\!]$. From this, with Cantor's diagonal argument, it is easy to derive that the set of such deterministic updating modes is an uncountable set. Therefore, in order to avoid confusions and misleadings, our following presentation separates voluntarily the deterministic updating modes from the non-deterministic ones. Deterministic updating modes are such that each configuration has a unique image, i.e. given a configuration, only one outgoing transition is possible. Non-deterministic updating modes are such that each configuration may admit several images, i.e. given a configuration, distinct outgoing transitions are possible.

6.2.3.4 Deterministic Updating Modes
6.2.3.4.1 Block-Sequential Updating Modes
Since the works of Robert (1980, 1986), the community working on Boolean networks, and more generally on automata networks, has paid particular attention to a specific family of deterministic and periodic updating modes, classically called *block-sequential updating modes* in the literature (Demongeot et al., 2008a,b, Aracena et al., 2009, Goles and Noual, 2010) and introduced as series-parallel updating modes by Robert. Informally, given a Boolean network f of dimension n, the idea of such modes is to partition the automata of $[\![n]\!]$ into disjoint blocks (or subsets), and to make automata of one block execute their updating functions synchronously (or in parallel) while blocks are iterated in series and periodically. Let E be an arbitrary finite set. A vector E' composed of subsets of E is an *ordered partition* of E if all the elements of E' are non-empty, pairwise disjoint, and if their union equals E. Such an updating mode $\mu = \text{bs}$ admissible for Boolean network f of dimension n is defined as an ordered partition $(W_1, \ldots, W_p)$ of $[\![n]\!]$, with $p \leq |[\![n]\!]|$.

6 Boolean Networks and Their Dynamics: The Impact of Updates

Definition 6.1 Let f be a Boolean network of dimension n and $\text{bs} = (W_1, \ldots, W_p)$ an ordered partition of $[\![n]\!]$. Dynamical system (f, bs) is defined by the transition graph $\mathcal{D}_{(f,\text{bs})} = (\mathbb{B}^n, \to_{(f,\text{bs})} \subseteq \mathbb{B}^n \times \mathbb{B}^n)$ where

$$\forall x, y \in \mathbb{B}^n, \ x \to_{(f,\text{bs})} y \iff y = \phi_{W_p} \circ \cdots \circ \phi_{W_1}(x)$$

The transitions operated in such dynamical systems are non-elementary, except in one case, when $\text{bs} = [\![n]\!]$, i.e. when all the transitions are synchronous. This latter case is certainly the most classical from the theoretical standpoint because the underlying dynamical system $(f, ([\![n]\!]))$ is directly defined by f such that for every $x \in \mathbb{B}^n$, its image is $\phi(x) = f(x)$. This is called the *synchronous* or *parallel* updating mode. Notice also that the number of possible block-sequential updating modes is exponential in the number of automata n, as it is given by the Fubini number whose recurrence formula is $\#_{\text{bs}}(n) = \sum_{i=0}^{n-1} \binom{n}{i} \#_{\text{bs}}(i)$, with $\#_{\text{bs}}(0) = 1$. As an illustration, consider Boolean network f of dimension 3 defined in Figure 6.1. Since it is composed of 3 automata, it admits 13 distinct block-sequential updating modes. If we compute the 13 underlying dynamical systems, it appears that 6 among them are different. Their respective transition graphs are depicted in Figure 6.2. Let us now focus on $\mathcal{D}_{(f,(\{1,3\},\{2\}))}$ pictured on the upright corner of that figure. We can see that the dynamical system $(f, (\{1, 3\}, \{2\}))$ has five different limit sets, four fixed points (001 ↻, 011 ↻, 100 ↻, and 110 ↻), which are not attractors, and one limit cycle of length 2 (000 ⇄ 101), which is an attractor whose basin of attraction is $\{010, 111\}$. The whole figure shows another interesting feature: the fixed points are preserved, regardless the block-sequential updating mode, which is not the case for the limit cycle. We will formally speak of this feature in Section 6.4.1. Block-sequential updating modes are definitely relevant from the mathematical standpoint. Their study has emphasized strong properties about the convergence of dynamical systems toward limit sets (Robert, 1986, Goles and Martínez, 1990), about their influence on these limit sets (Aracena et al., 2009, Elena, 2009, Goles and Salinas, 2010, Aracena et al., 2011, 2013), and we are still far from having elucidated each of their underlying dynamical consequences. They also have brought interesting knowledge in the framework of genetic regulation network modelling (Mendoza and Alvarez-Buylla, 1998, Aracena et al., 2006, Demongeot et al., 2010, Ruz et al., 2014). Nevertheless, the fact that they are defined by means of ordered partitions of the set of automata is a strong constraint. First, they are periodic. This is a quite convenient mathematical restriction as it allows us to leave the uncountable universe of deterministic updating modes and enter into a countable one as soon as period $p \in \mathbb{N}$ is

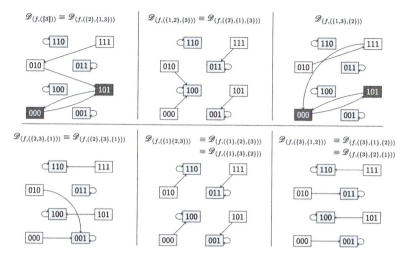

Figure 6.2 The six possible distinct block-sequential dynamics of Boolean network f defined in Figure 6.1 represented by their associated transition graphs $\mathcal{D}_{(f,bs)}$, where bs is among the 13 possible ordered partitions of $[\![3]\!]$.

determined and finite. Second, they make every automaton update its state exactly once during an updating period. The fact that every automaton updates its state at least once during a period is actually a good property because what would be the point of an automaton which never updates its state? Last but not least, the fact that it does only once per period prevents from observing peculiar phenomena. For instance, given a network, this makes it impossible for one of its subnetworks to act with its own clock, distinct from that of the rest of the network. Considering such biological phenomenology asks for conceiving more expressive and more relaxed deterministic updating modes in the framework of Boolean networks. Notice that more expressive deterministic periodic updating modes have never been studied in depth *per se*, even if they are evoked in Goles and Noual (2010) and Demongeot and Sené (2020).

6.2.3.4.2 Block-Parallel Updating Modes

A natural way to design new deterministic, periodic updating modes rests on the property above which says that each automaton needs to update its state at least once during a period. This leads to focus on periodic updating modes of arbitrary period $p \in \mathbb{N}$ defined as infinite periodic vectors $(W_1, \ldots, W_p, W_1, \ldots, W_p, \ldots)$ such that $\forall i \in \{1, \ldots, p\}, W_i \subseteq [\![n]\!]$ and $\bigcup_{i=1}^{p} W_i = [\![n]\!]$. For the sake of clarity, such updating modes are usually denoted by means of finite vectors such as $(W_i)_{i \in \mathbb{N}/p\mathbb{N}}$. From now on, given a dynamical system (f, μ), let us distinguish time steps from elementary time steps. There is one *time step* between configuration $x \in [\![n]\!]$ and another $y \in [\![n]\!]$ if $x \to_{(f,\mu)} y$. If this transition is non-elementary, as explained above, it can be decomposed into elementary ones and we say that there are p *elementary time steps* between x and y if $x \to_{(f,\mu)} y \iff x = x^0 \to_{\phi_{w_1}} x^1 \to_{\phi_{w_2}} \cdots \to_{\phi_{w_{p-1}}} x^{p-1} \to_{\phi_{w_p}} x^p = y$. These periodic updating modes are called *fair periodic updating modes* because for any elementary time step $t \in \mathbb{N}$, there exists $k \in \{1, \dots, p\}$ such that every automaton of $[\![n]\!]$ is updated in the elementary time interval $[t; t+k]$. Nevertheless, whilst the set of fair updating modes is indeed countable, it is too much big to hope studying it in depth. Indeed, for a network of dimension n and given a period p, let us argue that it is greater than the number of coverings of dimension p of $[\![n]\!]$, which means greater than $\sum_{k=0}^{n}(-1)^k \binom{n}{k}\binom{2^{n-k}-1}{p}$ (Comtet, 1966), because of the order on subsets and the subset repetition availability. To constrain this set of possible periodic updating modes, consider the dual of the set of block-sequential updating modes, that is the set of block-parallel updating modes. Informally, the idea of such modes is to partition automata of $[\![n]\!]$ into disjoint blocks and to make the automata of one block execute their updating functions sequentially while blocks are iterated in parallel. Formally, a *partitioned order* of $[\![n]\!]$ is a set $\{S_k\}_{1 \leq k \leq s}$, with $1 \leq s \leq |[\![n]\!]|$, such that:

- $\forall k \in \{1, \dots, s\}, \vec{S_k} \neq \emptyset$ is a vector of automata of $[\![n]\!]$ without repetitions;
- $\forall i \in [\![n]\!], i \in S_k \implies \forall \ell \in \{1, \dots, s\} \setminus \{k\}, i \notin S_\ell$;
- $\|_{k=1}^{s} S_k$ covers $[\![n]\!]$.

From this, we derive that a *block-parallel updating mode* $\mu = $ bp admissible for a Boolean network f of dimension n is a partitioned order $\{S_1, \dots, S_s\}$ of $[\![n]\!]$, with $s \leq |[\![n]\!]|$. Notice that the number of partitioned orders of a set equals that of its ordered partitions.

Proposition 6.1 Given a Boolean network f of dimension n, block-parallel updating modes admissible for f are fair periodic updating modes.

Proof: Given f and n, any block-parallel updating mode defined as a partitioned order $\{S_k\}_{1 \leq k \leq s}$ of $[\![n]\!]$ can be rewritten into a periodic vector $(W_\ell)_{1 \leq \ell \leq p}$, with period $p = \text{lcm}(|S_k|_{1 \leq k \leq s})$, of subsets of $[\![n]\!]$ of cardinal s, such that:

$$\forall \ell \in \{1, \dots p\}, W_\ell = \{i \in [\![n]\!] \mid \exists k, \exists j \in \{1, \dots, |S_k|\},$$
$$\exists m \in \mathbb{N}, (S_k)_j = i \text{ and } \ell = j + m|S_k|\}$$

Let us denote this rewriting by the non injective function rw such that $\text{rw}(\{S_k\}_{1\leq k\leq s}) = (W_\ell)_{1\leq \ell \leq p}$, and let us denote its right inverse by $\text{rw}_{\text{bp}}^{-1}$. The construction of rw implies notably that $\forall i \in [\![n]\!], \exists \ell \in \{1,\ldots,p\}, i \in W_\ell$. □

Definition 6.2 Let f be a Boolean network of dimension n and $\text{bp} = \{S_1,\ldots,S_s\}$ a partitioned order of $[\![n]\!]$. Dynamical system (f,bp) is defined by the transition graph $\mathcal{D}_{(f,\text{bp})} = (\mathbb{B}^n, \to_{(f,\text{bp})} \in \mathbb{B}^n \times \mathbb{B}^n)$ such that:

$$\forall x, y \in \mathbb{B}^n, \ x \to_{(f,\text{bp})} y \iff y = \phi_{W_p} \circ \cdots \circ \phi_{W_1}(x)$$

where $(W_\ell)_{1\leq \ell \leq p} = \text{rw}(\text{bp})$ and $p = \text{lcm}(|S_k|_{1\leq k\leq s})$.

As a first illustration, consider a Boolean network f of dimension 6 and the block-parallel updating mode $\text{bp} = \{(2,4,6),(5,1),(3)\}$. One can derive from the latter that automata 2, 4, and 6 are updated every three elementary time steps, automata 5 and 1 every two elementary time steps, and automaton 2 at each elementary time step. More precisely, automaton 2 (resp. 4 and 6) is updated at each elementary time step $t \in \mathbb{N}$ such that $t \equiv 0 \mod 3$ (resp. $t \equiv 1 \mod 3$, and $t \equiv 2 \mod 3$), automaton 5 (resp. 1) is updated at each elementary time step $t \in \mathbb{N}$ such that $t \equiv 0 \mod 2$ (resp. $t \equiv 1 \mod 2$). This corresponds exactly to the following periodic updating mode: $(\{2,3,5\},\{1,3,4\},\{3,5,6\},\{1,2,3\},\{3,4,5\},\{1,3,6\}) = \text{rw}(\text{bp})$. An insight of this rewriting is given in Figure 6.3. Now, given a configuration $x = (x_1, x_2, x_3, x_4, x_5, x_6)$, its global updating according to bp gives the following transition:

$$x^0 = (x_1,x_2,x_3,x_4,x_5,x_6) \to_{(f,\text{bp})} x^1$$
$$= \phi_{\{1,3,6\}} \circ \phi_{\{3,4,5\}} \circ \phi_{\{1,2,3\}} \circ \phi_{\{3,5,6\}} \circ \phi_{\{1,3,4\}} \circ \phi_{\{2,3,5\}}(x^0).$$

ϕ_2	ϕ_4	ϕ_6	ϕ_2	ϕ_4	ϕ_6	ϕ_2	ϕ_4	ϕ_6
ϕ_5	ϕ_1	ϕ_5	ϕ_1	ϕ_5	ϕ_1	ϕ_5	ϕ_1	ϕ_5
ϕ_3	ϕ_3	ϕ_3	ϕ_3	ϕ_3	ϕ_3	ϕ_3	ϕ_3	ϕ_3

0.0   0.1   0.5   1.0   1.1 t

First period   Second …

Figure 6.3 Insight of the rewriting of block-parallel updating mode $\text{bp} = \{(2,5,6),(5,1),(3)\}$ into a classical periodic updating mode form. This corresponds to the elementary time diagram of updating function executions. Two periods are pictured, the first one entirely, the second one partially. In the first period, the rectangles filled in light gray come from a reading in columns of the diagram and represent the subsets W_ℓ of three automata evolving in parallel. In the second period, the rectangles filled in gray come from a reading in row of the diagram and represent the sub-vectors S_k of automata evolving sequentially.

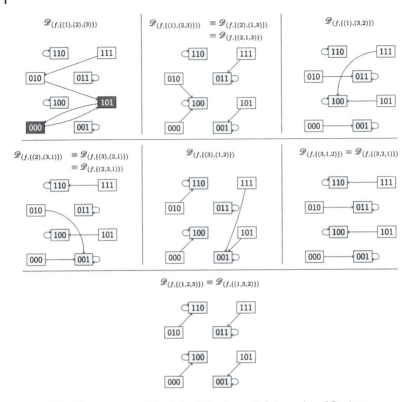

Figure 6.4 The seven possible distinct block-parallel dynamics of Boolean network f defined in Figure 6.1 represented by their associated transition graphs $\mathcal{D}_{(f,\text{bp})}$, where bp is amongst the 13 possible partitioned orders of $[\![3]\!]$.

As a second illustration, consider again Boolean network f of dimension 3 given in Figure 6.1. There exist 13 block-parallel updating modes admissible for it. Seven underlying dynamical systems are different which are depicted in Figure 6.4. Notice that five of them correspond to dynamical systems emerging from block-sequential updating modes given in Figure 6.2. What is interesting at this stage is that two of them are not: the one in the up right frame, and the one in the middle central frame. This shows that block-parallel updating modes can lead to dynamical systems not observable by means of block-sequential updating modes. Let us denote the set of block-parallel updating modes defined on $[\![n]\!]$ viewed as periodic modes by $\{\text{rw}(\text{bp}([\![n]\!]))\}$, and the set of block-parallel updating modes defined on $[\![n]\!]$ by $\{\text{bs}([\![n]\!])\}$. The following properties hold:

- $|\{\text{rw}(\text{bp}([\![n]\!]))\}| \leq |\{\text{bs}([\![n]\!])\}|$. The number of partitioned orders is upper-bounded by the number of ordered partitions since rewriting function rw is not injective. Indeed, for $n \geq 4$, there exist specific partitioned orders in which subvectors of automata lead to the same rewriting into vector of automata subsets. For instance, $\text{rw}(\{(1,2),(3,4)\}) = \text{rw}(\{(1,4),(3,2)\}) = (\{1,3\},\{2,4\})$.
- $\{\text{rw}(\text{bp}([\![n]\!]))\} \neq \{\text{bs}([\![n]\!])\}$, since updating repetitions in a period are allowed by block-parallel updating modes. Considering Boolean networks (and *a fortiori* automata networks), a consequence of this is that, given a network f, the set of its underlying block-parallel dynamical systems is not necessarily the set of its underlying block-sequential dynamical systems.
- $\{\text{rw}(\text{bp}([\![n]\!]))\} \cap \{\text{bs}([\![n]\!])\} \neq \emptyset$. Indeed, the parallel and the $n!$ sequential updating modes are both block-parallel and block-sequential updating modes.
- If bp is a partitioned order $\{S_k\}_{1 \leq k \leq s}$ such that $\forall k, \ell \in \{1, \ldots, s\}$, $|S_k| = |S_\ell| = c$, then rw(bp) is a block-sequential updating mode whose subset cardinals equal s and whose period is c. Conversely, if bs is an ordered partition $(W_k)_{1 \leq k \leq p}$ such that $\forall k, \ell \in \{1, \ldots, p\}$, $|W_k| = |W_\ell| = c$, then $\text{rw}_{\text{bp}}^{-1}(\text{bs})$ is a block-parallel updating mode of cardinal c whose sub-vector dimensions equal p. In other terms, rw is a bijection for such block-parallel and block-sequential updating modes sets and $\text{rw}_{\text{bp}}^{-1}$ is the (two-sided) inverse of rw in this case.

These properties are interesting *per se* since they emphasize that the block-parallel and block-sequential families of periodic updating modes are different whilst they share some properties. However, it would be of better interest to have families of updating modes which integrate both of them. To this end, let us introduce a generalization of block-parallel updating modes.

6.2.3.4.3 Toward Local-Clocks Updating Modes

Block-Parallel Generalization The general idea here consists in restarting from block-parallel updating modes and to extend them by allowing automata of distinct blocks to synchronize their updatings, thanks to the concept of *waiting delay*. To this end, consider the set $[\![n]\!] \cup \{0\}$, where element 0 is a fictitious automaton with no local function which will serve as a kind of waiting delay. Formally, a *partitioned order of $[\![n]\!]$ with delay* is a set $\{S_k\}_{1 \leq k \leq s}$, with $1 \leq s \leq n$, for which the following properties hold:

- $\forall k \in \{1, \ldots, s\}$, $\vec{S_k} \neq \vec{\emptyset}$ is a vector of automata of $[\![n]\!] \cup \{0\}$ without repetitions of elements of $[\![n]\!]$, and such that $\exists i \in [\![n]\!]$, $i \in S_k$;

- $\forall i \in [\![n]\!], i \in S_k \implies \forall \ell \in \{1, \ldots, s\} \setminus \{k\}, i \notin S_\ell$;
- $\|_{k=1}^{s} S_k$ covers $[\![n]\!]$;

From this, we derive that a *general block-parallel updating mode* $\mu = \mathsf{gbp}$ admissible for a Boolean network f of dimension n is a partitioned order $\{S_1, \ldots, S_s\}$ of $[\![n]\!]$ with delay, with $s \leq n$. This leads to the following proposition whose proof is similar to that of Proposition 6.1. In this case, however, the left inverse of rw is distinct from $\mathsf{rw}_{\mathsf{bp}}^{-1}$ since it needs to generate 0s in sub-vectors, and is denoted by $\mathsf{rw}_{\mathsf{gbp}}^{-1}$.

Proposition 6.2 Given a Boolean network f of dimension n, general block-parallel updating modes admissible for f are fair periodic updating modes.

Definition 6.3 Let f be a Boolean network of dimension n and $\mathsf{gbp} = \{S_1, \ldots, S_s\}$ a partitioned order of $[\![n]\!]$ with delay. Dynamical system (f, gbp) is defined by the transition graph $\mathcal{D}_{(f,\mathsf{gbp})} = (\mathbb{B}^n, \to_{(f,\mathsf{gbp})} \subseteq \mathbb{B}^n \times \mathbb{B}^n)$ such that:

$$\forall x, y \in \mathbb{B}^n, \quad x \to_{(f,lc)} y \iff y = \phi_{W_p} \circ \ldots \circ \phi_{W_1}(x)$$

where $(W_\ell)_{1 \leq \ell \leq p} = \mathsf{rw}(\mathsf{gbp})$ and $p = \mathsf{lcm}(|S_k|_{1 \leq k \leq s})$.

As a first illustration, consider a Boolean network f of dimension 4 and the general block-parallel updating mode $\mathsf{gbp} = \{(1, 0), (0, 2, 0), (3), (0, 4)\}$. One can derive from it that automaton 1 (resp. 2, 3, and 4) executes its updating function at elementary time step $t = 0$ (resp. $t = 1$, $t = 0$, and $t = 2$), and then every two (resp. three, one, two) elementary time steps. This corresponds exactly to the following periodic updating mode: $(\{1, 3\}, \{2, 3, 4\}, \{1, 3\}, \{3, 4\}, \{1, 2, 3\}, \{3, 4\})$. In other words, we have:

$$\mathsf{rw}(\mathsf{gbp}) = (\{1, 3\}, \{2, 3, 4\}, \{1, 3\}, \{3, 4\}, \{1, 2, 3\}, \{3, 4\}), \text{ and}$$

$$\mathsf{rw}_{\mathsf{gbp}}^{-1}((\{1, 3\}, \{2, 3, 4\}, \{1, 3\}, \{3, 4\}, \{1, 2, 3\}, \{3, 4\}))$$

$$= \mathsf{gbp}$$

An insight of this rewriting is presented in Figure 6.5. As a second illustration, consider once again Boolean network f of dimension 3 given in Figure 6.1, and the three following general block-parallel updating modes admissible for it:

- $\mathsf{gbp}_1 = \{(1, 0), (2), (3, 0, 0)\}$ is an updating mode of period 6 which makes automaton 1 (resp. 2, and 3) update its state at each two (resp. one, and three) elementary time steps from $t = 0$;

- $\text{gbp}_2 = \{(1,0,2),(3,0)\}$ is an updating mode of period 6 which makes automaton 1 (resp. 2, and 3) update its state at each three (resp. three, and two) elementary time steps from $t=0$ (resp. $t=2$, and $t=0$);
- $\text{gbp}_3 = \{(1,2),(3,0)\}$ is an updating mode of period 2 which makes automata 1 and 3 update their states at each two elementary time steps from $t=0$, and automaton 2 update its state at each two elementary time steps from $t=1$.

Their associated dynamical systems are depicted in Figure 6.6. First, remark that $\mathscr{D}_{(f,\{(1,0),(2),(3,0,0)\})} = \mathscr{D}_{(f,(\{1,2,3\},\{2\},\{1,2\},\{2,3\},\{1,2\},\{2\}))}$ is neither a block-sequential updating mode nor a block-parallel one. This emphasizes that general block-parallel updating modes can generate dynamics which are not observable thanks to block-sequential and block-parallel updating modes. Second, remark that $\mathscr{D}_{(f,\{(1,0,2),(3,0)\})} = \mathscr{D}_{(f,(\{1,3\},\emptyset,\{2,3\},\{1\},\{3\},\{2\}))}$ equals $\mathscr{D}_{(f,\{(3),(1,2)\})} = \mathscr{D}_{(f,(\{1,3\},\{1,2\}))}$. This emphasizes two distinct relevant features:

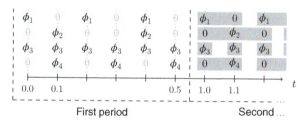

Figure 6.5 Insight of the rewriting of general block-parallel updating mode $\text{gbp} = \{(1,0),(0,2,0),(3),(0,4)\}$ into a classical periodic updating mode form. This corresponds to the elementary time diagram of updating functions executions. Two periods are pictured, the first one entirely, the second one partially. In the first period, the rectangles filled in light gray come from a reading in columns of the diagram and represent the subsets W_ℓ of automata of $[\![4]\!]$ evolving in parallel. In the second period, the rectangles filled in gray come from a reading in row of the diagram and represent the sub-vectors S_k of automata evolving sequentially.

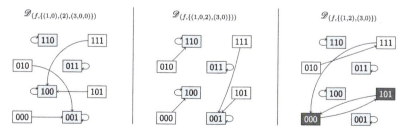

Figure 6.6 Three possible distinct general block-parallel dynamics of Boolean network f defined in Figure 6.1 represented by their associated transition graphs.

- The periodical form of a general block-parallel updating mode can be partially composed of empty subsets which correspond to elementary time steps where nothing happens. In this example, considering configuration $x = (x_1, x_2, x_3)$, its global updating according to gbp_2 gives the following transition:

$$x^0 = (x_1, x_2, x_3) \quad \to_{(f, \mathsf{gbp}_2)} \quad \phi_{\{2\}} \circ \phi_{\{3\}} \circ \phi_{\{1\}} \circ \phi_{\{2,3\}} \circ \phi_{\emptyset} \circ \phi_{\{1,3\}}$$
$$\iff x^0 = (x_1, x_2, x_3) \quad \to_{(f, \mu)} \quad \phi_{\{2\}} \circ \phi_{\{3\}} \circ \phi_{\{1\}} \circ \phi_{\{2,3\}} \circ \phi_{\{1,3\}}$$

- general block-parallel updating modes which are not block-parallel can generate the same dynamics directly observable thanks to block-parallel updating modes. Notice that it is also the case for block-sequential updating modes.

Third, remark that $\mathsf{gbp}_3 = \{(1, 2), (3, 0)\}$ is a block-sequential updating mode by definition since $\mathsf{rw}(\mathsf{gbp}_3) = (\{1, 3\}, \{2\})$. Actually, the general block-parallel updating modes of equal sub-vectors dimensions are either block-sequential updating modes or updating modes whose underlying dynamics equals a block-sequential dynamics (they can have empty subsets in their periodical form).

Local-Clocks Actually, the three previous classes of periodic updating modes "by blocks" are particular sub-classes of *local clocks updating modes* (Wilson, 2021, Wilson and Theyssier, 2021). The general idea underlying the local clocks updating modes rests on a local view, i.e. at the level of the automata of a network, not at the level of the network, of the scheduling of the updates over time. More formally, considering the time scale decomposed according to elementary transitions, given a Boolean network f of dimension n and a periodic updating mode μ, we say that μ is a local clocks updating mode if and only if, for each automaton $i \in [\![n]\!]$, there exist a local period p_i and an initial shift $\rho_i \in \{0, \ldots, m\}$ such that updating function ϕ_i is executed at each time step $t_i \in \mathbb{N}$ such that, for all $k \in \mathbb{N}$, $t_i = \rho_i + k p_i$. In other terms, each automaton has its own initial shift at which it enters in its own local updating period. From this, it is easy to see that block-sequential and (general) block-parallel updating modes respect this property. Indeed, for block-sequential ones of period p, all the automata execute their updating functions once per period, and thus, all have an updating period equal to p. Moreover, each automaton is initially shifted of a number of time steps equal to the number minus 1 of the subset it belongs to in the ordered partition. For (general) block-parallel ones, the local updating period of each i is the cardinal of the sub-vector it belongs to, and its initial shift corresponds to its position minus 1 in the sub-vector. Proposition 6.3 below derives directly from the definition of local clocks updating modes and Proposition 6.2.

Proposition 6.3 Given a Boolean network f of dimension n, local clocks updating modes admissible for f are fair periodic updating modes.

Nevertheless, it is important to notice that local clocks updating modes can lead to elementary dynamics which cannot be captured by the previous ones. This comes from the initial shifts which can force an updating transitory phase before entering into an updating periodic phase. As a consequence, they cannot be directly taken into account in general block-parallel updating modes since the latter do not allow any transitory phase. Indeed, by definition, any initial local shift generated by putting 0s at the beginning of sub-vectors of the partitioned order leads to increase the corresponding local period. As an illustration, consider any Boolean network f composed of 3 automata and the following local clocks updating mode defined by means of the two following vectors: $\rho = (3, 1, 0)$ and $p = (2, 3, 2)$, the initial local shift vector and the local period vector, respectively (see Figure 6.7). This updating mode generates a transitory phase of 3 time steps that cannot be interpreted in the framework of general block-parallel updating modes, and a periodic phase of periods of length 6 that can be. Indeed, the mode $\{(1,0), (0,2,0), (0,3)\} \equiv (\{1\}, \{2,3\}, \{1\}, \{3\}, \{1,2\}, \{3\})$ produces it. More generally, it can be observed that, when there exists at least one initial local shift greater than 1, local clocks updating modes induce a transitory phase of $\max_{i \in [\![n]\!]} (\rho_i) - 1$ time steps that cannot be captured by general block-parallel updating modes.

6.2.3.4.4 Updating Modes Induced by Boolean Networks with Memory

All the periodic updating modes discussed above, and more generally all the periodic updating modes leading to define dynamical systems whose transitions are the results of a global transition function, which is a composition of updating functions, have a particularity with respect to Boolean

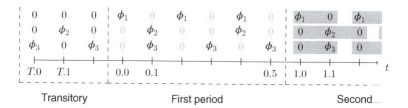

Figure 6.7 Elementary time diagram of updating functions executions under local clocks updating mode $lc = (\rho = (3, 1, 0), p = (2, 3, 2))$. The transitory phase and two periods are pictured, the first one entirely, the second one partially. In the first (resp. second) period, the rectangles filled in light gray (resp. gray) give the periodical reading (resp. the general block-parallel view) of the periodic phase of the updating mode.

networks and their possible dynamics. They are in some sense *context free*. Indeed, given a Boolean network f of dimension n and such an updating mode μ, configurations of $\mathbb{B}^n$, and thus automata of $[\![n]\!]$, evolve over time according to the schedule determined by μ, without taking into account anything else. The underlying dynamical systems are consequently memoryless. To go even further on deterministic updating modes, one could consider updating modes accounting for some context. The literature on Boolean networks recently put the emphasis on such modes through the gene protein Boolean networks model, introduced by Graudenzi and Serra (2009) and Graudenzi et al. (2011a,b). The general initial idea of their work rests on

- considering bipartite Boolean networks with two kinds of automata: a half of them models genes, the other half models their associated one-to-one proteins;
- considering that each protein has its own decay time that defines the number of time steps during, which it remains present in the cell after having been produced by the punctual expression of its associated gene.

Without entering into the details of this formalism, the study presented in Goles et al. (2020) shows that this new model of genetic regulation networks can be viewed as *Boolean networks with memory*, i.e. Boolean networks with an added delay vector so that an activated automaton remains activated during at least a number of time steps equal to its own delay. Formally, a *Boolean network with memory* f of dimension n is defined as a Boolean network f and a delay vector $d \in \mathbb{N}^n$.

Definition 6.4 Let f be a Boolean network of dimension n and $d = (d_1, \ldots, d_n) \in (\mathbb{N} \setminus \{0\})^n$ be a delay vector. The set of its configurations is defined as $X_{(f,d)} = \{(x, \delta) \in \mathbb{B}^n \times \mathbb{N}^n \mid \forall i \in [\![n]\!], \delta_i \in \{0, \ldots, d_i\}, x_i = 0 \iff \delta_i = 0, \text{ and } x_i = 1 \iff \delta_i \in \{1, \ldots, d_i\}\}$. Dynamical system $((f, d), \mathsf{p})$ is defined by the transition graph $\mathcal{D}_{((f,d),\mathsf{p})}$, where p represents the parallel updating mode such that $\mathsf{p} = ([\![n]\!])$, made of transitions based on updating function $\phi^\star : X_{(f,d)} \to X_{(f,d)}$ depending on the delays such that

$$\forall (x, \delta), (y, \delta') \in X_{(f,d)}, (x, \delta) \to_{((f,d),\mathsf{p})} (y, \delta') \iff (y, \delta') = \phi^\star_{[\![n]\!]}(x, \delta)$$

where

$$\forall i \in [\![n]\!], \phi^\star_{[\![n]\!]}(x, \delta)_i = (y_i, \delta'_i)$$

with

$$\delta'_i = \begin{cases} 0 & \text{if } f_i(x) = 0 \text{ and } \delta_i = 0, \\ \delta_i - 1 & \text{if } f_i(x) = 0 \text{ and } \delta_i > 0, \\ d_i & \text{if } f_i(x) = 1 \end{cases}$$

and

$$y_i = \begin{cases} 1 & \text{if } \delta'_i \geq 1, \\ f_i(x) & \text{if } \delta'_i = 0 \end{cases}$$

The previous definition shows that while Boolean networks with memory evolve globally according to the parallel updating mode, automata desynchronization is made possible locally, thanks to the delays whose evolution depends on the local context. In other terms, Boolean networks with memory are Boolean networks whose global dynamics depend on the local context of their automata at each time step. Nevertheless, notice that when $d = (1, \ldots, 1)$, the dynamics of Boolean network with firing memory (f, d) is nothing else but $\mathcal{D}_{(f,p)}$. Let us take once more Boolean network f of dimension 3 defined in Figure 6.1 as an example to illustrate the peculiarities and intricacies of the contextual updating modes at stake in Boolean networks with memory. To this end, let us combine f with delay vector $d = (2, 1, 1)$. For convenience and the sake of clarity, let us add a notation based on the specification of the set of configurations of (f, d), rather than considering configurations in $X_{(f,d)}$, which decouple the Boolean configurations of f from their possible associated delay configurations of $\mathbb{N}^n$; let us consider only the latter. So, for instance, delay configuration $\delta = (2, 0, 1)$ actually corresponds to configuration $(x = (1, 0, 1), d = (2, 0, 1))$ of (f, d). From this, let us abuse the notation and consider that the updating function $\phi^\star_{[\![n]\!]}$ maps the set of delay configurations to itself. Figure 6.8a depicts the deterministic dynamics of this Boolean network with memory (f, d). However, an interesting point is that when the deterministic dynamics is projected on the Boolean configurations of f by getting rid of local delays, then the obtained transition graph is not deterministic anymore, as pictured in Figure 6.8b. Indeed, for instance, configuration $(1, 1, 1)$ can reach both configurations $(0, 1, 0)$ and $(1, 1, 0)$. So, generally speaking, in some sense, working on an arbitrary Boolean network with memory (f, d) of dimension n by considering the projection of its dynamics on its partial configurations of $\mathbb{B}^n$ may involve considering the dynamics of f evolving according to a specific non-deterministic updating mode.

6.2.3.5 Non-deterministic Updating Modes
6.2.3.5.1 Asynchronous Updating Modes
The updating modes defined in the previous section enables specifying which automata should get updated simultaneously, possibly in a given sequence. The *asynchronous* updating mode considers *any* combination of automata to be updated simultaneously: it corresponds to the non-empty elementary transitions introduced in Section 6.2.3.2.

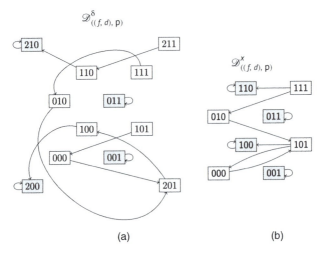

Figure 6.8 (a) Dynamics of Boolean network f with memory $d = (2, 1, 1)$ represented by its deterministic transition table on delay configurations and the associated transition graph $\mathscr{D}^{\delta}_{((f,d),p)}$; (b) projection of the dynamics of (f, d) on the underlying Boolean configurations, for any time step t in $\mathbb{N}$, and its associated transition graph $\mathscr{D}^{x}_{((f,d),p)}$.

Definition 6.5 Let f be a Boolean network of dimension n. Dynamical system (f, a) is defined by the transition graph $\mathscr{D}_{(f,a)} = (\mathbb{B}^n, \rightarrow_{(f,a)} \subseteq \mathbb{B}^n \times \mathbb{B}^n)$ where

$$\forall x, y \in \mathbb{B}^n, \ x \rightarrow_{(f,a)} y \iff \exists W \subseteq [\![n]\!], W \neq \emptyset, y = \phi_W(x)$$

The obtained dynamics are then non-deterministic. Remark that whenever k automata can change of state ($k = |\Delta(x, f(x))|$), then $2^k - 1$ transitions from x are generated by the asynchronous updating mode. From a modelling standpoint, the asynchronous updating mode aims at accounting for state changes that occur at different speed and whenever it is not possible to determine local clocks, due for instance to insufficient knowledge on the system or to an intrinsic stochasticity of the system. As we will detail later in this section, whereas the asynchronous updating mode captures all different time scales of state changes in Boolean automata, it is no longer complete when considering Boolean networks as abstraction of quantitative systems. Interestingly, this asynchronous updating mode has been barely considered so far when modelling biological systems with Boolean networks (where it is usually referred to as the *general asynchronous* updating mode). Indeed, in the modelling framework of René Thomas, one and only automaton can be updated in a transition, leading to the so-called *fully asynchronous* updating mode (usually referred to as asynchronous in the biological systems modelling community). Nevertheless, some applications consider updating modes that mix parallel and fully asynchronous transitions, thus giving (particular) trajectories of the asynchronous updating mode (Fauré et al., 2006, Das and Layek, 2016).

Definition 6.6 Let f be a Boolean network of dimension n. Dynamical system (f, fa) is defined by the transition graph $\mathcal{D}_{(f,\text{fa})} = (\mathbb{B}^n, \rightarrow_{(f,\text{fa})} \subseteq \mathbb{B}^n \times \mathbb{B}^n)$ where

$$\forall x, y \in \mathbb{B}^n, \ x \rightarrow_{(f,\text{fa})} y \iff \exists i \in [\![n]\!], y = \phi_i(x)$$

In Thomas (1991), Thomas justifies this modelling choice in the scope of gene regulatory networks with respect to practical observations on the delays for state changes of genes: "There is absolutely no reason why the time delays (…) should be equal. As a matter of fact, they are often very unequal. (…) This leads us to a fully asynchronous description, in which all the time delays are different in the absence of an accidental coincidence". Another practical advantage of the fully asynchronous updating mode is that there are at most n transitions from one configuration, simplifying greatly the representation of their dynamics. Figure 6.9 pictures the asynchronous and fully asynchronous dynamics of Boolean network f from Example 6.1.

6.2.3.5.2 Non-deterministic Updates

The updating modes considered so far correspond to the application and composition of elementary *deterministic updates* of configurations

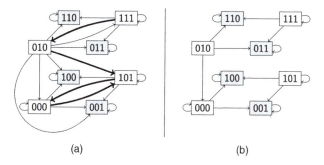

Figure 6.9 Asynchronous (a) and fully asynchronous (b) dynamics of Boolean network f defined in Figure 6.1 represented by their associated transition graphs $\mathcal{D}_{(f,a)}$ and $\mathcal{D}_{(f,\text{fa})}$. Non-loop transitions of the parallel updating mode are drawn in bold.

$\phi : \mathbb{B}^n \to \mathbb{B}^n$. As we have seen above, deterministic updates can generate non-deterministic updating modes by allowing different updates to be applied on the same configuration. The main exception is the case of Boolean networks with memory, where the dynamics has been defined at the end of Section 6.2.3.4 as the projection of a discrete parallel dynamics. Other updating modes have been defined in the literature, which cannot be expressed as the application nor the composition of elementary updates, such as the interval updating mode (Chatain et al., 2018, 2020) and the most permissive updating mode (Paulevé et al., 2020). Indeed, these latter can generate transitions that are neither elementary nor non-elementary transitions. In Paulevé and Sené (2021), we introduced an extension toward *non-deterministic updates*, with a unifying framework for expressing complex updating modes, including those mentioned above. Non-deterministic updates are functions mapping *sets of configurations*, i.e. of the form $\Phi : 2^{\mathbb{B}^n} \to 2^{\mathbb{B}^n}$. We define Φ as a map from sets of configurations to sets of configurations for enabling iterations and compositions of non-deterministic updates. Nevertheless, we assume that for any $X \subseteq \mathbb{B}^n$, $\Phi(X) = \bigcup_{x \in X} \Phi(\{x\})$: one can define Φ only from all singleton configuration sets. This restriction ensures that, for any $X \subseteq \mathbb{B}^n$, each configuration in the image set $y \in \Phi(X)$ can be computed from a singleton set $\{x\}$ for some $x \in \mathbb{B}^n$. In the following, we call such updates *set updates*. Starting from a singleton configuration set $\{x\}$, the iteration of set updates delineates the domains of configurations the system can evolve to. Thus, set updates naturally define transition relations between configurations:

Definition 6.7 Given a set update function Φ for Boolean networks of dimension n, the generated transition relation is given by $\varphi : (2^{\mathbb{B}^n} \to 2^{\mathbb{B}^n}) \to 2^{\mathbb{B}^n \times \mathbb{B}^n}$ with $\varphi(\Phi) = \{(x,y) \mid x \in \mathbb{B}^n, y \in \Phi(\{x\})\}$.

In contrast to the deterministic updates, non-deterministic updating modes can be characterized directly by set updates. Indeed, non-deterministic updating modes allow "superposing" alternative updates to generate different transitions from a single configuration x, although each of them is computed with a deterministic update. An example of this is one update ϕ, where $\phi(x) = y$, and another update ϕ', where $\phi'(x) = y' \neq y$. Now, let us imagine an updating mode superposing two set updates, Φ and Φ', where, for some configurations, $x \in \mathbb{B}^n$, $\Phi(\{x\}) \setminus \Phi'(\{x\}) \neq \emptyset$. One can then build a single set update Φ^* such that $\Phi^*(X) = \Phi(X) \cup \Phi'(X)$. It results that $\varphi(\Phi^*) = \varphi(\Phi) \cup \varphi(\Phi')$; thus, the updating mode can be assimilated to Φ^*. Remark that the limit sets of the generated dynamics $\varphi(\Phi)$ can be characterized as the $\subseteq$-smallest sets of configurations $X \subseteq \mathbb{B}^n$

such that $\Phi(X) = X$. As a first illustration of set updates and how they can characterize the updating modes, consider the following set update for Boolean networks of dimension n:

$$\Phi_a(X) = \{\phi_W(x) \mid x \in X, \emptyset \neq W \subseteq [\![n]\!]\}$$

This set update generates exactly all the non-empty elementary transitions: $\varphi(\Phi_a) = \to_a$. Thus, Φ_a characterizes the asynchronous updating mode. Similarly, let us now consider the following set update:

$$\Phi_{fa}(X) = \{\phi_i(x) \mid x \in X, i \in [\![n]\!]\}$$

Remark that $\varphi(\Phi_{fa}) = \to_{fa}$, i.e. Φ_{fa} characterizes the fully asynchronous updating mode.

6.2.3.5.3 Context-Dependent Updating Modes

Set updates enable specifying elementary updates that may depend on the state of automata in a given configuration. As an illustration, we show here an explicit characterization of the Boolean dynamics of Boolean networks with memory mentioned in the previous section. Recall from Definition 6.4 that a Boolean network with memory f of dimension n is parameterized with a delay vector $d \in \mathbb{N}^n$. Its configurations are the couples of binary and discrete configurations $X_{(f,d)}$ such that each automaton i having state 0 in the binary part has state 0 in the discrete part, and each automaton i having state 1 in the binary part has a state between 1 and d_i in the discrete part. First, let us define as $\mathsf{mem}(x)$ the set of all memory configurations that can be associated with binary configuration $x \in \mathbb{B}^n$, and conversely, let us denote by $\mathsf{bin}(\delta)$ the binary configuration corresponding to memory configuration $\delta \in \mathbb{N}^n$. Formally:

$$\forall x \in \mathbb{B}^n, \quad \mathsf{mem}(x) = \{\delta \in \mathbb{N}^n \mid x_i = 0 \iff \delta_i = 0,$$
$$x_i = 1 \iff \delta_i \in [\![d_i]\!]\},$$
$$\forall i \in [\![n]\!], \quad \mathsf{bin}(\delta)_i = \min\{\delta_i, 1\},$$
$$\forall x \in \mathbb{B}^n, \forall \delta \in \mathsf{mem}(x), \quad \mathsf{bin}(\delta) = x$$

It appears that $X_{(f,d)} = \{(\mathsf{bin}(\delta), \delta) \mid \delta \in \mathbb{N}^n, \forall i \in [\![n]\!], \delta_i \in \{0, \ldots, d_i\}\}$. Thus, one can reformulate the original definition by considering the deterministic parallel update of memory configurations $\delta \in \mathbb{N}^n$ and replacing x with $\mathsf{bin}(\delta)$: an automaton $i \in [\![n]\!]$ is set to state d_i whenever its local function f_i is evaluated to 1 on the corresponding binary configuration $\mathsf{bin}(\delta)$; otherwise, its state is decreased by 1, unless it is already 0. In particular, one can define the deterministic memory update $\phi_d^* : \mathbb{N}^n \to \mathbb{N}^n$

such that, for each $i \in [\![n]\!]$,

$$\phi_d^*(\delta)_i = \begin{cases} 0 & \text{if } f_i(\text{bin}(\delta)) = 0 \text{ and } \delta_i = 0, \\ \delta_i - 1 & \text{if } f_i(\text{bin}(\delta)) = 0 \text{ and } \delta_i \geq 0, \\ d_i & \text{if } f_i(\text{bin}(\delta)) = 1 \end{cases}$$

Let us now extend the above definitions to sets:

- $\forall X \subseteq \mathbb{B}^n$, $\text{MEM}(X) = \bigcup_{x \in X} \text{mem}(x)$;
- $\forall D \subseteq \mathbb{N}^n$, $\text{BIN}(D) = \{\text{bin}(\delta) \mid \delta \in D\}$; and
- $\forall D \subseteq \mathbb{N}^n$, $\Phi_d^*(D) = \{\phi_d^*(\delta) \mid \delta \in D\}$.

The memory set update can then be defined for any set of configurations $X \subseteq \mathbb{B}^n$ by first generating the set of corresponding memory configurations, then applying the deterministic update on them, and finally converting them back to binary configurations:

$$\Phi_d(X) = \text{BIN} \circ \Phi_d^* \circ \text{MEM}(X)$$

With this formulation, one can see that the memory updating mode, being the projection of configurations of Boolean networks with memory on their binary part, leads to non-deterministic dynamics. Indeed, whenever a configuration gets mapped to several possible memory configurations, and whenever for two of these configurations δ and δ', there is an automaton $i \in [\![n]\!]$, where $\phi_d(\delta)_i = 0$ and $\phi_d(\delta')_i \geq 1$. This can occur if and only if $d_i \geq 2$, $x_i = 1$, and $f_i(x) = 0$. Thus, the memory updating mode of Boolean networks can equivalently be parameterized by a set of automata $\overline{d} = \{i \in [\![n]\!] \mid d_i \geq 2\}$ and defined as the following set update:

$$\Phi_{\overline{d}}(X) = \Big\{ \phi_W(x) \mid x \in X, W \subseteq [\![n]\!],$$

$$W \supseteq \{i \in [\![n]\!] \mid i \notin \overline{d} \text{ or } f_i(x) = 1\} \Big\}$$

Remark that this definition no longer relies on delay configurations in $\mathbb{N}^n$. Overall, the memory updating mode of Boolean networks can be understood as a particular set of elementary transitions that depend on the configurations: the state changes from 1 to 0 of automata in $\overline{d}$ are applied asynchronously, whereas the other state changes are applied in parallel.

6.2.3.5.4 Beyond Elementary and Non-elementary Updates

Boolean networks are often employed as abstractions of a more detailed *concrete* model that would account for non-Boolean states, e.g. make them correspond to concentrations or copy numbers of biological molecules, or speed and delay of state changes of automata. Here, a Boolean model has the advantage of offering a simplified coarse-grained view of the concrete dynamics and requiring much less parameters. However, then, the question arises of how faithful are Boolean dynamics with respect to the quantitative dynamics from a formal standpoint. It has been recently underlined in Chatain et al. (2018) and Paulevé et al. (2020) that the elementary and non-elementary transitions of Boolean networks are not complete enough to capture particular quantitative trajectories. With a fixed logic, and starting from similar configurations, the quantitative system shows that an automaton can eventually get activated, whereas none of the elementary and non-elementary dynamics of the Boolean network can reproduce this behavior. Thus, recently, several updating modes generating transitions that are neither elementary nor non-elementary have been introduced, enabling to capture delays in the change of automata states (Chatain et al., 2018, 2020, Paulevé et al., 2020, Paulevé and Sené, 2021). They result in set updates Φ where, for some Boolean networks of dimension n and for some configurations $x \in \mathbb{B}^n$, there is $k \in \mathbb{N}$ such that there exists $y \in \Phi^k(\{x\})$, whereas $x \not\to_e^* y$. In this chapter, we focus on the *most permissive* updating mode of Boolean networks. "Permissive" refers to the transition relation it generates. It has been proven in Paulevé et al. (2020) that the most permissive updating mode captures *all* trajectories that can be achieved by any quantitative model in agreement with the Boolean network specification. We will come back more formally to this notion in Section 6.3.4. In Paulevé and Sené (2021), we also detail set update formalization of the *Interval* updating mode (Chatain et al., 2018, 2020) of Boolean networks, which accounts for a specific type of delay in the state updates that generate transitions being neither elementary nor non-elementary.

Most Permissive Updating Mode Consider the case whenever the state of an automaton i is used to compute the state of two distinct automata j and k and assume that i is increasing from 0. During its increase, there are times when i may be high enough to trigger a state change of j but not (yet) high enough for k. This can be illustrated on a concrete biological example, the so-called *Incoherent Feed-Forward Loop of type 3* (I3-FFL) (Mangan and Alon, 2003): a Boolean network f of dimension 3 with

$$f_1(x) = 1 \qquad f_2(x) = x_1 \qquad f_3(x) = \neg x_1 \wedge x_2.$$

Its influence graph $G(f)$ is as follows:

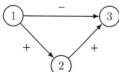

Starting from configuration $(0, 0, 0)$, the asynchronous updating mode predicts only the following non-reflexive transitions: $(0, 0, 0) \rightarrow_a (1, 0, 0) \rightarrow_a (1, 1, 0)$. However, it has been observed experimentally (Schaerli et al., 2014) and in quantitative models (Ishihara et al., 2005, Rodrigo and Elena, 2011) that depending on reaction kinetics, one can actually activate transiently automaton 3. Essentially, the idea is that during the increase of the state of automaton 1, there is a period of time where automaton 1 is high enough so that automaton 2 can consider it active, i.e. $x_1 = 1$, but automaton 3 still considers it inactive, i.e. $x_1 = 0$. Then, the state of automaton 2 can increase and so does the state of automaton 3. Leaving the Boolean network framework, one could model such a behavior using 3 ordered states for automaton 1: $0, \frac{1}{2}, 1$. Then, automaton 2 can be updated to state 1 whenever $x_1 \geq \frac{1}{2}$ and automaton 3 whenever $(x_1 < 1) \wedge (x_2 = 1)$. In other words, when automaton 1 is mild active, automaton 2 can become active, and then automaton 3 as well, etc., until automaton 1 becomes fully active, inhibiting automaton 3. However, this activation of automaton 3 cannot be predicted at the Boolean level by combining the deterministic updates defined so far, whereas the logic encoded by f is correct. Without introducing any parameter, the most permissive updating mode captures these additional dynamics by accounting for *all* possible threshold orderings and all updates that can happen over a switch of a Boolean state. In some sense, the most permissive updating mode abstracts both the quantitative domain of automata and the duration of state changes. Their original definition (Paulevé et al., 2020) is based on the introduction of pseudo *dynamical* states, namely, increasing and decreasing. An automaton can change from 0 to increasing whenever it can interpret the state of the other automata so that its local function is satisfied. Once in increasing state, it can change to the state 1 without any condition, or to the decreasing state whenever it can interpret the state of other automata so that its local function is not satisfied. Whenever an automaton is in a dynamical state, the automata can freely interpret its state as either 0 or 1. The most permissive updating mode can equivalently be expressed in a more standard way by means of composition of set updates (Paulevé and Sené, 2021). This definition relies on the notion of sub-hypercubes. A sub-hypercube of dimension n can be specified by a vector $h \in \{0, 1, *\}^n$ where components of the vectors having value $*$ are *free*: the vertices of the sub-hypercube are all the binary vectors $x \in \mathbb{B}^n$ such that for each dimension $i \in [\![n]\!]$, either $h_i = *$ or $h_i = x_i$. Thus, if we project a sub-hypercube over the d dimensions that are free, we obtain the hypercube of dimension d. For instance, the sub-hypercube for $h = (0, *, *)$ has four vertices: $(0, 0, 0), (0, 0, 1), (0, 1, 0)$, and $(0, 1, 1)$. Given a configuration x, the computation of the next configurations according to the most permissive updating mode is done in two stages. A first stage consists in computing all the elementary updates of a single automaton and then widening the resulting set. The *widening* is defined using function $\nabla : 2^{\mathbb{B}^n} \to 2^{\mathbb{B}^n}$, which computes the vertices of the smallest sub-hypercube containing the given set of configurations. For instance, $\nabla(\{(0,0,1),(0,1,1)\}) = \{(0,0,1),(0,1,1)\}$, and $\nabla(\{(0,0,1),(0,1,0)\}) = \{(0,0,0),(0,0,1),(0,1,0),(0,1,1)\}$. Given a set of automata W, the *widening set update* $\Phi_{W,\nabla} : 2^{\mathbb{B}^n} \to 2^{\mathbb{B}^n}$ applies this operator on the results of the elementary set update, or equivalently with the fully asynchronous set update, on the automata of W. This widening is re-iterated until a fixed point is reached. Then, a *narrowing* $\Lambda_W : 2^{\mathbb{B}^n} \to 2^{\mathbb{B}^n}$ filters the computed configurations X to retain only those where the states of automata in W can be computed with f from X.

Definition 6.8 The *most permissive set update* Φ_{MP} of a Boolean network of dimension n is given by

$$\Phi_{\text{MP}}(X) = \bigcup_{W \subseteq \llbracket n \rrbracket} \Lambda_W \circ \Phi^{\omega}_{W,\nabla}(X)$$

where for any $X \subseteq \mathbb{B}^n$ and any $W \subseteq \llbracket n \rrbracket$:

$$\nabla(X) = \prod_{i=1}^{n} \{x_i \mid x \in X\}$$
$$\Phi_{W,\nabla}(X) = \nabla(X \cup \{\phi_i(x) \mid x \in X, i \in W\}),$$
$$\Lambda_W(X) = \{x \in X \mid \forall i \in W, \exists y \in X, \; x_i = f_i(y)\}$$

and where $\Phi^{\omega}_{W,\nabla}(X)$ denotes the iteration of $\Phi_{W,\nabla}$ until reaching a fixed point (remark that for any $k \in \mathbb{N}$, $\Phi^{k}_{W,\nabla}(X) \subseteq \Phi^{k+1}_{W,\nabla}(X) \subseteq \mathbb{B}^n$).

Example 6.2 Figure 6.10 shows the dynamics generated from the configuration 000 by the asynchronous and most permissive updating mode on the Boolean network of the I3-FFL defined page 33. By instantiating Definition 6.8, we obtain:

$$\Phi_{\{1,2,3\},\nabla}(\{(0,0,0)\}) = \nabla(\{(0,0,0),(1,0,0)\})$$
$$= \{(0,0,0),(1,0,0)\},$$
$$\Phi^{2}_{\{1,2,3\},\nabla}(\{(0,0,0)\}) = \nabla(\{(0,0,0),(1,0,0)\} \cup \{(1,1,0)\})$$
$$= \{(0,0,0),(1,0,0),(0,1,0),(1,1,0)\},$$
$$\Phi^{3}_{\{1,2,3\},\nabla}(\{(0,0,0)\}) = \nabla(\{(0,0,0),(1,0,0),(0,1,0),(1,1,0)\}$$
$$\cup \{(0,1,1)\}) = \mathbb{B}^n,$$
$$\Lambda_{\{1,2,3\}}(\mathbb{B}^n) = \{(1,0,0),(1,0,1),(1,1,0),(1,1,1)\}$$

Thus, $(1,1,1) \in \Phi_{\text{MP}}(\{(0,0,0)\})$, whereas $(0,0,0) \not\to^{*}_{\text{e}} (1,1,1)$, i.e. configuration $(1,1,1)$ is reachable from $(0,0,0)$ using the most permissive updating mode, whereas there are no elementary (and non-elementary) paths between these two configurations.

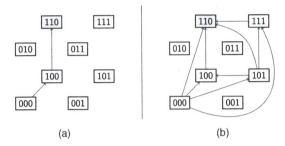

Figure 6.10 (a) Asynchronous and (b) most permissive dynamics *reachable from the configuration* $(0,0,0)$ for the Boolean network f of the I3-FFL defined page 33. For the sake of readability, loops are omitted. In the two examples, there are loops on each configuration involved in the displayed transitions.

Finally, let us conclude by listing some basic properties of the most permissive updating mode:

1. It preserves the fixed points of f: for any configuration $x \in \mathbb{B}^n$, $f(x) = x$ if and only if $\Phi_{MP}(x) = \{x\}$.
2. It subsumes elementary transitions: $\to_e \, \subseteq \, \varphi(\Phi_{MP})$.
3. Its transition relation is reflexive and transitive: $\Phi_{MP} = \Phi_{MP}^2$.
4. (by 2 and 3) Its transition relation subsumes non-elementary transitions: $\to_e^* \, \subseteq \, \varphi(\Phi_{MP})$.
5. (by 4 and the example) There exist Boolean networks such that the most permissive transition relation is strictly larger than non-elementary transitions, i.e. there exist $x, y \in \mathbb{B}^n$ such that $y \in \Phi_{MP}(\{x\})$ but $x \not\to_e^* y$.

6.3 Biological Case Studies

As discussed in the introduction of this chapter and in Section 6.2.1, Boolean network features have been particularly appreciated to apprehend regulation network modelling from both theoretical and applied standpoints. As explained, this comes notably from their setting simplicity and their abstraction level, which allow us to focus on the natural computations operated by the modeled systems and thus deduce or predict properties of the latter. In general, computational models of biological systems are appealing for bringing potential explanations to observed phenomena and for performing experiments *in silico*, which may help prioritizing wet lab experiments (Eduati et al., 2020). Biological processes involve a myriad of features, such as the shape of cells, the shape of molecules, their location, their movement, the way they interact, the way they fold, etc. Thus, to be manageable and keep some interpretability of the results, a model of a biological process is first a harsh selection of the biological features. A model that is able to reproduce an observed phenomenon thus supports an hypothesis according to which the selected features and the granularity of their representation are sufficient to explain the phenomenon. With this perspective, Boolean networks offer a very high abstraction of the system, focusing on the structure of the causal influences between the components of the network and without considering explicitly quantitative features such as the chronological time and quantities of molecules. The Boolean network setting specifies the logic of automata state changes, whereas the updating mode specifies the logic of the orchestration of state changes over time, potentially abstracting features related to time and quantities. Moreover, because of their discrete and finite nature, one can obtain an exhaustive assessment of their trajectories and limit sets that are very often theoretical representatives of real observable physiological characteristics. This provides a powerful tool to formally demonstrate that some behaviors are impossible to reproduce with a given Boolean network and updating mode.In this section, we summarize a selection of case studies for which the updating mode plays a central role:

- the robustness of predictions of limit sets and trajectories leading to them when varying the updating mode with the modelling of the floral morphogenesis of *A. thaliana* (Section 6.3.1);
- the definition of custom updating modes to reflect time constraints and accurately reproduce sequences of state changes within the cell cycle (Section 6.3.2) and for biological rhythms synchronization (Section 6.3.3); and
- the formal analysis of the absence of trajectories whenever the Boolean network is considered as an abstraction of a quantitative system, by relying on the most permissive updating mode and illustrated on an incoherent feed-forward loop system (Section 6.3.4).

6.3.1 Floral Morphogenesis of *A. thaliana*

The floral development of *A. thaliana* is among the model organisms the most studied in vegetal biology since the early 1900s and the first plant whose nuclear genome was sequenced (The Arabidopsis Genome Initiative, 2000). The first model based on Boolean networks introduced in the literature for the genetic network of the floral development of this plant comes from Mendoza and Alvarez-Buylla (1998). It is based on a threshold Boolean network, namely, a Boolean network whose local functions are threshold functions, composed of 12 automata, each one corresponding to a particular gene or gene complex acting on the control of the floral morphogenesis. In this seminal paper, the dynamics of this network is studied according to a specific block-sequential updating mode. This mode was chosen arguing its pertinence according to the activation time of the genes in the flowering and the flower morphogenesis processes. One year later, together with Thieffry, the same authors transformed this model into a classical Boolean network, i.e. with no threshold local functions, and used the method developed by Thomas (1991) to study its fully asynchronous dynamics (Mendoza et al., 1999). These two distinct choices, for both the mathematical model at stake and the updating mode, led the authors to highlight that the underlying dynamical systems converge toward six fixed points. More interestingly, among these six fixed points, four correspond to the four floral organs of the *A. thaliana*: sepals, petals, carpels, and stamens; the two remaining fixed points correspond to (i) inflorescence cells of the plant, which do not belong to its flowers, and to (ii) an expression pattern never observed in wild-type plants until now, which could be produced experimentally. These works emphasized that qualitative modelling with Boolean networks is helpful in the sense that their limit sets may capture real biological structures, such as organs, tissues, or cellular types, as evoked by Delbrück (1949). However, from a more theoretical standpoint, further work needs to be done to achieve a deep understanding of the network itself. Demongeot et al. (2010), basing themselves on the initial threshold Boolean network model, wanted to understand the network in depth. First, they highlighted that there exist updating modes, e.g. the parallel updating mode, for which the underlying dynamical systems admit not only the former fixed points but also limit cycles. Then, an analysis of the local functions led them to construct a simplified/compressed network with the same asymptotic dynamics and to show that the asymptotic richness depends mainly on five genes that are parts of two distinct symmetric strongly connected components, which can actually be considered as the dynamical complexity engines of the network. This actually explained the possible existence of limit cycles by applying a result of (Goles and Olivos, 1981). The natural question then was: now we can state the phenomenological sense of the six fixed points, what is the biological likeliness of the limit cycles? The answer to this question was given by the asynchronous dynamics, i.e. all the possible elementary transitions, of the model that highlights their unlikeliness by showing that the probabilities to enter and stay into an asymptotic oscillation is very weak. Eventually, these previous results were complemented and refined in 2020 by an analysis of the dynamics of this network when modeled by a

Boolean network with memory (Goles et al., 2020). Finally, we can notice that after two decades, the scientific community eventually gets a sharp understanding of this simple network modelling the floral development of *A. thaliana* from both theoretical and applied standpoints, thanks notably to the particular attention paid to updating modes. This highlights the importance of considering a consistent updating mode depending on the question addressed and that any updating mode may be *a priori* relevant.

6.3.2 Cell Cycle

Together with the modelling of cell differentiation and reprogramming processes that occupy a vast part of Boolean network models in biology, the modelling of cellular rhythms and oscillators is another prominent application of Boolean networks. In such systems, we expect that the sustained oscillatory behaviors are captured by the Boolean network as non-singleton limit sets. The system is usually assumed to be in a configuration of this limit set. One of the most studied biological oscillator is the cell cycle. The cell cycle refers to the successive divisions of a cell, which goes through a specific and well-characterized sequence of events, including the duplication of its genome (the so-called S phase, for synthesis) and its mitosis (M phase). Boolean network models of cell cycles usually focus on reproducing accurately this sequence of events that can be observed by the activity of specific proteins, as well as its control and coupling with other biological processes. The cell cycle involves numerous genes and proteins and is a tightly regulated process. It is also expected that the chronological time, notably related to the speed of different interactions, plays a crucial role in this regulation. With this latter consideration, an accurate modelling using Boolean networks seems challenging. The Boolean model of (Fauré et al., 2006) is a nice demonstration of how updating modes enable abstracting time adequately and obtain a realistic dynamical model. Their model gathers well-known influences between key regulators of the cell cycle, which have been extensively studied with quantitative models. These regulators involve different types of proteins (including cyclins, transcription factors, complexes, and inhibitors), the activity of some of them being well-known markers for the different phases. In total, the model comprises 10 automata. In their study, Fauré et al. first compared the limit cycles obtained with the parallel and fully asynchronous updating modes when the input automaton of the network (*CycD*) is activated. The parallel updating mode leads to one limit cycle between 7 configurations, which matches with the available data and previous quantitative models. However, many state changes are compressed into a single transition, making difficult a finer analysis on the possible sequences of events during the cell cycle. The fully asynchronous updating mode then enables results in a single limit cycle of 112 configurations, which contains well-detailed descriptions of the sequences of events of the cell cycle. However, the limit cycle contains many spurious trajectories, notably with shortcuts avoiding key checkpoints of the cell cycle. In order to refine the Boolean dynamics, the authors proposed a custom updating mode that can be seen related to the Memory Boolean networks and block-sequential updating modes studied in this chapter. The main principle is to group the state changes of automata into different priority classes and further split these priority classes into automata that are updated according to the fully asynchronous updating mode and other ones that are updated with the parallel updating mode. The priority classes allow treating more accurately the fast and slow biological processes. These classes are then refined further by identifying state changes that are controlled by the same regulatory process and thus should happen simultaneously. The obtained dynamics consists then in a selection of elementary transitions that depend on the state changes of automata. In the scope of the cell cycle model, this translates into the prediction of a limit cycle of 18 configurations with a finer description of sequences of cell cycle events and for which the different trajectories cannot be discriminated with the current knowledge of the biological process.

6.3.3 Vegetal and Animal Zeitgebers

Despite the fundamental role of updating modes discussed so far in this section related to biological case studies, regardless of the fact that they are block-sequential, fully asynchronous, or even asynchronous, and the large spectrum of questions they can help us answer, it turns out that they do not necessarily own the relevant properties to capture some biological phenomenological intricacies. Indeed, while they allow in some sense to modulate the internal clocks of a regulation process modelled by a network, they do it by considering the network as a whole, with no distinction between its own components. In Demongeot and Sené (2020), the authors focused on Zeitgebers. Zeitgebers are classically defined in biology and medicine as exogenous cues that synchronize endogenous biological rhythms. In the framework of genetic regulation network modelling, considering Zeitgebers as sorts of timers of genetic or physiological origin, somehow external to the very functional components of the network, they highlight that the latter cannot be modeled by means of the until now classically studied updating modes mentioned above, for the reasons evoked. Among the best known examples of a Zeitgeber here, namely, a subnetwork having its own clock that synchronizes the dynamics of the whole network that contains it, is certainly the role played by genes PER and TIM, which work together to control the circadian rhythm of *Drosophila melanogaster* (Hardin et al., 1990, Sehgal et al., 1994, Goldbeter, 1995). However, using threshold Boolean networks both as a genetic network model and as a neurophysiologic network model, Demongeot and Sené (2020) showed that some deterministic and periodic updating modes belonging to the local clocks family, namely, the block-parallel updating modes, are good candidates to capture synchronization phenomena, in the sense that they are expressive enough, which is essential for the biological modelling standpoint, without being too expressive, which is important for the mathematical and computational standpoint in order to stay away from a combinatorial explosion argument. More precisely, they emphasized schematic models taking into account the existence of specific genetic regulation subnetworks acting as Zeitgebers in two distinct contexts:

- the genetic control of plant growth – they proposed a model of vegetal growth by considering the regulatory relations between the circadian clocks genes CCA1 and TOC1 and abstract genes associated with auxin flows corresponding to the localized expressions of auxin (Thellier et al., 2004, Covington and Harmer, 2007, Bendix et al., 2015). In this model, the sub-network composed of CCA1 and TOC1 plays a timer role, which synchronizes the regulation of the other component acting as the functional auxin part of the network serving the plant growth, thanks to a specific block-parallel updating mode. The dynamical role of this timer is to produce a regular scheme of growth, which seems to coincide pretty much with the quasi-perfect morphogenesis of plants such as *Araucaria araucana*.
- the cardiorespiratory pace – they proposed a model of the clock governing the cardiorespiratory regulation at the neurophysiological level. Four components are considered, which are derived from Beauchaine (2001), Dergacheva et al. (2010), Moraes et al. (2014), and Demongeot et al. (2020): inspiratory and expiratory neurons that play the role of a timer synchronizing the sinoatrial node and the baroreceptor of the heart. Together with an adapted block-parallel updating mode, the underlying network behaves in two phases as it can be observed in nature. Nevertheless, even if the timer modelling seems to comply with the biological assumptions and observations, it is not exactly the case of the cardiac activity modelling, which paves the way to further refinements.

6.3.4 Abstraction of Quantitative Models

In many applications, the Boolean modelling of biological systems abstracts both the quantitative time of the interactions and the quantitative state of the interacting automata. This is notably the case when we consider gene regulatory networks, where the activation of a gene is generally assumed to depend on a sufficient amount of transcription factors. Although this dependency is often assumed to be non-linear and modeled as a threshold function, it is nevertheless a quantitative process. Indeed, differences of thresholds between the different gene activations may play an important role in several biological processes. The Boolean abstraction may also hinder the validation of the model with respect to observations of the biological system. The current experimental observations mostly rely on counting the population of different molecules either cell-per-cell (single-cell measurement technologies), or in a mixture of cells. Actually, the activity of genes is often linked to the amount of their transcripts and is thus a quantitative feature. A qualitative interpretation of such data then relies again on delineating thresholds to determine the Boolean state of automata. As illustrated by the previous case studies, the updating mode of Boolean networks is effectively employed to model features related to the time or speed of different Boolean processes. In these case studies, the updating modes actually select specific elementary paths of the Boolean network dynamics, avoiding predicting spurious transitions. In a way, they assume that the asynchronous dynamics (generating elementary transitions) gives a boundary on the admissible paths or equivalently on the admissible non-elementary transitions: if there is no elementary path from a configuration x to a configuration y, then none of the updating modes considered in the above case studies can generate transitions that would connect these two configurations. Given the abstraction level imposed by Boolean networks for modelling biological systems, one may wonder whether the assumption that the elementary and non-elementary transitions capture its state changes holds in general, i.e. if there exist cases of systems for which the asynchronous Boolean network misses transitions. The answer to this question heavily depends on the type of systems we want to model. In this section, we focus on quantitative systems, such as multivalued networks (Thomas and d'Ari, 1990) or ordinary differential equations (Glass and Kauffman, 1973, de Jong, 2002). From a formal standpoint, this question relates to the correctness of the abstract interpretation of a quantitative system by a Boolean network. The incoherent feed-forward loop system I3-FFL developed page 33 is a simple counter-example showing the incompleteness of elementary and non-elementary transitions of

Boolean networks. As discussed above, several quantitative models and experimental studies show that the output automaton of the system can be transiently activated from the configuration where all automata are inactive. However, the asynchronous dynamics shows no such paths. From a model validation perspective, this is critical, as the Boolean network would likely be considered as incoherent with the data, whereas the logic it encodes is correct. Indeed, the observed behavior can be recovered using the most permissive updating mode without any additional parameter. Paulevé et al. (2020) have demonstrated that the most permissive updating mode leads to a complete and minimal abstraction of state changes of any quantitative model being a *refinement* of the Boolean network. In other words, if there is no path between two configurations in the most permissive dynamics, then no quantitative refinement of the model can create this path. The definition of a *refinement* is obviously key. Let us first give a general definition of a quantitative model: a quantitative model F can be defined as a function mapping discrete or continuous configurations to the derivative of the state of automata. Then, F is a refinement of f if and only if the derivative of automaton i is strictly positive (resp. negative) in a given quantitative configuration z only if there is a binarization $\tilde{z}$ of z so that $f_i(\tilde{z}) = 1$ (resp. 0). The binarization has to map the quantitative 0 to the Boolean 0, and the maximal quantitative state, e.g. with multivalued networks, to the Boolean 1; otherwise, any Boolean value can be considered. One may remark that the updating mode that enables the transition from any configuration to any other ones verifies the completeness criteria. However, in addition to being non-minimal, such dynamics would make impossible any interesting prediction from the model. As we mentioned several times, Boolean networks are employed to model differentiation processes, where the system can have multiple limit behaviors. Predicting which attractors the system may reach, and how to control this reachability, is a prominent application. However, these predictions rely on the absence of paths between certain configurations of the Boolean network. Thus, an updating mode generating too many spurious paths between configurations would largely hinder such predictions. Paulevé et al. (2020) have shown on several case studies that the most permissive updating mode predicts the same reachable attractors and control as identified in previous studies with the fully asynchronous updating mode. Thus, the most permissive updating mode brings a formal abstraction of quantitative systems, with strong guarantees of capturing possible stage changes while being stringent enough to capture differentiation processes. Moreover, as we well detail in Section 6.4.2, the analysis of the most permissive dynamics requires a much lower computational cost than elementary and non-elementary updating modes, enabling addressing genome-scale models with several thousands of automata (Paulevé et al., 2020).

6.4 Fundamental Knowledge

Beyond their wide use in the context of qualitative modelling of biological networks, which has led to numerous significant advances about genetic regulation, Boolean networks is being studied *per se* from the theoretical (mathematical and computational) standpoint, notably because we are still far from understanding their underlying properties in depth. For instance, despite the real progresses realized these last two decades, the way that information is transmitted along the automata and the ability to produce this or that global dynamical behavior depending on local interactions are still not well understood. Trying to answer this kind of questions needs to focus on distinct discrete structures (e.g. influence graphs, sets of local functions, transition graphs, and updating modes) and to relate them by means of methods and tools coming from different areas of mathematics and computer science such as combinatorics, dynamical system theory, algorithms, computational complexity theory, and computability theory. In this section, in order to give an insight into some relevant fundamental problems on Boolean networks (and more generally on automata networks), some seminal results obtained are presented following two research lines: the links between static (syntactic) and dynamical (semantic) properties of Boolean networks and the inherent complexity of some classical problems underlying Boolean network studies.

6.4.1 Structural Properties and Attractors

Given an influence graph or a Boolean network, knowing the underlying possible asymptotical dynamics is a problem that is known to be complex from the computational standpoint (see Section 6.4.2 below). Nevertheless, past studies gave very general and important results on the subject. This subsection aims to present some of these results while emphasizing links with the role of updating modes.

6.4.1.1 Fixed Points Stability

The result presented hereafter is certainly the most classical of discrete and finite dynamical systems. It concerns fixed points and their relative stability depending on updating modes and holds for any kind of finite dynamical system. More precisely, Theorem 6.1 states that, considering any Boolean network f, the parallel fixed points of f are necessarily fixed points of (f, μ), for any updating mode μ, which governs the evolution of f by organizing the executions of associated updating functions over time.

Theorem 6.1 *(Folklore) Let $f : X \to X$ be an automata network whose configurations are elements of X. Let π be the parallel updating mode. Let $FP(f) = FP(f, \pi)$ be the set of fixed points of dynamical system (f, π). Then we have, for any μ defining an organization of updating functions over time:*

$$FP(f) \subseteq FP(f, \mu).$$

Proof: Consider a Boolean network f of dimension n, and $x \in \mathbb{B}^n$ a fixed point of (f, π). Since $x = (x_1, \ldots, x_n)$ is a fixed point of (f, π), $x_1 = f_1(x)$ (resp. $x_2 = f_2(x), \ldots, x_n = f_n(x)$) is a fixed point of local function f_1 (resp. $f_2, \ldots, f_n$). So, since x is actually a vector composed of the fixed points of the n local functions, regardless of the way an updating mode μ updates states from x, it cannot change them, by definition. In other words, $\forall W \subseteq [\![n]\!], \phi_W(x) = x$, which is the expected result. $\square$

A first remark is that the reciprocal of this theorem does not hold. Indeed, there exists an automata network $f : X \to X$ and pertinent updating mode μ such that $FP(f, \mu) \not\subseteq FP(f)$. For instance, consider the Boolean network of dimension 3 defined as $f(x) = (f_1(x) = x_3, f_2(x) = x_1, f_3(x) = x_2)$ whose associated influence graph is a cycle of length 3 whose every arc is positively signed. If this Boolean network evolves according to the parallel updating modes, then it admits two fixed points such that $FP(f) = \{(0,0,0), (1,1,1)\}$. Now, let $\mu = (\{2,3\}, \{1,3\}, \{1,2\})$ be a pertinent periodic updating mode. Notice that μ is not a local clocks updating mode. Then, it is easy to compute the dynamics of (f, μ) and to conclude that this dynamical system has four fixed points and is such that $FP(f, \mu) = \{(0,0,0), (0,1,0), (1,0,1), (1,1,1)\}$. As a consequence, even if playing with updating modes cannot lead to destroy fixed points according to Theorem 6.1, it can lead to create fixed points. Although it has never been addressed in depth, this fixed point generation by updating modes is very interesting and certainly deserves further analyses. A second remark is that we could imagine other kinds of updating modes creating transitions that are not the result of local function executions, such as transitions imposed by some kind of external events/behaviors to the underlying network itself. In that case, there would be no guarantee to conserve the "canonical" fixed points of f.

6.4.1.2 Feedback Cycles as Engines of Dynamical Complexity

The previous result related to fixed points stability emphasizes a strong intrinsic dynamical property of finite dynamical systems and thus of

Boolean networks. Of course, such a property is very important and pertinent. However, we have to keep in mind that, in the context of modelling by means of Boolean networks, in most situations, the inputs of the problems that are to be addressed are either Boolean networks, i.e. their definitions as collections of local functions, or influence graphs (or even communication graphs in which the interactions are not necessarily effective). As a consequence, it is natural to ask for finding/characterizing structural properties on these latter objects, which induce relevant properties about their underlying possible dynamical systems. The literature offers us several results relating structural and dynamical properties of Boolean networks. Among the most classical ones is the theorem of (Robert, 1980) stating the crucial role of feedback cycles in the influence graphs of automata networks for them to have a non-trivial dynamics. Theorem 6.2 below is the adaptation to Boolean networks of the seminal theorem of Robert and states that the presence of a feedback cycle in the influence graph of a Boolean network f is a necessary condition for f to admit several limit sets. From a purely theoretical standpoint, any updating mode that is mathematically correct is reasonable, but, if we consider updates in a context of modelling, some constraints need to be taken in account. In this framework, consider the following definition: an updating mode is called *reasonable* if its application leads every local function to be executed infinitely often.[1]

Theorem 6.2 *(Robert (1980))* *Let f be a Boolean network of dimension n and $\mathcal{G}_f = (\llbracket n \rrbracket, E_f)$ its associated influence graph. If $\mathcal{G}_f$ is acyclic, then for any reasonable updating mode μ, every configuration of dynamical system (f, μ) converges toward a unique attractor $y \in \mathbb{B}^n$, which is a fixed point.*

Proof: If $\mathcal{G}_f = (\llbracket n \rrbracket, E_f)$ is acyclic, then it is a directed acyclic graph (DAG) and, as a consequence, (i) at least one of its vertices is a source, i.e. a vertex with in-degree 0, and (ii) it can be structured into layers defined recursively such that the layer of depth 0 contains all the sources, and the layer of depth k contains all the vertices that have at least one in-neighbor belonging to the layer of depth $k - 1$. Notice that as soon as a vertex i is a source, i has no incoming influence. Thus, the state of its corresponding automaton does not depend on the state of any automaton in $\llbracket n \rrbracket$, which means that its local function is necessarily constant, namely, $f_i(x) = 0$ or $f_i(x) = 1$. Moreover, since $\mathcal{G}_f$ is a DAG, it has at least one topological

---

1 All the updating modes discussed in this chapter are reasonable. Notice that, obviously, the best example of a non-pertinent updating mode is (∅), which does not update any state. Any Boolean network evolving according to this updating mode admits 2^n fixed points because each of its configurations is stable by definition.

ordering, i.e. a total ordering of its vertices such that for each arc $(i,j) \in E_f$, i comes before j in the ordering. Let L be the topological ordering of $\mathcal{G}_f$ obtained by applying a variation of the Kahn's algorithm (Kahn, 1962), where the initial set of sources is structured into a queue where vertices are ordered both increasingly depending on the depths of reachable layers and lexicographically according to $[\![n]\!]$ and reordered similarly when a vertex is removed from the queue. In other terms, L is the unique topological ordering that takes into account both the layered structure of $\mathcal{G}_f$ and the lexicographical order on $[\![n]\!]$. Consider now that L is organized according to the layers of $\mathcal{G}_f$ such that $L = (L_0, \ldots, L_\ell)$, where L_0 is the restriction of L to the sources of $\mathcal{G}_f$, and L_k is the restriction of L to the automata whose states only depend on automata that belong to layers between L_0 and L_{k-1}.

Now, let us consider any initial configuration $x \in \mathbb{B}^n$ and time step $t_0 \in \mathbb{N}$ after which all the automata of L_0 have updated their state for the first time. Since the updating mode μ is reasonable, t_0 exists. Since every automaton of L_0 is a source in $\mathcal{G}_f$, we have: for all $i \in L_0$, $x_i^{t_0} = f_i(x)$ and will remain fixed, i.e. $\forall t_k \geq t_0 \in \mathbb{N}, x_i^{t_k} = 0$ or $x_i^{t_k} = 1$. Consider now time step $t_1 > t_0$ after which all the automata of L_1 have updated their state for the first time. Again, for the same reason, t_1 exists. By definition, these automata only depend on the automata of L_0, whose states are fixed since $t_1 > t_0$. As a result, from time step t_1, the state of every automaton of L_1 will remain fixed because it has updated its state by executing its local function whose variables are fixed. From this, by an induction on the layer depth of L and by applying the same reasoning on every $(L_k)_{2 \geq k \geq \ell}$, there exists time step t_ℓ after which the states of all automata of $[\![n]\!]$ remain fixed. Let $x' = f^{t_\ell}(x) \in \mathbb{B}^n$. Since x' is the image of any configuration after t_ℓ time steps, it is the unique fixed point of (f, μ). □

Despite the apparent simplicity of its statement and of its proof that can easily be extended to finite dynamical systems, Theorem 6.2 is certainly one of the most important theorems of dynamical system theory (remark that generalizations of it can be found in Shih and Dong (2005), Richard (2015)). Indeed, it emphasizes that feedback cycles are actually sorts of fundamental engines to create asymptotic diversity. To go beyond, it is a formalization and a concretization of the key principle of cybernetics (Wiener, 1948), the *circular causality*.

6.4.1.3 About Signed Feedback Cycles

The previous result of Robert (1980) and its extensions highlight the major role played by feedback cycles on the asymptotic behavioral complexity of Boolean networks. At the same period, Thomas mentioned the importance of distinguishing cycles according to their nature. Let $\mathscr{C} = (\llbracket n \rrbracket, E)$ be a (signed) cycle of length n, where $E \subseteq \llbracket n \rrbracket \times \{-, +\} \times \llbracket n \rrbracket$. With no loss of generality, consider that the arcs of $\mathscr{C}$ are ordered according to the lexicographical order on $\llbracket n \rrbracket$ such that $E = \{(1, 2), (2, 3), \ldots, (n, 1)\}$. Now, let us consider that $\mathscr{C}$ is the influence graph of a Boolean network f defined on $\llbracket n \rrbracket$ and rename it as $\mathscr{C}_f$. Obviously, since $\mathscr{C}_f$ is a cycle, the arity of every local function of f equals 1, so that:

$$\forall i \in \llbracket n \rrbracket \setminus \{1\}, f_i(x) = \begin{cases} \neg x_{i-1} \\ x_{i-1} \end{cases} \quad \text{and} \quad f_1(x) = \begin{cases} \neg x_n \\ x_n \end{cases}.$$

The definition below discriminates Boolean network cycles depending on the signs of their arcs.

Definition 6.9 Let $\mathscr{C}_f = (\llbracket n \rrbracket, E_f)$ be a cycle that represents the influence graph of a Boolean network f. $\mathscr{C}_f$ is said to be a *positive cycle* (resp. a *negative cycle*) if the number of negative arcs that compose it is even (resp. odd).

Let us give an insight into understand the trajectory of a Boolean cycle configuration of dimension n: consider one automaton of the Boolean cycle as special in the sense that it is the unique observation point from which an observer can see the evolution of the dynamics of the cycle and denote this automaton by $i \in \llbracket n \rrbracket$; now, imagine that at time step $t \in \mathbb{N}$, the state of i is $x_i^t = b$, and imagine that the cycle evolves according to the parallel updating mode. Then, information b will travel along the cycle as follows: at time step $t + 1$, b (resp. $\neg b$) becomes the state of automaton $i + 1$, except of course in the case where $i = n$ in which b becomes the state of automaton 1, if arc $(i, i+1)$ (or if arc $(n, 1)$ when relevant) is positive (resp. negative); at each following time step, the information keeps traveling along the cycle, until the time step $t + n$ is reached. At this time step, if the observer evaluates the state of i, then:

$$x_i^{t+n} = \begin{cases} b & \text{if the cycle is positive} \\ \neg b & \text{if the cycle is negative} \end{cases}.$$

Indeed, only a negative cycle can negate the initial information after one complete round, thanks to the odd number of its negative arcs. At the beginning of the 1980s, Thomas proposed two general rules related to these two types of cycles (Thomas, 1981):

1. The presence of a positive cycle in the influence graph of a dynamical system is necessary for this system to have several fixed points.
2. The presence of a negative cycle in the influence graph of a dynamical system is necessary for this system to have a limit cycle.

These two rules led to numerous studies aiming to prove their validity in different contexts. In particular, they gave rise to several theorems in both discrete and continuous settings. Although they are not the purpose of this chapter, we invite the readers to see Snoussi (1998), Gouzé (1998), Cinquin and Demongeot (2002), and Soulé (2003, 2006), which deal with these theorems in the continuous case. Let us come back now to the discrete case on which our focus is. Notice that the first rule was proven in more or less general frameworks, see:

- Aracena et al. (2004) and Aracena (2008) for the parallel Boolean case;
- Remy et al. (2008) for the fully asynchronous Boolean case; and
- Richard and Comet (2007) and Richard (2009) for the fully asynchronous discrete case.

Following these lines while focusing on the Boolean case, in Noual (2012b) and Sené (2012), the authors generalized the existing results to the asynchronous updating mode. Remind that, given a Boolean network, the asynchronous updating mode is a non-deterministic mode that authorizes every possible transition inducing subsets of automata of this network, i.e. all the transitions made possible by any deterministic updating modes as well as the fully asynchronous updating mode. This extension to the asynchronous case is in consequence quite pertinent. Indeed, it induces the validity of the first Thomas' rule for a very large set of updating modes (namely, an infinite one actually).

Theorem 6.3 *(Noual (2012b))* *Let f be a Boolean network of dimension n, $\mathcal{G}_f = (\llbracket n \rrbracket, E_f)$ its associated influence graph, and let a be the asynchronous updating mode. If (f, a) has several fixed points, then there is a positive cycle in $\mathcal{G}_f$.*

Iden of the proof: The general idea of the reasoning rests on a proof by contradiction. Indeed, consider that $\mathcal{G}_f$ does not contain any positive cycle. Then, $\mathcal{G}_f$ is either acyclic or it admits at least one negative cycle so that, if it admits several ones, these cannot induce a positive cycle by hypothesis. Let us admit that $\mathcal{G}_f$ is acyclic. In this case, Theorem 6.2 applies and (f, a) has only one fixed point. Now, let us consider the other case. Without entering into the details here, the idea of the related proof is to show that a negative cycle Boolean network can never suppress all the local instabilities, an automaton i being unstable in a configuration x if and only if $f_i(x) \neq x_i$, and a local instability in x being the number of unstable automata of x. In particular, if a configuration has zero local instability, then it is a fixed point. Summarizing, given any configuration $x \in \mathbb{B}^n$, there exists no asynchronous trajectory $x \to^*_{(f,a)} y$ such that y is a fixed point. This can be proven by focusing on the structure of the transition graph $\mathcal{D}_{(f',a)}$ of a cyclic negative Boolean network f' of dimension n'. Indeed, $\mathcal{D}_{(f',a)}$ is a layered graph in which each layer contains configurations of equivalent instability level, which allows us to show that a configuration instability can be an odd number between 1 and n'. In other words, this induces that negative cycles cannot generate fixed points. As a consequence, the presence of a positive cycle in $\mathcal{G}_f$ is necessary for (f,a) to admit several fixed points. □

Concerning now the second rule about negative cycles, notice that it has been proven in Richard (2010) for the discrete fully asynchronous case. While we are not going to enter into details here, an interesting fact concerning this rule is that it has also been generalized to the asynchronous case by Noual (2012b) and Sené (2012). However, contrary to the first rule developed above, the fact that it holds under the asynchronous updating mode does not induce that it does for all deterministic modes because of the very nature of limit cycles and their high sensitivity to synchronism (Goles and Salinas, 2010, Demongeot et al., 2012, Noual and Sené, 2018). Notably, the dynamics in parallel of any positive cycle of length greater than 1 admits limit cycles. Thomas' rules, together with Robert's theorem, put the highlight on the remarkable role of feedback cycles. Nevertheless, the results presented until now do not explain in depth the dynamics of positive and negative cycles *per se*. That is in particular why studies dedicated to cycles have been led. Remy et al. (2003) proved that, under the fully asynchronous updating mode hypothesis, positive (resp. negative) cycles of length n admit exactly two fixed points x and $\bar{x}$ (resp. one limit cycle of length $2n$) as attractors. Demongeot et al. (2012) then gave a complete combinatorial and dynamical characterization of cycle dynamics in parallel, which extends to every block-sequential updating mode by the main result of (Goles and Noual, 2010) that shows that computing a block-sequential cycle dynamics can be reduced to computing the parallel dynamics of a smaller cycle of same sign. A synthesis of these results on cycles can be found in Demongeot et al. (2021). Then, works focused on cycle tangential intersections to make a first step in the direction of comprehending how combinations of cycles operate in more complex networks (Noual, 2012a,b, Alcolei et al., 2016). The previous results succeeded in giving deep knowledge on Boolean cycles and their tangential intersections. These results also suggested that cycle intersections would be key structural elements to reduce drastically the degrees of freedom of Boolean networks and their ability to have too many attractors, which seems not to be likely empirically in biology for instance. Nevertheless, while acquiring knowledge (even complete) about these specific patterns forms a mandatory first step before going further, it is far from being sufficient to comprehend globally those of more complex networks, possibly composed of numerous intersections of cycles of different kinds. In this framework, the second part of Hilbert's XVIth problem (Hilbert, 1902) asking for the maximal number of limit cycles and their relative sizes in polynomial vector fields gives rise to addressing problems that are discrete variations of it: perform deep analyses of the structure of finite Boolean dynamical systems and their asymptotic combinatorics. The question of counting the maximal number of fixed points was addressed in Aracena et al. (2017) through the following question: "Given an influence graph $G = (V, E)$, with $|V| = n$, what is maxFP$(G) = \max(\{\text{card}(\text{FP}(G,f)) \mid f : \mathbb{B}^n \to \mathbb{B}^n\})$?", where maxFP$(G)$ is the maximum number of fixed points that a Boolean network f admitting G as influence graph can have. For the sake of clarity here, notice that the input of this problem is not a Boolean network but an influence graph which, by definition, can admit several distinct Boolean networks. Furthermore, the authors chose the parallel updating mode as the reference updating mode in their study, resting on the fixed points stability theorem (see Theorem 6.1), in order to consider only the very fixed points of f and those that could be generated by specific modes, as if the latter were some kinds of side effects. They obtain, among other results, the following theorem.

Theorem 6.4 *(Aracena et al. (2017))* *Let $G = (V, E)$ be an influence graph, $\nu(G)$ be the maximum number of disjoint cycles of G (also known as its packing number), and $\tau^+(G)$ be the minimum number of vertices meeting every positive cycle of G (also known as its positive feedback vertex set). Then:*

$$\nu(G) + 1 \leq \text{maxFP}(G) \leq 2^{\tau^+(G)}$$

This result paves the way to refinements, in particular of the upper bound. Indeed, here, we can see that only positive cycles are considered in this latter bound. To refine it, we could, for instance, take into account the role of negative cycles, in the sense that Demongeot et al. (2012) showed that their intersections with positive cycles tends to maintain the convergence toward fixed points while reducing it. However, although this seems to be a natural and promising research track, the fact remains that obtaining such refinements will be highly difficult since the impact of negative cycles is still very weakly understood because of their versatility. Eventually, remark that similar questions could be tackled about limit cycles. However, these limit sets are peculiar in the sense that, contrary to fixed points that are rather stable depending on updating modes, limit cycles are very sensitive to updating modes. So, for the sake of generality, such studies would necessarily ask to be parameterized with updating modes, which obviously adds to the difficulty of dealing with limit cycles counting. An approach that could lead to make progress on this would consist for instance in changing the viewing angle by focusing more on analyzing the complexity of counting through pertinent decision problems.

6.4.2 Computational Complexity

Given a Boolean network f of dimension n and an updating mode μ, we focus on the computational complexity of determining basic dynamical properties related to fixed points, reachability between configurations, and limit sets. All these properties reduce to simple graph properties of transition graph $\mathcal{D}_{(f,\mu)}$ as discussed in Section 6.2.3.3. However, this graph is of exponential size with n, both in terms of number of nodes (configurations) and arcs (transitions). Moreover, recall that updating modes such as the asynchronous one can generate an exponential number of outgoing transitions from a single configuration, leading thus to a doubly exponential number of arcs in the transition graph. However, this is only an upper bound on the actual complexity of computing the desired dynamical properties. In what follows, we give an overview of the complexity of deciding dynamical properties of Boolean networks. As usual, these properties are expressed as decision problems, i.e. expecting a yes or no answer. The complexity will be expressed in function of the number of dimensions n, the size of Boolean network f, and additional inputs of the problem, typically configurations, when relevant. Boolean network f is assumed to be given in a symbolic form, for instance expressed using propositional logic as done along this chapter (in contrast with a truth table). Moreover, we assume that given a configuration x, evaluating $f(x)$ is linear in the size of f. Let us first recall the bases of computational complexity classes (Papadimitriou, 1995, Arora and Barak, 2009): the P class (resp. NP class, PSPACE class) is formed by the decision problems that can be solved by algorithms running in polynomial time (resp. in polynomial time with non-deterministic choices, in polynomial space) according to the size of its inputs. We know that P $\subseteq$ NP $\subseteq$ PSPACE, where "$\subseteq$" can be understood as "simpler". A problem is harder than another problem if the latter can be "reduced" to the former (intuitively, it is a particular case of the former problem). In general, different types of reductions can be considered. In this chapter, we essentially rely on polynomial-time reductions. A problem is hard for a given complexity class if it is among the hardest problems of this class. A problem is complete for a given complexity class if it belongs to and is among the hardest problems of this class. The famous SAT problem of determining if a formula expressed in propositional logic (essentially Boolean variables and logic connectors) has a satisfying solution is NP-complete. It is not known yet if NP = PSPACE, but in practice, NP-complete problems are much more tractable than the PSPACE-complete ones, by several orders of magnitude. Hereafter, we also refer to the coNP class, delimiting the problems for which finding a counter-example is in NP, and to the P^{NP} and $coNP^{coNP}$ classes, where A^B denotes the problems that can be solved with complexity A assuming problems of class B can be solved in one instruction (oracle); remark that $P^P = P$ and $NP^P = NP$. These complexity classes belong to the polynomial hierarchy and are subject to the following properties: $NP \subseteq P^{NP}$ and $coNP \subseteq P^{NP} \subseteq coNP^{coNP} \subseteq PSPACE$. Together with the details of the complexity results, we also give pointers to software tools that implement the corresponding dynamical analyses. However, before entering into the details, let us present one very recent result that is certainly the most general in the framework of automata networks, and thus of Boolean networks, because it emphasizes that all relevant problems we may want to address in the context of Boolean networks are actually complex. In the 1950s, Rice (1953) showed that *any non-trivial property of the function computed by a Turing machine is undecidable*, namely, no algorithm leading to a correct positive or negative answer can be constructed. This result is a major result in computer science and more precisely in computability theory as it generalizes for instance the undecidability of the halting problem (Turing, 1936). In Gamard et al. (2021), among others, the authors suggested a meta-theorem for automata networks stating that *any non-trivial property of the function computed by an automata network admits a high computational complexity*, namely, is (co)NP-hard at least. This means notably that we are currently not able to conceive an efficient algorithm to decide if such properties are true or false.

6.4.2.1 Existence of a Fixed Point

The fixed points are a particular type of limit sets consisting of a single configuration. They form one of the most basic and important characteristics of Boolean networks. Most modelling studies rely on the computation of the fixed points to offer a first validation of the model with respect to expected stable behaviors of the system.

Proposition 6.4 Given a Boolean network f of dimension n, deciding if there exists $x \in \mathbb{B}^n$ such that $f(x) = x$ is NP-complete.

The belonging of this problem to NP comes directly from the complexity of evaluating $f(x)$, assumed to be in polynomial time. The completeness can be proven by reductions from different NP-complete problems, such as the SAT problem or the PARTITION problem (Alon, 1985, Floréen and Orponen, 1989, Formenti et al., 2014). Notice that the problem is formulated independently of the updating mode: the fixed points considered here are fixed points of the n-dimensional Boolean function f. By the folklore theorem (Theorem 6.1), the fixed points of f are fixed points of the dynamical system (f, μ). For many of the defined updating modes, including parallel, asynchronous, fully asynchronous, block-sequential, block-parallel, and most permissive, the fixed points of f correspond exactly to the fixed point of the dynamical system. As pointed out in Section 6.4.1.1, there exist updating modes that can generate fixed points that are not fixed points for f. In that case, the complexity of deciding the existence of the fixed point may be different and will likely depend on the definition of the updating mode. However, in the case whenever computing a transition is in polynomial time, we can already settle that it is at most in NP. Then, determining the existence of a fixed point being either of f or of the dynamical system is NP-complete. Following the link between the influence graph of a Boolean network and its fixed points discussed in Section 6.4.1.3, the complexity of deciding the existence of a Boolean network having a fixed given influence graph and a given minimal number of fixed point has been characterized by Bridoux et al. (2019). Notably, deciding whether there exists a Boolean network having a given influence graph and having at least one fixed point is in P; if having $k \geq 2$ fixed points is NP-complete. The former case comes from the result that one can build a Boolean network having at least one fixed point if and only if each non-singleton strongly connected component of the input influence graph contains at least one positive cycle, which can be decided in polynomial time. The latter result is done by reduction from the 3-SAT problem.

6.4.2.1.1 Software Tools

The resolution of the decision problem of the existence of a fixed point can be directly encoded as a SAT problem. When a fixed point exists, the resolution of the SAT problem comes with an instance of configuration x so that $f(x) = x$. Then, one can re-iterate with the decision of the existence of a fixed point different from x, and so on, in order to obtain an explicit listing of the fixed points of the Boolean network, as implemented in the software PINT (Paulevé, 2017). In practice, the identification of a single fixed point by SAT solving is scalable to networks with several hundreds of thousands of automata. The number of iterations of SAT solving being linear with the number of fixed points, the complete and explicit listing of fixed points can only be obtained when there is a limited number of them; otherwise, we usually perform a partial enumeration. Another approach for listing fixed points relies on data structures such as decision diagrams as done in the software GINSIM (Naldi et al., 2007, 2018a). There, the set of fixed points is encoded symbolically, and their enumeration is done by enumerating particular paths within the diagram structure.

6.4.2.2 Reachability Between Configurations

The reachability problem consists in deciding whether there exists a trajectory leading from a given configuration x to a given configuration y with a given updating mode μ. Such properties are employed to predict future states that may be observed in the future in the modeled system. Let us first focus on the classical parallel, asynchronous, and fully asynchronous updating modes. As there is at most 2^n configurations to explore, the problem can be solved in a polynomial space: consider a non-deterministic simulation algorithm that iteratively applies a transition modifying the current configuration of the network. If there exists a trajectory from x to y, then there is at least one execution of the algorithm that will encounter the configuration y after at most $2^n - 1$ transitions. Thus, it is sufficient to store the current configuration of the network and a counter that is incremented after each transition. Considering a binary encoding, this counter requires n bits. Thus, the non- deterministic simulation algorithm requires a linear space and is thus in PSPACE. With the parallel updating mode, the PSPACE-hardness derives by reduction from the reachability problem in reaction systems, a subclass of synchronous Boolean networks (Dennunzio et al., 2019). With fully asynchronous and asynchronous updating modes, the PSPACE-hardness derives by reduction from the reachability problem in synchronous Boolean networks. Indeed, similarly to cellular automata (Nakamura, 1981), one can define a Boolean network g so that asynchronous and fully asynchronous dynamics give reachability relations that are equivalent with the synchronous dynamics of f (Paulevé et al., 2020).

Proposition 6.5 Given a Boolean network f of dimension n, an updating mode μ, and two configurations $x, y \in \mathbb{B}^n$, deciding if $x \xrightarrow[(f,\mu)]{*} y$ is PSPACE-complete with $\mu \in \{\mathsf{p}, \mathsf{fa}, \mathsf{a}\}$.

Regarding the sequential updating modes, it is interesting to remark that one can encode the parallel dynamics of a Boolean network f of dimension n as a sequential dynamics of a Boolean network g in at most polynomial time. One naive approach is to double the number of automata, the first set storing the next state of each original automaton and the second set storing its state before the update. The first set is updated first, and then the second set copies the states of the first set of automata: define g with $2n$ dimensions such that for all $x, x' \in \mathbb{B}^n$, for each $i \in [\![n]\!]$, $g_i(xx') = f_i(x')$ and $g_{n+i}(xx') = x_i$. Then, for all $x, y \in \mathbb{B}^n$, $x \to_{(f,p)} y$ if and only if $xx \to_{(f,(1,\cdots,2n))} yy$. Note that Bridoux et al. (2017), Bridoux (2018, 2019) proposed a more efficient encoding requiring less than $2n$ automata. Thus, the reachability problem with the sequential updating mode is harder than with the parallel updating mode. Because it lies in PSPACE, it is therefore PSPACE-complete as well. Finally, because the parallel updating mode is a particular case of the block-sequential and block-parallel, the complexity result applies for these updating modes as well and any other generalization of them. As demonstrated in Paulevé et al. (2020), the case of the permissive updating mode is quite different. Following Definition 6.8, the computation of the configurations succeeding configuration x is done by a two-step process repeated for each subset of automata W. First, it will compute the smallest sub-hypercube h that contains x and such that for each dimension $i \in W$, either $h_i = *$ or for each vertex z of the sub-hypercube, $f_i(z) = h_i = x_i$. This computation can be performed by the following algorithm, where $c(h)$ denotes the set of vertices of the sub-hypercube h:

```
1 h := x
2 repeat |W| times
3     for each i ∈ W such that h_i ≠ *:
4         if ∃z ∈ c(h) such that f_i(z) ≠ x_i:
5             h_i := *
```

The second step consists in filtering the obtained sub-hypercube by removing states of automata in W, which cannot be computed from any vertex of h. From the computation above, these states correspond to the states of automata of W in x:

```
6 for each i ∈ W such that h_i = *:
7     if not (∃z ∈ c(h) such that f_i(z) = x_i) :
8         h_i := 1 - x_i
```

Then, y is reachable from x if and only if y is a vertex of such an obtained sub-hypercube, for at least one subset W. Let us analyze the complexity of this algorithm. First, with a fixed W, remark that the computation of h relies on several tests of the form "$\exists z \in c(h) : f_i(z) = b$", with $b \in \mathbb{B}$. This is exactly the SAT problem. Thus, in the general case, such decisions are NP-complete, and in the case whenever f is locally monotone (thus f_i is monotone), such decisions are in P. Then, given the configuration y, it is key to remark that there is no need to compute the sub-hypercubes for all the possible subsets W. Indeed, let us focus on the case whenever for a fixed W, y is not a vertex of the computed sub-hypercube. Two cases arise: (i) y is not a vertex of the sub-hypercube before the filtering: then, it is also the case for all the subsets of W; (ii) y is no longer a vertex of the sub-hypercube after the filtering: it means that there is a subset $D \subseteq W$ of automata that has been filtered and such that $\forall i \in D$, $x_i = y_i$. Then, y is not a vertex of all the sub-hypercubes computed with W minus a strict subset of D. Thus, the overall procedure starts with $W = [\![n]\!]$ and repeatedly removes the subset D until either y is a vertex of the sub-hypercube (and it is thus reachable from x) or is not a vertex of the sub-hypercube before filtering (and is thus not reachable from x).

Proposition 6.6 Given a Boolean network f of dimension n, and two configurations $x, y \in \mathbb{B}^n$, deciding if $x \to^*_{(f,\text{MP})} y$ is in P if f is locally monotone; otherwise, it is in P^{NP}.

6.4.2.2.1 Software Tools

For the synchronous, fully asynchronous, and asynchronous updating modes, the verification of reachability properties is usually tackled by generic tools related to the model checking of discrete and finite dynamical systems: the reachability property being expressed in temporal logics such as CTL (Clarke and Emerson, 1981). Tools such as GINSIM (Naldi et al., 2018a) enable exporting Boolean networks to files in suitable formats for model checkers such as NuSMV (Cimatti et al., 2002). The verification of reachability properties can take advantage of static analysis to reduce the transitions to explore, as offered by the tool PINT (Paulevé, 2017), enhancing the scalability of the computation in many cases (Paulevé, 2018). The verification of reachability properties using the most permissive updating mode is implemented in the tool MPBN (Paulevé et al., 2020). In practice, because of the differences in complexity, the verification of reachability properties using asynchronous updating modes scales up to networks with around one hundred automata; using the most permissive updating mode, it scales up in the order of hundreds of thousands of automata (Paulevé, 2020).

6.4.2.3 Limit Configurations

The limit configurations and reachable limit configurations are among the most analyzed dynamical properties of Boolean networks when modelling biological systems. Limit configurations generalize fixed points by taking into account non-singleton limit sets. Combined with reachability analysis, they delineate the possible long-term behaviors of the network from a given initial configuration. In this section, we focus on the following decision problem: *Given a configuration, does it belong to a limit set of the Boolean network with the given updating mode?* The advantage of such a formulation is that the size of the input is independent of that of the limit set. Furthermore, it makes a direct link with the problem of identification and enumeration of the limit sets. Moreover, we do not explicitly address the reachability of the limit configurations from a given initial configuration. However, this comes naturally by combining with the decision problem discussed in the previous section. How can we determine that a configuration x is a limit configuration? This problem is actually tied to the reachability problem: x is a limit configuration if and only if it is reachable from any configuration y reachable from x (with the updating mode μ). Equivalently, x is not a limit configuration if and only if there exists a configuration y reachable from x but x is not reachable from y. Using the same reasoning as for reachability, one can deduce that this problem is in coNPSPACE = PSPACE for the parallel, fully asynchronous, and asynchronous updating modes. As for the reachability, the completeness can be derived by reduction from the same problem in synchronous reaction systems demonstrated to be PSPACE-complete in Dennunzio et al. (2019). Then, using the reduction of synchronous Boolean networks to fully asynchronous and asynchronous updating modes, we obtain that the decision problem of limit configurations is PSPACE-complete.

Proposition 6.7 Given a Boolean network f of dimension n and a configuration $x \in \mathbb{B}^n$, deciding if x belongs to a limit set of f with the updating mode $\mu \in \{p, fa, a\}$ is PSPACE-complete.

With the exact same arguments as for the reachability problem, this complexity also applies to the sequential updating modes and any generalization of them. The case of the most permissive updating mode is again much more specific (Paulevé et al., 2020). Indeed, it appears that the limit sets of the most permissive dynamics have a particular shape: they always correspond to particular sub-hypercubes, namely, to the smallest sub-hypercubes that are closed by f, also known as *minimal trap spaces* (Klarner et al., 2015). A sub-hypercube h is closed by f if and only if for each of its vertices x, $f(x)$ is also a vertex of the hypercube. Fixed points are a particular case of minimal trap spaces where the sub-hypercubes have dimension 0 (all the automata have a fixed state). This property comes from the fact that whenever two configurations lying a diagonal are reachable from each other with the most permissive updating mode, then so are the adjacent vertices. In other words,

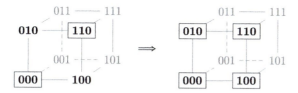

Figure 6.11 Illustration of the property of limit sets with the most permissive updating mode, where boxed configurations belong to a same limit set: whenever two configurations belong to the same limit set, so do all the configurations of the smallest sub-hybercube that contains both of them.

and without loss of generality, let us consider that configurations $x = a00b$ and $y = a11b$, where a and b are binary vectors, are reachable from each other. This implies that there exists a set of automata W such that x and y belong to the sub-hypercube h computed according to the previous section from x or y or equivalently belong to $\Lambda_W \circ \Phi^\omega_{W,V}(\{x\}) = \Lambda_W \circ \Phi^\omega_{W,V}(\{y\})$, following the notations of Definition 6.8. Thus, because of the sub-hypercube structure, both $a01b$ and $a10b$ belong to h and are thus reachable from x and y, and conversely. This is illustrated by Figure 6.11.

Proposition 6.8 $A \subseteq \mathbb{B}^n$ is a limit set of f with the most permissive updating mode if and only if A forms a minimal sub-hypercube closed by f.

Thus, determining if a configuration x belongs to a limit set with the most permissive updating mode boils down to determining the existence of a minimal sub-hypercube h that contains x and is closed by f. Consider IS-NOT-CLOSED(f, h) the problem of deciding if the sub-hypercube h is not closed by f: it is equivalent to deciding if there exists an automaton $i \in [\![n]\!]$ with $h_i \neq *$ and a vertex z of h such that $f_i(z) \neq h_i$, which is NP-complete in general, and P whenever f is locally monotone. Then, the complementary problem IS-CLOSED(f, h) is in coNP in the general case and in P in the locally monotone case. Now, consider IS-NOT-MINIMAL(f, h) the problem of deciding if the sub-hypercube h closed by f is not minimal: it can be solved by deciding wherever there exists a sub-hypercube h' that is strictly included in h and which is closed by f, which is at most NP$^{\text{IS-CLOSED}}$. Thus, the complementary problem IS-MINIMAL(f, h) is in coNP$^{\text{IS-CLOSED}}$, i.e. coNP$^{\text{coNP}} = \Pi^P_2$ in the general case and coNP in the locally monotone case.

Theorem 6.5 Given a Boolean network f of dimension n and a configuration $x \in \mathbb{B}^n$, deciding if x is a limit configuration of f with the most permissive updating mode is in coNP whenever f is locally monotone, and in Π^P_2 otherwise.

6.4.2.3.1 Software Tools

Similarly to the analysis of fixed points, the software tools usually focus on the enumeration of the limit sets of a given Boolean network and a given updating mode. The case of the parallel and fully asynchronous updating modes is implemented in several software tools such as BoolSim (Garg et al., 2008) (parallel and fully asynchronous), BNS (Dubrova and Teslenko, 2011) (parallel), and Cabean (Su and Pang, 2020) (fully asynchronous). They rely either on SAT solving (BNS) or on symbolic representations of the reachable configurations with decision diagrams. To tackle the potential combinatorial explosion of the number of configurations within a limit set, the methods output them as unions of sub-hypercubes. This representation comes quite directly when using decision diagrams, for instance. In practice, the scalability is roughly similar to the reachability analysis, i.e. in the order of the hundreds of automata. In the case of the most permissive updating mode, because the configurations of the limit sets correspond to specific sub-hypercubes, their identification can be reduced to the subset-minimal solutions of an Answer-set programming problem (Baral, 2003) as done by the tool mpbn (Paulevé et al., 2020). In practice, this approach can be applied to networks with several hundreds of thousands of components, at least for a partial enumeration of their limit sets when there are too many of them for an explicit enumeration (Paulevé, 2020).

6.5 Conclusion

6.5.1 Updating Modes and Time

In this chapter, we have developed some elements related to Boolean networks and their use as models of biological complex systems, in particular as models of genetic regulation networks. More precisely, our purpose has been focused on the updating modes, by underlining that Boolean network dynamics strongly depend on the manner automata execute their local functions. If we consider these updating modes as means of representation of the relations that automata maintain with *time*, countless questions about the nature and consideration of time raise. These questions are obviously relevant in mathematics and computer science (notice that one main problematic in this discipline concerns synchronism and asynchronism), but they are all the more pertinent in the context of theoretical biology, insofar as the answers that we could bring in this context would possibly increase the knowledge of the way time flows at different scales of living organisms. When a real interacting system $\mathscr{S}$ is modeled by a Boolean network, or by an automata network f by extension, we generally consider that every state of $\mathcal{S}$ can be associated with one (or more) configuration(s) of f, up to a specific encoding. Moreover, we classically think of these systems as collections of entities that interact with each other *over time*. As a consequence, the very concept of a transition from a configuration x to a configuration y in the dynamics of f implicitly integrates a notion of time that places x before y, which leads us to make an association between a trajectory and the intuitive concept of temporal flow. However, what can be the semantics of a transition actually? In this paragraph, we propose three distinct visions and are aware of the non-exhaustiveness of the latter.

6.5.1.1 Modelling Durations

When we study automata networks and the dynamical systems they can model, we associate them with transition graphs that are either deterministic or non-deterministic. In this latter case, they can be either non-stochastic or stochastic, i.e. complemented with a measure of probability in order to weight each of their transitions. In the general case, such systems induce a time domain that is either continuous or discrete (in this chapter, the time domain is $\mathbb{N}$). In this framework, a natural abstraction consists in seeing the trajectories as the temporal flow. However, the coherence with physical time would imply that the discrete transitions embed the concept of duration and represent all the same duration. This corresponds precisely to a real-time discretization. However, whenever an automata network is considered as a model of a real system, this vision can be perceived as unrealistic. This is why some studies increase the concept of transition by adding that of duration, so that the transitions can be associated with distinct durations. They are then labeled by the amount of time they are supposed to last or, more precisely, by the duration of the event(s) that they are supposed to represent (Thomas, 1983, Bernot et al., 2004, Siebert and Bockmayr, 2008). In addition to the questions presented in the following paragraphs, this choice of modelling the duration of a temporal flow raises specific theoretical questions, including how long does a network take to reach an asymptotic behavior in the best case, in the worst case, or on average? what is the probability that the network passes asymptotically by this or that configuration?... However, integrating this concept of chronometric time (through durations) within a fundamentally discrete formalism raises a problem extremely difficult to solve. In particular, in Noual and Sené (2011), the authors emphasized how this problem leads naturally, almost necessarily, to prefer a continuous framework. In order to avoid this problem, and thus not to give the discrete framework more capacities that it has while retaining its intrinsic advantages, the following paragraphs shows that we can choose to model time differently.

6.5.1.2 Modelling Precedence

A second way of conceiving time with discrete dynamical systems consists in no longer considering the trajectories as witnesses of the durations of a temporal flow but as representatives of a relation of precedence. They are then nothing else but sequences of events without duration. In this case, if transitions $x \to^* y$ and $x' \to^* y'$ are both achievable at the same time, then this vision makes it possible to give meaning that $x \to^* y$ may last longer than $x' \to^* y'$, under certain conditions, and conversely that $x' \to^* y'$ may last longer than $x \to^* y$, under other conditions. Thus, the different behaviors of an automata network can occur at different time scales without new information being specified to distinguish them. The modeled time then deviates slightly from the notion of real time to become a *logical time*, which requires less of knowledge on the transitions between states of the modeled real system. The transitions and trajectories are then only sequences of events that follow each other without integrating any notion of duration. More precisely, a transition $x \to y$ does not hold information about the real process it models but only expresses the result of a possible observation of the system at a certain time step returning that the system is in state y, knowing that the previous observation of the system returned the state x. As a result, the time that transitions and trajectories take cannot be measured, only the number of events and their succession can be. This is the classical approach of time that is considered when we work on deterministic or non-deterministic discrete dynamical systems. Typical questions highlighting this conception of time are the following (Melliti et al., 2013): how many steps does a network take to reach limit sets? and Is such behavior always observed after such event?...

6.5.1.3 Modelling Causality

Another view we can have of time relates to the causality between events, where the notion of duration of transitions is replaced by a relation of causes and consequences. Causality essentially refers to the fact that an event can only be triggered once a particular condition is met, and triggering an event may preempt other *concurrent* events. This modelling aims at revealing a fine-grained dependency structure between transitions and events. Then, precedence becomes an emerging property, as causal relations rule which events *must* appear before others, which events can never appear after others, and which sets of events can be interleaved freely. Causal modelling focuses on the *minimal* conditions for triggering a certain event or observe a given behavior, which relates to determining the prime implications for transitions. Questions related to this context are typically about the existence or accessibility properties: is this transition possible?, is a configuration that checks such properties reachable?...as well as for the control: which modification of the model is sufficient to make a transition possible or impossible? Notice that, in this context, the transition graph hinders important parts of this information, as it forgets the local functions that generated it. The concurrency theory offers conceptual tools, including event structures for reasoning efficiently on the causality between events that can be naturally applied to Boolean networks with their deterministic and non-deterministic updating modes (Chatain et al., 2020).

6.5.2 Toward an Updating Mode Hierarchy

Echoing with different modelling hypotheses and specifications of time, we presented along this chapter a large range of updating modes for Boolean networks. In this chapter, we endeavored to define them with a unifying mathematical framework of deterministic and non-deterministic updates. The relationship between the different updating modes is a fundamental question that aims at bringing a structure between these numerous dynamics. Let us say that an updating mode μ is *(weakly) simulated* by an updating mode μ', noted $\mu \leqslant \mu'$, if and only if, for any Boolean network f and any pair (x, y) of its configurations, if there exists a path from x to y in the dynamics generated by μ, then there exists a path from x to y in the dynamics generated by μ'. Formally:

$$\mu \leqslant \mu' \iff \forall f, \forall x, \forall y, \ x \to^*_{(f,\mu)} y \implies x \to^*_{(f,\mu')} y.$$

A class of updating modes C is then weakly simulated by a class of updating modes C', if for any updating mode of the former class, there exists an updating mode of the latter class that weakly simulates it:

$$C \leqslant C' \iff \forall \mu \in C, \exists \mu' \in C', \ \mu \leqslant \mu'.$$

The obtained hierarchy is depicted in Figure 6.12.

6.5.2.1 Software Tools

In the past 20 years, numerous software tools for modelling and analyzing dynamics of Boolean networks have been developed. The CoLoMoTo interactive notebook (Naldi et al., 2018b) provides a distribution of many of these tools promoting their accessibility and the writing and publishing of reproducible computational analysis of Boolean networks. Based on this software environment, we designed interactive notebooks for reproducing the examples given in this chapter with the different updating modes. They are available at https://github.com/pauleve/updating-modes-notebooks (archived at https://doi.org/10.5281/zenodo.6966830), and can

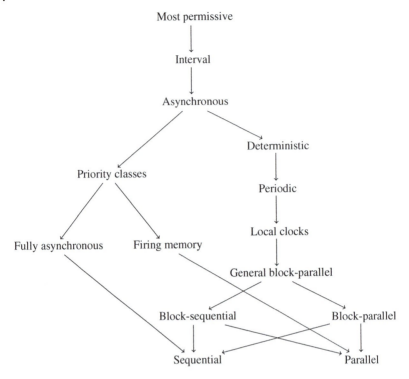

Figure 6.12 Weak simulation relation between the classes of updating modes mentioned in this chapter with any fixed Boolean network.

be easily executed and re-used for analyzing different Boolean networks and implement custom updating modes.

6.5.3 Opening on Intrinsic Simulations

The hierarchy presented in Figure 6.12 shows the relations between the paths generated by the different updating modes on any fixed Boolean network. Another direction is to consider simulation relations between the dynamical systems, i.e. Boolean networks coupled with an updating mode. A perspective of the work presented in this chapter may focus on simulations of Boolean networks evolving with non-deterministic updates by Boolean networks evolving with deterministic updates. A first natural way is by following a classical determinization of the dynamics. Indeed, one can encode any set of configurations in $\mathbb{B}^n$ as one configuration in $\mathbb{B}^{2^n}$. Let us consider such an encoding $c : 2^{\mathbb{B}^n} \to \mathbb{B}^{2^n}$, where for all $x \in \mathbb{B}^n$,

$c(X)_x = 1$ if $x \in X$, otherwise $c(X)_x = 0$ (we slightly abuse notations here by specifying a vector index by its binary representation). Now, it is clear that for any set update $\Phi : 2^{\mathbb{B}^n} \to 2^{\mathbb{B}^n}$ of a Boolean network f of dimension n, one can define a Boolean network g such that for all sets of configurations $X \subseteq \mathbb{B}^n$, $g(c(X)) = c(\Phi(X))$. This encoding is complete in the sense that any transition generated by Φ is simulated in (g, p). However, these simulations are nothing else but a brute-force encoding in which we get rid of the transition relation by increasing exponentially the state space. Moreover, with this deterministic encoding, the structure of the transition relation of $(f, \mu = \Phi)$ is lost, which makes much more difficult characterizing dynamical features of (f, μ) such as its limit sets for instance. Actually, a fundamental matter here lies in the concept of simulation at stake: we are interested in intrinsic simulations that go far beyond the classical concepts of encoding or simulation. Indeed, intrinsic simulations aim at conserving dynamical structures in addition to operated computations. So, one of the first questions to answer would consist in defining formally different kinds of intrinsic simulations. Nevertheless, firstly, consider the following intrinsic simulation: a dynamical system (f, μ) simulates another (g, μ') if $\mathcal{D}_{(g,\mu')}$ is a sub-graph of $\mathcal{D}_{(f,\mu)}$. With this rather simple definition, it is direct to state that, with a and $\overline{d}$ the asynchronous and memory updating modes, respectively, for any Boolean network f, (f, a) simulates $(f, \overline{d})$. Some natural questions related to Boolean networks updated with memory are as follows:

- Are there Boolean networks whose dynamics obtained according to $\overline{d}$ remains deterministic, whatever $\overline{d}$?
- If so, what are their properties and what are the equivalent deterministic updating modes?

To go further, consider the most permissive updating mode. It is direct that (f, μ) does not simulate (f, MP), except for very particular f. Let us now consider a more general intrinsic simulation: a dynamical system (f, μ) simulates another (g, μ') if $\mathcal{D}_{(g,\mu')}$ is a graph obtained from $\mathcal{D}_{(f,\mu)}$ thanks to edge deletions and vertex shortcuts. A lot of promising questions arise from this, in particular related to $\overline{d}$ and MP updating modes, among which for instance:

- Let per be a deterministic periodic updating mode. How can $(f, \overline{d})$ be simulated by (g, per)? The answer is known for per = p (Goles et al., 2020), but it seems pertinent to find a generalization to deterministic periodic updating modes, and even more general deterministic updating modes.
- Intuitively, any (f, MP) might be simulated by (g, a), where f and g are Boolean networks and the dimension of g is greater than that of f. However, how many automata need to be added to g depending on the dimension of f?

All answers, even partial or negative, will bring a better understanding of updating modes and Boolean networks, which would lead to pertinent further development in both Boolean network theory and their application in systems biology.

Acknowledgments

This work was primarily funded by our salaries as French State agents and secondarily by Agence Nationale pour la Recherche (ANR) in the scope of the projects ANR-18-CE40-0002 FANs (SS) and ANR-20-CE45-0001 BNeDiction (LP).

References

Wassim Abou-Jaoudé, Pedro T. Monteiro, Aurélien Naldi, Maximilien Grandclaudon, Vassili Soumelis, Claudine Chaouiya, and Denis Thieffry. Model checking to assess T-helper cell plasticity. *Frontiers in Bioengineering and Biotechnology*, 2: 86, 2015. doi: https://doi.org/10.3389/fbioe.2014.00086.

Wassim Abou-Jaoudé, Denis Thieffry, and Jérôme Feret. Formal derivation of qualitative dynamical models from biochemical networks. *Biosystems*, 149: 70–112, 2016. doi: https://doi.org/10.1016/j.biosystems.2016.09.001.

Jamil Ahmad, Olivier Roux, Gilles Bernot, Jean-Paul Comet, and Adrien Richard. Analysing formal models of genetic regulatory networks with delays. *International Journal of Bioinformatics Research and Applications*, 4: 240–262, 2008. doi: https://doi.org/10.1504/IJBRA.2008.019573.

Aurore Alcolei, Kévin Perrot, and Sylvain Sené. On the flora of asynchronous locally non-monotonic Boolean automata networks. In *Proceedings of the International Workshop on Static Analysis and Systems Biology*, volume 326 of *Electronic Notes in Theoretical Computer Science (ENTCS)*, pages 3–25. Elsevier, 2016. doi: https://doi.org/10.1016/j.entcs.2016.09.016.

Noga Alon. Asynchronous threshold networks. *Graphs and Combinatorics*, 1: 305–310, 1985. doi: https://doi.org/10.1007/BF02582959.

Julio Aracena. Maximum number of fixed points in regulatory Boolean networks. *Bulletin of Mathematical Biology*, 70: 1398–1409, 2008. doi: https://doi.org/10.1007/s11538-008-9304-7.

Julio Aracena, Jacques Demongeot, and Eric Goles. Positive and negative circuits in discrete neural networks. *IEEE Transactions on Neural Networks*, 15: 77–83, 2004. doi: https://doi.org/10.1109/TNN.2003.821555.

Julio Aracena, Mauricio González, Alejandro Zu niga, Marco A. Mendez, and Verónica Cambiazo. Regulatory network for cell shape changes during

Drosophila ventral furrow formation. *Journal of Theoretical Biology*, 239: 49–62, 2006. doi: https://doi.org/10.1016/j.jtbi.2005.07.011.

Julio Aracena, Eric Goles, Andres Moreira, and Lilian Salinas. On the robustness of update schedules in Boolean networks. *Biosystems*, 97: 1–8, 2009. doi: https://doi.org/10.1016/j.biosystems.2009.03.006.

Julio Aracena, Eric Fanchon, Marco Montalva, and Mathilde Noual. Combinatorics on update digraphs in Boolean networks. *Discrete Applied Mathematics*, 159: 401–409, 2011. doi: https://doi.org/10.1016/j.dam.2010.10.010.

Julio Aracena, Luis Gómez, and Lilian Salinas. Limit cycles and update digraphs in Boolean networks. *Discrete Applied Mathematics*, 161: 1–12, 2013. doi: https://doi.org/10.1016/j.dam.2012.07.003.

Julio Aracena, Adrien Richard, and Lilian Salinas. Number of fixed points and disjoint cycles in monotone Boolean networks. *SIAM Journal on Discrete Mathematics*, 31: 1702–1725, 2017. doi: https://doi.org/10.1137/16M1060868.

Sanjeev Arora and Boaz Barak. *Computational Complexity: A Modern Approach*. Cambridge University Press, 2009.

Chitta Baral. *Knowledge Representation, Reasoning and Declarative Problem Solving*. Cambridge University Press, 2003.

Theodore Beauchaine. Vagal tone, development, and Gray's motivational theory: toward an integrated model of autonomic nervous system functioning in psychopathology. *Development and Psychopathology*, 13: 183–214, 2001. doi: https://doi.org/10.1017/s0954579401002012.

Claire Bendix, Carine M. Marshall, and Frank G. Harmon. Circadian clock genes universally control key agricultural traits. *Molecular Plants*, 8: 1135–1152, 2015. doi: https://doi.org/10.1016/j.molp.2015.03.003.

Arndt Benecke. Chromatin code, local non-equilibrium dynamics, and the emergence of transcription regulatory programs. *The European Physical Journal E*, 19: 353–366, 2006. doi: https://doi.org/10.1140/epje/i2005-10068-8.

Gilles Bernot, Jean-Paul Comet, Adrien Richard, and Janine Guespin. Application of formal methods to biological regulatory networks: extending Thomas' asynchronous logical approach with temporal logic. *Journal of Theoretical Biology*, 229: 339–347, 2004. doi: https://doi.org/10.1016/j.jtbi.2004.04.003.

Florian Bridoux. Sequentialization and procedural complexity in automata networks. arXiv:1803.00438, 2018.

Florian Bridoux. *Simulations intrinsèques et complexités dans les réseaux d'automates*. PhD thesis, Université d'Aix-Marseille, 2019.

Florian Bridoux, Pierre Guillon, Kévin Perrot, Sylvain Sené, and Guillaume Theyssier. On the cost of simulating a parallel Boolean automata network with a block-sequential one. In *Proceedings of the International Conference on Theory and Applications of Models of Computation*, volume 10185 of *Lecture Notes in Computer Science (LNCS)*, pages 112–128. Springer, 2017. doi: https://doi.org/10.1007/978-3-319-55911-7_9.

Florian Bridoux, Nicolas Durbec, Kévin Perrot, and Adrien Richard. Complexity of maximum fixed point problem in Boolean networks. In *Proceedings of the Conference on Computability in Europe*, volume 11558 of *Lecture Notes in Computer Science (LNCS)*, pages 132–143. Springer, 2019. doi: https://doi.org/10.1007/978-3-030-22996-2_12.

Thomas Chatain, Stefan Haar, and Loïc Paulevé. Boolean networks: beyond generalized asynchronicity. In *Proceedings of International Workshop on Cellular Automata and Discrete Complex Systems*, volume 10975 of *Lecture Note in Computer Science (LNCS)*, pages 29–42. Springer, 2018. doi: https://doi.org/10.1007/978-3-319-92675-9_3.

Thomas Chatain, Stefan Haar, Juraj Kolčák, Loïc Paulevé, and Aalok Thakkar. Concurrency in Boolean networks. *Natural Computing*, 19: 91–109, 2020. doi: https://doi.org/10.1007/s11047-019-09748-4.

Alonso Church. A set of postulates for the foundation of logic. *Annals of Mathematics*, 33: 346–366, 1932. doi: https://doi.org/10.2307/1968337.

Alonso Church. An unsolvable problem of elementary number theory. *American Journal of Mathematics*, 58: 345–363, 1936. doi: https://doi.org/10.2307/2371045.

Alessandro Cimatti, Edmund Clarke, Enrico Giunchiglia, Fausto Giunchiglia, Marco Pistore, Marco Roveri, Roberto Sebastiani, and Armando Tacchella. NuSMV 2: An ppenSource tool for symbolic model checking. In *Proceedings of the International Conference on Computer Aided Verification*, volume 2404 of *Lecture Notes in Computer Science (LNCS)*, pages 359–364. Springer, 2002. doi: https://doi.org/10.1007/3-540-45657-0_29.

Olivier Cinquin and Jacques Demongeot. Positive and negative feedback: strinking a balance between necessary antagonists. *Journal of Theoretical Biology*, 216: 229–241, 2002. doi: https://doi.org/10.1006/jtbi.2002.2544.

Edmund M. Clarke and E. Allen Emerson. Design and synthesis of synchronization skeletons using branching-time temporal logic. In *Proceedings of the Workshop on Logic of Programs*, volume 131 of *Lecture Notes in Computer Science (LNCS)*, pages 52–71. Springer, 1981. doi: https://doi.org/10.1007/BFb0025774.

David P. A. Cohen, Loredana Martignetti, Sylvie Robine, Emmanuel Barillot, Andrei Zinovyev, and Laurence Calzone. Mathematical modelling of molecular pathways enabling tumour cell invasion and migration. *PLoS Computational Biology*, 11: e1004571, 2015. doi: https://doi.org/10.1371/journal.pcbi.1004571.

Samuel Collombet, Chris van Oevelen, Jose Luis Sardina Ortega, Wassim Abou-Jaoudé, Bruno Di Stefano, Morgane Thomas-Chollier, Thomas Graf, and Denis Thieffry. Logical modeling of lymphoid and myeloid cell specification and transdifferentiation. *Proceedings of the National Academy of Sciences of the United States of America*, 114: 5792–5799, 2017. doi: https://doi.org/10.1073/pnas.1610622114.

Louis Comtet. Recouvrements, bases de filtre et topologies d'un ensemble fini. *Comptes rendus de l'Académie des Sciences – Série A*, 262: 1091–1094, 1966.

Michel Cosnard and Jacques Demongeot. On the definitions of attractors. In *Proceedings of the International Symposium on Iteration Theory and its Functional Equations*, volume 1163 of *Lecture Notes in Mathematics*, pages 23–31. Springer, 1985. doi: https://doi.org/10.1007/BFb0076414.

Patrick Cousot and Radhia Cousot. Abstract interpretation: a unified lattice model for static analysis of programs by construction or approximation of fixpoints. In *Proceedings of the ACM SIGACT-SIGPLAN Symposium on Principles of programming languages*, pages 238–252. Association for Computing Machinery, 1977. doi: https://doi.org/10.1145/512950.512973.

Michael F. Covington and Stacey L. Harmer. The circadian clock regulates auxin signaling and responses in *Arabidopsis*. *PLoS Biology*, 5: e222, 2007. doi: https://doi.org/10.1371/journal.pbio.0050222.

Yves Crama and Peter L. Hammer. *Boolean Functions: Theory, Algorithms, and Applications*, volume 142 of *Encyclopedia of Mathematics and Its Applications*. Cambridge University Press, 2011.

Haimabati Das and Ritwik Kumar Layek. Estimation of delays in generalized asynchronous Boolean networks. *Molecular BioSystems*, 12: 3098–3110, 2016. doi: https://doi.org/10.1039/C6MB00276E.

Maria I. Davidich and Stefan Bornholdt. Boolean network model predicts cell cycle sequence of fission yeast. *PLoS One*, 3: e1672, 2008. doi: https://doi.org/10.1371/journal.pone.0001672.

Hidde de Jong. Modeling and simulation of genetic regulatory systems: a literature review. *Journal of Computational Biology*, 9: 67–103, 2002. doi: https://doi.org/10.1089/10665270252833208.

Max Delbrück. Génétique du bactériophage. In *Unités biologiques douées de continuité génétique*, volume 8 of *Colloques internationaux du CNRS*, pages 91–103, 1949.

Jacques Demongeot and Sylvain Sené. About block-parallel Boolean networks: a position paper. *Natural Computing*, 19: 5–13, 2020. doi: https://doi.org/10.1007/s11047-019-09779-x.

Jacques Demongeot, Adrien Elena, and Sylvain Sené. Robustness in regulatory networks: a multi-disciplinary approach. *Acta Biotheoretica*, 56: 27–49, 2008a. doi: https://doi.org/10.1007/s10441-008-9029-x.

Jacques Demongeot, Christelle Jézéquel, and Sylvain Sené. Boundary conditions and phase transitions in neural networks. Theoretical results. *Neural Nerworks*, 21: 971–979, 2008b. doi: https://doi.org/10.1016/j.neunet.2008.04.003.

Jacques Demongeot, Eric Goles, Michel Morvan, Mathilde Noual, and Sylvain Sené. Attraction basins as gauges of the robustness against boundary conditions in biological complex systems. *PLoS One*, 5: e11793, 2010. doi: https://doi.org/10.1371/journal.pone.0011793.

Jacques Demongeot, Mathilde Noual, and Sylvain Sené. Combinatorics of Boolean automata circuits dynamics. *Discrete Applied Mathematics*, 160: 398–415, 2012. doi: https://doi.org/10.1016/j.dam.2011.11.005.

Jacques Demongeot, Dan Istrate, Hajer Khlaifi, Lucile Mégret, Carla Taramasco, and René Thomas. From conservative to dissipative non-linear differential systems. An application to the cardio-respiratory regulation. *Discrete & Continuous Dynamical Systems – Series S*, 13: 2121–2134, 2020. doi: https://doi.org/10.3934/dcdss.2020181.

Jacques Demongeot, Tarek Melliti, Mathilde Noual, Damien Regnault, and Sylvain Sené. On Boolean automata isolated cycles and tangential double-cycles dynamics. In *Automata and Complexity: Essays presented to Eric Goles on the occasion of his 70th birthday*. Emergence, Complexity and Computation, vol 42. Springer, 2021. doi: https://doi.org/10.1007/978-3-030-92551-2_11

Alberto Dennunzio, Enrico Formenti, Luca Manzoni, and Antonio E. Porreca. Complexity of the dynamics of reaction systems. *Information and Computation*, 267: 96–109, 2019. doi: https://doi.org/10.1016/j.ic.2019.03.006.

Olga Dergacheva, Kathleen J. Griffioen, Robert A. Neff, and David Mendelowitz. Respiratory modulation of premotor cardiac vagal neurons in the brainstem. *Respiratory Physiology & Neurobiology*, 174: 102–110, 2010. doi: https://doi.org/10.1016/j.resp.2010.05.005.

Elena Dubrova and Maxim Teslenko. A SAT-based algorithm for finding attractors in synchronous Boolean networks. *IEEE/ACM Transactions on Computational Biology and Bioinformatics*, 8: 1393–1399, 2011. doi: https://doi.org/10.1109/tcbb.2010.20.

Federica Eduati, Patricia Jaaks, Jessica Wappler, Thorsten Cramer, Christoph A. Merten, Mathew J. Garnett, and Julio Saez-Rodriguez. Patient-specific logic models of signaling pathways from screenings on cancer biopsies to prioritize personalized combination therapies. *Molecular Systems Biology*, 16: e8664, 2020. doi: https://doi.org/10.15252/msb.20188664.

Adrien Elena. *Robustesse des réseaux d'automates booléens à seuil aux modes d'itération. Application à la modélisation des réseaux de régulation génétique*. PhD thesis, Université Grenoble 1 – Joseph Fourier, 2009.

François Fages and Sylvain Soliman. Abstract interpretation and types for systems biology. *Theoretical Computer Science*, 403: 52–70, 2008. doi: https://doi.org/10.1016/j.tcs.2008.04.024.

Nazim Fatès. A guided tour of asynchronous cellular automata. In *Proceedings of the International Workshop on Cellular Automata and Discrete Complex Systems*, volume 8155 of *Lecture Notes in Computer Science (LNCS)*, pages 15–30. Springer, 2013. doi: https://doi.org/10.1007/978-3-642-40867-0_2.

Aurélien Fauré, Adrien Naldi, Claudine Chaouiya, and Denis Thieffry. Dynamical analysis of a generic Boolean model for the control of the mammalian cell cycle. *Bioinformatics*, 22: e124–e131, 2006. doi: https://doi.org/10.1093/bioinformatics/btl210.

Patrik Floréen and Pekka Orponen. On the computational complexity of analyzing Hopfield nets. *Complex Systems*, 3: 577–587, 1989.

Enrico Formenti, Luca Manzoni, and Antonio E. Porreca. Fixed points and attractors of reaction systems. In *Proceedings of the Conference on Computability in Europe*, volume 8493 of *Lecture Notes in Computer Science (LNCS)*, pages 194–203. Springer, 2014. doi: https://doi.org/10.1007/978-3-319-08019-2_20.

Guilhem Gamard, Pierre Guillon, Kévin Perrot, and Guillaume Theyssier. Rice-like theorems for automata networks. In *Proceedings of the Symposium of the Theoretical Aspects of Computer Science*, volume 187 of *Leibniz International Proceedings in Informatics (LIPIcs)*, pages 32:1–32:17. Schloss Dagstuhl, 2021. doi: https://doi.org/10.4230/LIPIcs.STACS.2021.32.

Abhishek Garg, Alessandro Di Cara, Ioannis Xenarios, Luis Mendoza, and Giovanni De Micheli. Synchronous versus asynchronous modeling of gene regulatory networks. *Bioinformatics*, 24: 1917–1925, 2008. doi: https://doi.org/10.1093/bioinformatics/btn336.

Carlos Gershenson. Updating schemes in random Boolean networks: do they really matter? In *Proceedings of the International Conference on Simulation and Synthesis of Living Systems*, pages 238–243. MIT Press, 2004. doi: https://doi.org/10.7551/mitpress/1429.003.0040.

Leon Glass and Stuart A. Kauffman. Logical analysis of continuous, non-linear biochemical control networks. *Journal of Theoretical Biology*, 39: 103–129, 1973. doi: https://doi.org/10.1016/0022-5193(73)90208-7.

Albert Goldbeter. A model for circadian oscillations in the *Drosophila* period protein (PER). *Proceedings of the Royal Society of London B: Biological Sciences*, 261: 319–324, 1995. doi: https://doi.org/10.1098/rspb.1995.0153.

Eric Goles and Servet Martínez. *Neural and Automata Networks: Dynamical Behavior and Applications*, volume 58 of *Mathematics and Its Applications*. Kluwer Academic Publishers, 1990.

Eric Goles and Mathilde Noual. Block-sequential update schedules and Boolean automata circuits. In *Proceedings of the International Workshop on Cellular Automata and Discrete Complex Systems*, pages 41–50. Discrete Mathematics and Theoretical Computer Science, 2010.

Eric Goles and Jorge Olivos. Comportement périodique des fonctions à seuil binaires et applications. *Discrete Applied Mathematics*, 3: 93–105, 1981. doi: https://doi.org/10.1016/0166-218X(81)90034-2.

Eric Goles and Lilian Salinas. Sequential operator for filtering cycles in Boolean networks. *Advances in Applied Mathematics*, 45: 346–358, 2010. doi: https://doi.org/10.1016/j.aam.2010.03.002.

Eric Goles, Fabiola Lobos, Gonzalo A. Ruz, and Sylvain Sené. Attractor landscapes in Boolean networks with firing memory: a theoretical study applied to genetic networks. *Natural Computing*, 19: 295–319, 2020. doi: https://doi.org/10.1007/s11047-020-09789-0.

Jean-Luc Gouzé. Positive and negative circuits in dynamical systems. *Journal of Biological Systems*, 6: 11–15, 1998. doi: https://doi.org/10.1142/S0218339098000054.

Alex Graudenzi and Roberto Serra. A new model of genetic networks: the gene protein Boolean network. In *Proceedings of the Workshop italiano su vita artificiale e calcolo evolutivo*, pages 283–291. World Scientific, 2009. doi: https://doi.org/10.1142/9789814287456_0025.

Alex Graudenzi, Roberto Serra, Marco Villani, Annamaria Colacci, and Stuart A. Kauffman. Robustness analysis of a Boolean model of gene regulatory network with memory. *Journal of Computational Biology*, 18: 559–577, 2011a. doi: https://doi.org/10.1089/cmb.2010.0224.

Alex Graudenzi, Roberto Serra, Marco Villani, Chiara Damiani, Annamaria Colacci, and Stuart A. Kauffman. Dynamical properties of a Boolean model of gene regulatory network with memory. *Journal of Computational Biology*, 18: 1291–1303, 2011b. doi: https://doi.org/10.1089/cmb.2010.0069.

Paul E. Hardin, Jeffrey C. Hall, and Michael Rosbah. Feedback of the *Drosophila* period gene product on circadian cycling of its messenger RNA levels. *Nature*, 343: 536–540, 1990. doi: https://doi.org/10.1038/343536a0.

Inman Harvey and Terry Bossomaier. Time out of joint: attractors in asynchronous random Boolean networks. In *Proceedings of the European Conference on Artificial Life*, pages 67–75. MIT Press, 1997.

Ben Hesper and Paulien Hogeweg. Bioinformatica: een werkconcept. *Kameleon*, 1: 18–29, 1970.

David Hilbert. Mathematical problems. *Bulletin of the American Mathematical Society*, 8: 437–479, 1902.

Paulien Hogeweg. Simulating the growth of cellular forms. *Simulation*, 31: 90–96, 1978. doi: https://doi.org/10.1177/003754977803100305.

Paulien Hogeweg and Ben Hesper. Interactive instruction on population interactions. *Computers in Biology and Medicine*, 8: 319–327, 1978. doi: https://doi.org/10.1016/0010-4825(78)90032-X.

Shuji Ishihara, Koichi Fujimoto, and Tatsuo Shibata. Cross talking of network motifs in gene regulation that generates temporal pulses and spatial stripes. *Genes to Cells*, 10: 1025–1038, 2005. doi: https://doi.org/10.1111/j.1365-2443.2005.00897.x.

Arthur Kahn. Topological sorting of large networks. *Communications of the ACM*, 5: 558–562, 1962. doi: https://doi.org/10.1145/368996.369025.

Stuart A. Kauffman. Metabolic stability and epigenesis in randomly constructed genetic nets. *Journal of Theoretical Biology*, 22: 437–467, 1969a. doi: https://doi.org/10.1016/0022-5193(69)90015-0.

Stuart A. Kauffman. Homeostasis and differentiation in random genetic control networks. *Nature*, 224: 177–178, 1969b. doi: https://doi.org/10.1038/224177a0.

Stuart A. Kauffman. Gene regulation networks: a theory for their global structures and behaviors. In *Current Topics in Developmental Biology* (Eds. A.A. Moscona and Alberto Monroy), volume 6, pages 145–181. Elsevier, 1971.

Stuart A. Kauffman, Carsten Peterson, Björn Samuelsson, and Carl Troein. Random Boolean network models and the yeast transcriptional network. *Proceedings of the National Academy of Sciences of the United States of America*, 100: 14796–14799, 2003. doi: https://doi.org/10.1073/pnas.2036429100.

Marcelle Kaufman, Jacques Urbain, and René Thomas. Towards a logical analysis of the immune response. *Journal of Theoretical Biology*, 114: 527–561, 1985. doi: https://doi.org/10.1016/S0022-5193(85)80042-4.

Hannes Klarner, Alexander Bockmayr, and Heike Siebert. Computing maximal and minimal trap spaces of Boolean networks. *Natural Computing*, 14: 535–544, 2015. doi: https://doi.org/10.1007/s11047-015-9520-7.

Stephen C. Kleene. General recursive functions of natural numbers. *Mathematische Annalen*, 112: 727–742, 1936a. doi: https://doi.org/10.1007/BF01565439.

Stephen C. Kleene. λ-definability and recursiveness. *Duke Mathematical Journal*, 2: 340–353, 1936b. doi: https://doi.org/10.1215/S0012-7094-36-00227-2.

Scott A. Mangan and Uri Alon. Structure and function of the feed-forward loop network motif. *Proceedings of the National Academy of Sciences of the United*

States of America, 100: 11980–11985, 2003. doi: https://doi.org/10.1073/pnas.2133841100.

Warren S. McCulloch and Walter Pitts. A logical calculus of the ideas immanent in nervous activity. *Journal of Mathematical Biophysics*, 5: 115–133, 1943. doi: https://doi.org/10.1007/BF02478259.

Karim Mekhail and Danesh Moazed. The nuclear envelope in genome organization, expression and stability. *Nature Reviews Molecular Cell Biology*, 11: 317–328, 2010. doi: https://doi.org/10.1038/nrm2894.

Tarek Melliti, Damien Regnault, Adrien Richard, and Sylvain Sené. On the convergence of Boolean automata networks without negative cycles. In *Proceedings of International Workshop on Cellular Automata and Discrete Complex Systems*, volume 8155 of *Lecture Notes in Computer Science (LNCS)*, pages 124–138. Springer, 2013. doi: https://doi.org/10.1007/978-3-642-40867-0_9.

Luis Mendoza. A network model for the control of the differentiation process in Th cells. *BioSystems*, 84: 101–114, 2006. doi: https://doi.org/10.1016/j.biosystems.2005.10.004.

Luis Mendoza and Elena R. Alvarez-Buylla. Dynamics of the genetic regulatory network for *Arabidopsis thaliana* flower morphogenesis. *Journal of Theoretical Biology*, 193: 307–319, 1998. doi: https://doi.org/10.1006/jtbi.1998.0701.

Luis Mendoza, Denis Thieffry, and Elena R. Alvarez-Buylla. Genetic control of flower morphogenesis in Arabidopsis thaliana: a logical analysis. *Bioinformatics*, 15: 593–606, 1999. doi: https://doi.org/10.1093/bioinformatics/15.7.593.

John Milnor. On the concept of attractor. *Communications in Mathematical Physics*, 99: 177–185, 1985. doi: https://doi.org/10.1007/BF01212280.

Davi J. A. Moraes, Benedito H. Machado, and Daniel B. Zoccal. Coupling of respiratory and sympathetic activities in rats submitted to chronic intermittent hypoxia. *Progress in Brain Research*, 212: 25–38, 2014. doi: https://doi.org/10.1016/B978-0-444-63488-7.00002-1.

Katsuhiko Nakamura. Synchronous to asynchronous transformation of polyautomata. *Journal of Comuter and System Sciences*, 23: 22–37, 1981. doi: https://doi.org/10.1016/0022-0000(81)90003-9.

Aurélien Naldi, Denis Thieffry, and Claudine Chaouiya. Decision diagrams for the representation and analysis of logical models of genetic networks. In *Proceedings of the International Conference on Computational Methods in Systems Biology*, volume 4695 of *Lecture Notes in Computer Science (LNCS)*, pages 233–247. Springer, 2007. doi: https://doi.org/10.1007/978-3-540-75140-3_16.

Aurélien Naldi, Jorge Carneiro, Claudine Chaouiya, and Denis Thieffry. Diversity and plasticity of Th cell types predicted from regulatory network modelling. *PLoS Computational Biology*, 6: e1000912, 2010. doi: https://doi.org/10.1371/journal.pcbi.1000912.

Aurélien Naldi, Céline Hernandez, Wassim Abou-Jaoudé, Pedro T. Monteiro, Claudine Chaouiya, and Denis Thieffry. Logical modeling and analysis of cellular regulatory networks with GINsim 3.0. *Frontiers in Physiology*, 9: 646, 2018a. doi: https://doi.org/10.3389/fphys.2018.00646.

Aurélien Naldi, Céline Hernandez, Nicolas Levy, Gautier Stoll, Pedro T. Monteiro, Claudine Chaouiya, Tomáš Helikar, Andrei Zinovyev, Laurence Calzone, Sarah Cohen-Boulakia, Denis Thieffry, and Loïc Paulevé. The CoLoMoTo interactive notebook: accessible and reproducible computational analyses for qualitative biological networks. *Frontiers in Physiology*, 9: 680, 2018b. doi: https://doi.org/10.3389/fphys.2018.00680. URL http://colomoto.org/notebook.

Mathilde Noual. Dynamics of circuits and intersecting circuits. In *Proceedings of the International Conference on Language and Automata Theory and Applications*, volume 7183 of *Lecture Notes in Computer Science (LNCS)*, pages 433–444. Springer, 2012a. doi: https://doi.org/10.1007/978-3-642-28332-1_37.

Mathilde Noual. *Updating automata networks*. PhD thesis, École normale supérieure de Lyon, 2012b.

Mathilde Noual and Sylvain Sené. Towards a theory of modelling with Boolean automata networks – I. Theorisation and observations. arXiv:1111.2077, 2011.

Mathilde Noual and Sylvain Sené. Synchronism versus asynchronism in monotonic Boolean automata networks. *Natural Computing*, 17: 393–402, 2018. doi: https://doi.org/10.1007/s11047-016-9608-8.

Christos H. Papadimitriou. *Computational Complexity*. Addison-Wesley, 1995.

Loïc Paulevé. Pint: a static analyzer for transient dynamics of qualitative networks with IPython interface. In *Proceedings of the International Conference on Computational Methods in Systems Biology*, volume 10545 of *Lecture Notes in Computer Science (LNCS)*, pages 309–316. Springer, 2017. doi: https://doi.org/10.1007/978-3-319-67471-1_20.

Loïc Paulevé. Reduction of qualitative models of biological networks for transient dynamics analysis. *IEEE/ACM Transactions on Computational Biology and Bioinformatics*, 15: 1167–1179, 2018. doi: https://doi.org/10.1109/TCBB.2017.2749225.

Loïc Paulevé. Notebooks demonstrating Most Permissive Boolean Networks. 2020. doi: https://doi.org/10.5281/zenodo.3936123.

Loïc Paulevé and Sylvain Sené. Non-deterministic updates of Boolean networks. In *Proceedings of the International Workshop on Cellular Automata and Discrete Complex Systems*, volume 90 of *OpenAccess Series in Informatic (OASIcs)*, pages 10:1–10:16. Schloss Dagstuhl, 2021. doi: https://doi.org/10.4230/OASIcs.AUTOMATA.2021.10

Loïc Paulevé, Juraj Kolčák, Thomas Chatain, and Stefan Haar. Reconciling qualitative, abstract, and scalable modeling of biological networks. *Nature Communications*, 11: 4256, 2020. doi: https://doi.org/10.1038/s41467-020-18112-5.

Élisabeth Remy, Brigitte Mossé, Claudine Chaouiya, and Denis Thieffry. A description of dynamical graphs associated to elementary regulatory circuits. *Bioinformatics*, 19 (Suppl. 2): ii172–ii178, 2003. doi: https://doi.org/10.1093/bioinformatics/btg1075.

Élisabeth Remy, Paul Ruet, and Denis Thieffry. Graphic requirement for multistability and attractive cycles in a Boolean dynamical framework. *Advances in Applied Mathematics*, 41: 335–350, 2008. doi: https://doi.org/10.1016/j.aam.2007.11.003.

Henry Gordon Rice. Classes of recursively enumerable sets and their decision problems. *Transactions of the American Mathematical Society*, 74: 358–366, 1953.

Adrien Richard. Positive circuits and maximal number of fixed points in discrete dynamical systems. *Discrete Applied Mathematics*, 157: 3281–3288, 2009. doi: https://doi.org/10.1016/j.dam.2009.06.017.

Adrien Richard. Fixed point theorems for Boolean networks expressed in terms of forbidden subnetworks. *Theoretical Computer Science*, 583: 1–26, 2015. doi: https://doi.org/10.1016/j.tcs.2015.03.038.

Adrien Richard. Negative circuits and sustained oscillations in asynchronous automata networks. *Advances in Applied Mathematics*, 44: 378–392, 2010. doi: https://doi.org/10.1016/j.aam.2009.11.011.

Adrien Richard and Jean-Paul Comet. Necessary conditions for multistationarity in discrete dynamical systems. *Discrete Applied Mathematics*, 155: 2403–2413, 2007. doi: https://doi.org/10.1016/j.dam.2007.04.019.

François Robert. Itérations sur des ensembles finis et automates cellulaires contractants. *Linear Algebra and its Applications*, 29: 393–412, 1980. doi: https://doi.org/10.1016/0024-3795(80)90251-7.

François Robert. *Discrete Iterations: A Metric Study*, volume 6 of *Springer Series in Computational Mathematics*. Springer, 1986.

Guillermo Rodrigo and Santiago F. Elena. Structural discrimination of robustness in transcriptional feedforward loops for pattern formation. *PLoS One*, 6: e16904, 2011. doi: https://doi.org/10.1371/journal.pone.0016904.

Gonzalo A. Ruz, Eric Goles, M. Montalva, and Gary B. Fogel. Dynamical and topological robustness of the mammalian cell cycle network: a reverse engineering approach. *Biosystems*, 115: 23–32, 2014. doi: https://doi.org/10.1016/j.biosystems.2013.10.007.

Yolanda Schaerli, Andreea Munteanu, Magüi Gili, James Cotterell, James Sharpe, and Mark Isalan. A unified design space of synthetic stripe-forming networks. *Nature Communications*, 5: 4905, 2014. doi: https://doi.org/10.1038/ncomms5905.

Amita Sehgal, Jeffrey L. Price, Bernice Man, and Michael W. Young. Loss of circadian behavioral rhythms and per RNA oscillations in the Drosophila mutant timeless. *Science*, 263: 1603–1606, 1994. doi: https://doi.org/10.1126/science.8128246.

Sylvain Sené. Sur la bio-informatique des réseaux d'automates. Université d'Évry – Val d'Essonne, 2012.

Mau-Hsiang Shih and Jian-Lang Dong. A combinatorial analogue of the Jacobian problem in automata networks. *Advances in Applied Mathematics*, 34: 30–46, 2005. doi: https://doi.org/10.1016/j.aam.2004.06.002.

Heike Siebert and Alexander Bockmayr. Temporal constraints in the logical analysis of regulatory networks. *Theoretical Computer Science*, 391: 258–275, 2008. doi: https://doi.org/10.1016/j.tcs.2007.11.010.

El Houssine Snoussi. Necessary conditions for multistationarity and stable periodicity. *Journal of Biological Systems*, 6: 3–9, 1998. doi: https://doi.org/10.1142/S0218339098000042.

Christophe Soulé. Graphical requirements for multistationarity. *ComPlexUs*, 1: 123–133, 2003. doi: https://doi.org/10.1159/000076100.

Christophe Soulé. Mathematical approaches to differentiation and gene regulation. *Comptes Rendus – Biologies*, 329: 13–20, 2006. doi: https://doi.org/10.1016/j.crvi.2005.10.002.

Cui Su and Jun Pang. CABEAN: a software for the control of asynchronous Boolean networks. *Bioinformatics*, 37: 879–881, 2020. doi: https://doi.org/10.1093/bioinformatics/btaa752.

The Arabidopsis Genome Initiative. Analysis of the genome sequence of the flowering plant *Arabidopsis thaliana*. *Nature*, 408: 796–815, 2000. doi: https://doi.org/10.1038/35048692.

Michel Thellier, Jacques Demongeot, Vic Norris, Janine Guespin, Camille Ripoll, and René Thomas. A logical (discrete) formulation for the storage and recall of environmental signals in plants. *Plant Biology*, 6: 590–597, 2004. doi: https://doi.org/10.1055/s-2004-821090.

René Thom. *Structural Stability and Morphogenesis*. W. A. Benjamin, Inc., 1975.

René Thomas. Boolean formalization of genetic control circuits. *Journal of Theoretical Biology*, 42: 563–585, 1973. doi: https://doi.org/10.1016/0022-5193(73)90247-6.

René Thomas. On the relation between the logical structure of systems and their ability to generate multiple steady states or sustained oscillations. In *Numerical methods in the study of critical phenomena*, volume 9 of *Springer Series in Synergetics*, pages 180–193. 1981. doi: https://doi.org/10.1007/978-3-642-81703-8_24.

René Thomas. Logical vs. continuous description of systems comprising feedback loops: the relation between time delays and parameters. *Studies in Physical and Theoretical Chemistry*, 28: 307–321, 1983.

René Thomas. Regulatory networks seen as asynchronous automata: a logical description. *Journal of Theoretical Biology*, 153: 1–23, 1991. doi: https://doi.org/10.1016/S0022-5193(05)80350-9.

René Thomas and Richard d'Ari. *Biological Feedback*. CRC Press, 1990.

René Thomas and Marcelle Kaufman. Multistationarity, the basis of cell differentiation and memory. II. Logical analysis of regulatory networks in terms of feedback circuits. *Chaos: An Interdisciplinary Journal of Nonlinear Science*, 11: 180–195, 2001. doi: https://doi.org/10.1063/1.1349893.

Alan M. Turing. On computable numbers, with an application to the entscheidungsproblem. *Proceedings of the London Mathematical Society*, 42: 230–265, 1936. doi: https://doi.org/10.1112/plms/s2-42.1.230.

John von Neumann. *Theory of Self-reproducing Automata*. University of Illinois Press, 1966. Edited and completed by A. W. Burks.

Norbert Wiener. *Cybernetics: Or Control and Communication in the Animal and the Machine*. Hermann & Cie, Paris, and John Wiley & Sons, New York, 1948.

Martín Ríos Wilson. *On automata networks dynamics: an approach based on computational complexity theory*. PhD thesis, Universidad de Chile and Université d'Aix-Marseille, 2021.

Martín Ríos Wilson and Guillaume Theyssier. On symmetry versus asynchronism: at the edge of universality in automata networks. arXiv:2105.08356, 2021.

David J. Wooten, Jorge G. T. Za nudo, David Murrugarra, Austin M. Perry, Anna Dongari-Bagtzoglou, Reinhard Laubenbacher, Clarissa J. Nobile, and Réka Albert. Mathematical modeling of the *Candida albicans* yeast to hyphal transition reveals novel control strategies. *PLoS Computational Biology*, 17: e1008690, 2021. doi: https://doi.org/10.1371/journal.pcbi.1008690.

Jorge G. T. Za nudo, Maurizio Scaltriti, and Réka Albert. A network modeling approach to elucidate drug resistance mechanisms and predict combinatorial drug treatments in breast cancer. *Cancer Convergence*, 1: 5, 2017. doi: https://doi.org/10.1186/s41236-017-0007-6.

7

Analyzing Long-Term Dynamics of Biological Networks With Answer Set Programming

Emna Ben Abdallah[1], Maxime Folschette[2], and Morgan Magnin[3]

[1] *Independent Researcher, Nantes, France*
[2] *Univ. Lille, CNRS, Centrale Lille, UMR 9189 CRIStAL, Lille, France*
[3] *Centrale Nantes, Université de Nantes, CNRS, LS2N, Nantes, France*

7.1 Introduction

Recent advances in molecular biology have made it possible to produce comprehensive genome maps of many living organisms. However, these static maps – made popular in the case of humans in particular through the "Human Genome Project" (Collins et al., 2003) – are not sufficient to account for the intrinsic complexity of living things. It is indeed necessary to consider genetic phenomena with regard to the full complexity of the dynamics of their interactions, either internal or with their environment (for example, the stress represented by an external component such as light or the ingestion of a protein).

Meanwhile, the development of biotechnologies such as DNA chips and the rise of high-throughput sequencing techniques (such as RNAseq) facilitate the production of time series data corresponding to the expression of several thousand genes. The question then arises of the meaning to be given to this profusion of information. One of the main current challenges in systems biology is thus to integrate these high-throughput data and to infer the associated genetic regulatory networks. As the interactions occur at different scales (genes, proteins, biochemical components, cells, etc.), there is a need to develop integrative methods that can formally and automatically learn biological data to facilitate understanding, at a systemic level, of the phenomena involved.

In fact, the modelling of biological regulatory mechanisms can be divided into two main trends. The first, quantitative, is based on ordinary differential equations involving the quantitative expression of interacting components

Systems Biology Modelling and Analysis: Formal Bioinformatics Methods and Tools,
First Edition. Edited by Elisabetta De Maria.
© 2023 John Wiley & Sons, Inc. Published 2023 by John Wiley & Sons, Inc.

(see Chapter 9). However, these equations are generally non-linear, which prevents their analytical resolution. In addition, the biological data obtained experimentally are generally relatively noisy, which means that they must be filtered efficiently. Conversely, the second tendency consists in addressing the problem using a discrete methodology. In other words, the expression of each component is assessed according to several (usually two, sometimes three or four) qualitative levels and no longer quantitatively. Even though such discrete modelling might be considered a less faithful abstraction of reality, it has been shown to be effective in addressing many qualitative biological questions (such as understanding the interactions between several components of a biological system or determining the reachability of a state).

In this chapter, the emphasis is placed on regulatory processes at the genetic level. We choose here to abstract certain mechanisms, e.g. the ones at the molecular level, which would require specifying in detail the behavior of each biological element such as proteins, plasmids, etc.

Difficulties encountered in analysis and prediction can be considered to fall into one of the following four broad categories:

- Parameter identification: discrete models, such as synchronous Boolean networks introduced by Stuart Kauffman (1969) or asynchronous multivalued networks introduced by René Thomas (1973), do not only require information about the network topology and the type of influences – activation or inhibition – between genes but also on the respective strength of each interaction (see Chapter 6). This type of information is needed to determine, for example, the predominant influence when a gene is the single target of two influences with opposite effects. As we mentioned previously, such information is represented either at the core of a network node update function or in a dedicated parameterization of the model. In either cases, this information can be deduced from experimental biological observations by automatic methods such as model-checking or constraint programming.
- Model inference: broader than the identification of parameters, the problem of inference also encompasses the determination of the overall structure of the network. Considering time series data that can be converted (once discretized) into a state-transition system, it may be important – to facilitate data analysis – to build a discrete model compatible with the available data.
- The identification of stable states and attractors, key properties of most biological systems, which are tackled in this chapter.
- Model control, which is the ability to control a model so that it exhibits a desired behavior or, on the contrary, guarantees that undesirable

behaviors are avoided. This issue is closely linked to the analysis of the aforementioned key properties. This is the most recent, but also the most difficult, research topic in the field of systems biology, closely linked to the problem of gene therapy.

Answer Set Programming (ASP) is a logic programming paradigm that has proven to be useful in all these four challenges. We will here focus on the ones that seem, to us at least, the most representative of ASP merits, especially the full identification of attractors.

7.2 State of the Art

7.2.1 Qualitative Modelling of Biological Systems

In the last decades, the emergence of a wide range of new technologies has made it possible to produce a massive amount of biological data (genomics, proteomics, etc.). This leads to considerable developments in systems biology, which takes profit from this data. In order to understand the nature of a cellular function or more broadly a living biological system (healthy or diseased), it is indeed essential to study not only the individual properties of cellular components but also their interactions. The behavior and functionalities of the cells emerge from such networks of interactions.

Considering this paradigm, the long-term behavior of a regulatory network's dynamics is of specific interest (Wuensche, 1998). Indeed, at any moment, a system may fall into a *trap domain*, which is a part of its dynamics that cannot be escaped. While evolving, the system may eventually fall into new and smaller trap domains, thus reducing its possible future behaviors by making previous states no longer reachable. This phenomenon depends on biological disruptions or other complex phenomena. Such outline has been interpreted as distinct responses of the organism, such as differentiating into distinct cell types in multicellular organisms (Huang et al., 2005).

Moreover, when refining a model of a living system, one way to remove inconsistencies or to predict missing information in biological models consists in comparing the attractors of the model with the experimentally observed long-term behavior. For instance, the model of the cellular development of *Drosophila melanogaster* was described using Boolean networks and their attractors (González et al., 2008, Albert and Othmer, 2003).

As mentioned in the previous sections, the modelling of biological regulatory mechanisms can be divided into two main categories: based on either ordinary differential equations or discrete values. Even if discrete modelling can be roughly seen as a less faithful abstraction, it has proven to be effective in addressing many biological questions, such as understanding how biological systems evolve or determining whether certain states are reachable. In fact, as data in biology are often more qualitative than quantitative, it is natural that an alternative view to differential equations started to emerge in the late 1960s. This qualitative modelling is based on the following principle: the expression of a component can be encoded with a Boolean variable. This corresponds to the fact that a component (e.g. a gene) is "on" (i.e. the regulations it controls are expressed) or "off." The relations between Boolean variables are governed by activation or inhibition relations, respectively, represented by positive or negative signed arcs. Various kinds of mathematical models implementing these precepts have been proposed for the modelling of Biological Regulatory Networks (BRNs). The two main families of such models have been Stuart Kauffman's synchronous Boolean networks (Kauffman, 1969) on the one hand and René Thomas' asynchronous networks (Thomas, 1991) on the other hand (see Chapter 6).

In this chapter, the focus is on a subclass of automata networks, called Asynchronous Automata Networks (AAN) (Folschette et al., 2015, Paulevé, 2016a), which is an extension of a previous framework called Process Hitting (Paulevé et al., 2014), and is convenient to model BRNs. This qualitative approach is based on three key concepts:

- Biological components (e.g. genes) are abstracted in the form of *automata*. The different local states (that are not restricted to Boolean values) of each automaton correspond to the different discrete qualitative levels of the components represented by the automaton.
- Interaction between biological components is modeled in an atomic way by local transitions on the automata, where each one is conditioned by a set of required local states in different automata and can modify the local state of a unique automaton.
- In the modelling task, such a representation makes it possible to build the largest possible dynamics and then to proceed by successive refinements to restrict the possible behaviors (Paulevé et al., 2011).

To sum up, AANs allow us to have multiple requirements for a local transition to occur but *not* to synchronize several local transitions in different automata. In this sense, they are considered *Asynchronous Automata Networks*, although in the following, we define and apply both the asynchronous and synchronous semantics to such networks. Here, the synchronous semantics allows us to fire all local transitions at once but not to add information about specific transitions being synchronized. Depending on the chosen semantics (asynchronous or synchronous), AANs especially encompass the Boolean frameworks of René Thomas and Stuart Kauffman.

Qualitative frameworks have received substantial attention because of their capacity to capture the switching behavior of genetic or biological processes and therefore the study of their long-term behavior. This adds up to the choice of a qualitative representation for the identification of trap domains. In such a qualitative framework, a minimal trap domain can take two different forms: it can be either a *stable state*, also called *fixed point* or *steady-state*, which is one state in which the system does not evolve anymore, or an *attractor*, which is a minimal set of states that cannot be escaped and thus loops indefinitely.

7.2.2 Identifying Attractors: A Major Challenge

The computational problem of finding all attractors in a BRN is difficult. Even the simpler problem of deciding whether the system has a stable state, which can be seen as the smallest kind of attractor, is NP-hard (Zhang et al., 2007). Based on this, many studies have proven that computing attractors in BRNs is also a NP-hard problem (Klemm and Bornholdt, 2005, Akutsu et al., 2012). Although some methods exist with a lesser complexity, consisting for instance in randomly selecting an initial state and following a long-enough trajectory, hoping to eventually finding an attractor, they are not exhaustive and may miss some (hard to reach) attractors.

Therefore, in the absence of more efficient exhaustive methods, it is still relevant to develop an approach to resolve the original NP-hard problem of attractor identification. Such an approach consists in exhaustively examining all possible states of a network, along with all possible paths from each of these states. Obviously, this brute force method is very time- and memory-consuming: 2^n initial states have to be considered for a Boolean model with n nodes, and multivalued networks raise this value even more. Furthermore, a sufficient number of computations have to be performed to ensure that all trajectories have been explored and all attractors are found. This high complexity justifies the use of a tool able to tackle such hard problems.

The simplest way to detect attractors is to enumerate all the possible states and to run a simulation from each one until an attractor is reached (Somogyi and Greller, 2001). This method ensures that all attractors are detected but it has an exponential time complexity; therefore, its applicability is highly restricted by the network size.

Regarding Boolean networks only, algorithms for detecting attractors have been extensively studied in the literature. Irons (2006) proposes to analyze partial states in order to discard potential attractors more efficiently. This method improves the efficiency from exponential time to polynomial time for a subset of biological Boolean models that is highly dependent on the topology of the underlying network (in terms of indegree, outdegree, and update functions). Another method, called GenYsis (Garg et al., 2007), starts from one (randomly selected) initial state and detects attractors by computing the successor and predecessor states of this initial state. It works well for small Boolean networks but becomes inefficient for large Boolean networks.

More generally, the efficiency and scalability of attractor detection techniques are further improved with the integration of two techniques. This first is based on Binary Decision Diagrams (BDD), a compact data structure for representing Boolean functions. In Zhao et al. (2014), algorithms have been based on the reduced-order binary decision diagram (ROBDD) data structure, which further speeds up the computation time of attractor detection. These BDD-based solutions only work for BRNs of a hundred of nodes and also suffer from the infamous state explosion problem, as the size of the BDD depends both on the regulatory functions and the number of nodes in the BRN. The other technique consists in representing the attractor enumeration problem as a satisfiability (SAT) problem such as in Dubrova and Teslenko (2011). The main idea is inspired by SAT-based bounded model-checking: the transition relation of the BRN is unfolded into a bounded number of steps in order to construct a propositional formula, which encodes attractors and which is then solved by a SAT solver. In every step, a new variable is required to represent a state of a node in the BRN. It is clear that the efficiency of these algorithms largely depends on the number of unfolding steps and the number of nodes in the BRN. The method presented below is also inspired from this bounded model-checking approach.

In Mushthofa et al. (2014), the authors separated the rules that describe the network (the nodes and their interactions: activation or inhibition) from the rules that define its dynamics (for instance, a gene will be activated in the next state if all its activators are active or when at least one of its activators is active at the current state). This allows us to obtain more flexible simulations, and the authors are also chose to use the declarative paradigm ASP (Baral, 2003) in order to have more liberty in the expression of evolution rules. They illustrated that specifying large networks with rather complicated behaviors becomes cumbersome and error prone in paradigms such as SAT, whereas this is much less the case in a declarative approach such as theirs.

7.2.3 Answer Set Programming for Systems Biology

More recently, ASP has been recently used successfully in systems biology on two different challenges. The first one consists in the synthesis of discrete models from a different background knowledge. In Chevalier et al. (2019), the authors focus on Boolean networks and exhibit an ASP-based method to determine the Boolean function from information about the attractors (either stable states or trap spaces). They show the scalability of their approach, which is able to tackle networks up to 50 nodes with an in-degree of 15 per node, and advocate for the associated benefits by addressing the cell differentiation process in the central nervous system. The efficient matching between biological data given as an input and a family of admissible models is really one of the cornerstones that motivates for the use of ASP. In Lemos et al. (2019), the authors introduce a method to revise models, encoded as Boolean logical regulatory graphs, from time series data. Such work stems from the need to be able to revise the existing Boolean models that may become inconsistent when new data is made available. The authors define various atomic revision operations, e.g. negation of a regulator (changing a regulator from an inhibitor to an activator, and conversely), operator substitution (turning a Boolean operator from AND to OR, and conversely), and removal of a regulator (removing all occurrences of a given regulator from a logical function). They show how such a method can be applied to biological case studies by applying this approach to five models with associated time series containing a dozen of nodes, with a number of time steps varying from 10 to 13.

The second main challenge – which is explored in detail in this chapter – consists in the identification of attractors in biological regulator networks. In Abdallah et al. (2017), the authors proposed a first version of the approach and algorithms that will be explained in more detail below. Meanwhile, other authors got interest in such a problem. In Khaled and Benhamou (2020), the authors search and enumerate attractors in asynchronous Boolean networks that have the form of a circuit using ASP by introducing a new resolution semantics that does not use the usual negation by failure. Instead, they rely on a weak version of the negation by failure that allows them to provide an enumeration approach preventing the simulation of the underlying networks. This approach targets networks with a given structure (all nodes must have an incoming edge and some results have been obtained only for cyclic interaction graphs). The methodology detailed in this chapter is different in the sense that it is iterative and without any assumption on the topology of the interaction graph.

7.2.4 Enumerating Attractors of a Biological Model Using Answer Set Programming

Our goal in this chapter is to present two methods to enumerate minimal trap spaces of a BRN modeled in AAN: (i) finding all stable states, and (ii) enumerating all attractors up to a given length n. A *stable state*, also called *fixed point*, is a state with no more possible dynamical evolution. An *attractor* is a minimal trap domain, which is not a fixed point; in other words, it is a minimal set of at least two states that cannot be dynamically escaped. Here, the *length* refers to the number of dynamical transitions required to cover all states of the attractor, which can be equal or higher than the number of states if there are complex dynamical branchings. This notion of length is required as it gives a higher bound to the dynamical exploration performed and makes this method fall under bounded model checking.

We focus on two widespread non-deterministic semantics: asynchronous and synchronous. It has to be noted that the stable states are the same in both dynamics (Klarner et al., 2015). Although we do not prove it, we claim that our method is also correct for the enumeration of attractors in the asynchronous dynamics and returns all attractors up to the given length. Regarding the synchronous dynamics, we also claim that out method is correct for *simple attractors*, that is, attractors made of exactly one loop; however, it might miss some *complex attractors* that are compositions of several loops. We use ASP to perform these aforementioned enumerations.

The originality of our approach is to consider AAN models with ASP. Within the AAN formalism, interactions are modeled as automata transitions instead of generic influences. This allows us to define finer influences between components than in formalisms based on parameters or evolution functions. For instance, an AAN model can contain non-monotonic influences or be built as a union of models to check several dynamics simultaneously (Paulevé et al., 2011). As mentioned in the Introduction of this chapter, AANs are more expressive than the modellings of René Thomas (Thomas, 1973) or Kauffman (Kauffman, 1969), which also means that the models in these formalisms can be expressed and analyzed as AANs, including their multivalued iterations. Moreover, automata-like transitions happen to be easily represented in ASP. Finally, as ASP is a programming paradigm, this would theoretically allow a much wider expression power in the description of update semantics (generalized, block-parallel, with memory, with priorities, etc.), although these are not tackled here.

Although some ideas of this approach were previously introduced in Abdallah et al. (2015) and Abdallah et al. (2017), the method presented in this chapter has been improved. Thanks to the possibility now offered by the ASP solver Clingo of interfacing with scripts, we propose an approach using Python to perform post-filtering or pre-filtering to enumerate all attractors up to a given length without yielding many spurious solutions as in the previous iterations. This conjunction of ASP and Python illustrates the benefits of such a combination to tackle efficient enumeration of answer sets. Given these changes, we have extended our previous benchmarks to reflect the changes made on the method and its implementation. As such, we summarize our results in updated case studies that illustrate the benefits of our approach.

This chapter is organized as follows: After discussing the possible uses of ASP in systems biology in the current section, Section 7.3 briefly presents the ASP framework that is used later. Then, Section 7.4 presents the main definitions related to the AAN formalism and the specific properties on which we will focus later on, that is, stable states and attractors. Section 7.5 details our approach to define an AAN model using ASP rules and enumerate all stable states or attractors in such a model. The merits of such an approach are then emphasized in Section 7.6, where we give benchmarks of our methods on several models of different sizes (up to 100 components). Finally, Section 7.7 concludes and gives some perspectives to this work.

7.3 Basic Notions of Answer Set Programming

ASP is a purely declarative language based on the stable model semantics of logic programming (Baral, 2003). As illustrated by the literature study proposed in the previous section, ASP has proved to be efficient to address highly computational problems.

In this section, we will introduce the basic elements of ASP syntax and way of modelling problems. We will focus here on the parts that will be necessary to us to model biological systems in the following parts of this chapter.

7.3.1 Syntax and Rules

An answer set *program* is a finite set of *rules* of the form:

$$a_0 \leftarrow a_1, \ldots, a_m, \text{not } a_{m+1}, \ldots, \text{not } a_n \tag{7.1}$$

where $n \geq m \geq 0$, a_0 is an *atom* or $\perp$, all $a_1, \ldots, a_n$ are atoms, and the symbol "*not*" denotes *negation as failure*. The intuitive reading of such a rule is that whenever $a_1, \ldots, a_m$ are known to be true and there is no evidence for any of the negated atoms $a_{m+1}, \ldots, a_n$ to be true, then a_0 has to be true as well. An atom or a negated atom is also called a *literal*. The atom on the left of the arrow is called the *head* of the rule while the set of literals on the right is called its *body*. The order of the rules in the program, and of literals in a body, does not change its meaning.

Some special rules are noteworthy. A rule where $m = n = 0$ is called a *fact* and is useful to represent data because the head atom a_0 is always true. It is often written without the central arrow, as in rule (7.2). On the other hand, a rule where $n > 0$ and $a_0 = \bot$ is called a *constraint*. Because $\bot$ can never become true, if the body of a constraint is true, this invalidates the whole solution. Constraints are thus useful to filter out unwanted solutions when there are several candidate solutions. The symbol $\bot$ is usually omitted in a constraint, as in rule (7.3).

$$a_0 \tag{7.2}$$

$$\leftarrow a_1, \ldots, a_m, \text{not } a_{m+1}, \ldots, \text{not } a_n \tag{7.3}$$

In the ASP paradigm, the search of solutions consists in computing the answer sets of a given program. Intuitively, the answer sets of a program are the minimal sets of atoms (in terms of set inclusion) that respect all rules of this program. Note that for a given program, there might be no, one, or several answer sets.

Formally, an answer set for a given program is defined by Gelfond and Lifschitz (1988) as follows. An *interpretation I* is a finite set of propositional atoms. A rule r as given in (7.1) is *true under I* if and only if:

$$\{a_1, \ldots, a_m\} \subseteq I \wedge \{a_{m+1}, \ldots, a_n\} \cap I = \emptyset \Rightarrow a_0 \in I$$

An interpretation I is a *model* of a program P if each rule $r \in P$ is true under I. Finally, let P^I be the program obtained from P by deleting all rules that contain a negated atom appearing in I and deleting all negated atoms from the remaining rules. I is an *answer set* (also called *solution* in the following) of P if I is a minimal model of P^I.

Example 7.1 Consider the following program:

$$a$$
$$b \leftarrow a, \text{not } c$$
$$c \leftarrow b, \text{not } a$$

The atom a is given in a fact and is thus always true. The atoms b and c depend on the presence of the other atoms. The unique solution (answer set) of this program is $\{a, b\}$. Indeed, it is the only minimal set of atoms that respects all rules.

Example 7.2 Programs can yield no answer set, one answer set, or several answer sets. For example, the following program

$$b \leftarrow \text{not } c$$
$$c \leftarrow \text{not } b$$

produces two answer sets: $\{b\}$ and $\{c\}$. Indeed, the absence of c makes b true, and conversely, the absence of b makes c true.

Cardinality constructs, depicted by a set of atoms in curly braces, are another way to obtain multiple answer sets. The most usual way of using a cardinality is in the head:

$$l \{q_1, \ldots, q_k\} u \leftarrow a_1, \ldots, a_m, \text{not } a_{m+1}, \ldots, \text{not } a_n$$

where $k \geq 0$, l is an integer and u is an integer or the infinity (∞). Such a cardinality means that, under the condition that the body is satisfied, the answer set X must contain at least l and at most u atoms from the set $\{q_1, \ldots, q_m\}$, or in other words: $l \leq |\{q_1, \ldots, q_m\} \cap X| \leq u$ where $\cap$ is the symbol of set intersection and $|\cdot|$ denotes the cardinality of a set. We note that several answer sets may match this definition, as there may be numerous solutions X to this equation. If they are not explicitly given, l defaults to 0 and u defaults to ∞.

Example 7.3 Using a cardinality construct, the program of Example 7.2 can be summed up into the following program containing only one cardinality fact:

$$1 \{b, c\} 1$$

7.3.2 Predicates

Atoms in ASP are expressed as *predicates* with an arity; that is, they can apply to *terms* (also called *arguments*). For instance, let us consider the following program:

$$\text{parentOf(jenny, charles)} \tag{7.4}$$

$$\text{parentOf(mary, jenny)} \tag{7.5}$$

$$\text{grandparentOf}(X, Z) \leftarrow \text{parentOf}(X, Y), \text{parentOf}(Y, Z) \tag{7.6}$$

This program first contains two facts in lines (7.4) and (7.5), stating that Jenny is a parent of Charles and Mary is a parent of Jenny. This is expressed by predicate "parentOf" with arity 2 (it has two arguments) and three constants "jenny", "charles", and "mary.". Then, line (7.6) expresses that a grandparent of someone is the parent of their parent. This is expressed by predicate "grandparentOf" also of arity 2, but this time using variables (X, Y, and Z). A variable can stand for any existing atom in the program. By convention, constant names start with a low letter or are surrounded by double quotes, and variable names start with a capital letter. Moreover, variables have a scope limited to the rule, meaning that if X was used in another rule, it would be considered another variable.

Solving an ASP program as explained above requires that it contains no variable; for this, a *grounding* step is first required, consisting in the removal of all free variables by replacing them by possible constants while preserving the meaning of the program. In the example above, the grounding step produces the following variable-free program, where X, Y, and Z are replaced by all possible constants:

parentOf(jenny, charles).   parentOf(mary, jenny).

grandparentOf(mary, mary) ← parentOf(mary, mary),
    parentOf(mary, mary).

grandparentOf(mary, charles) ← parentOf(mary, mary),
    parentOf(mary, charles).

grandparentOf(mary, charles) ← parentOf(mary, jenny),
    parentOf(jenny, charles).

grandparentOf(mary, jenny) ← parentOf(mary, charles),
    parentOf(charles, jenny).

... (23 more grounded rules) ...

The two first rules in the grounded rules above are simply the facts; all the others come from the replacing of all variables in rule (7.6). Solving then consists in iteratively removing the rules whose body is false. The answer set is the set of head predicates of the remaining rules, that is:

{ parentOf(jenny, charles),
parentOf(mary, jenny),
grandparentOf(mary, charles) }.

Finally, predicates are very useful in cardinality constructs as defined in the previous section: instead of explicitly giving all atoms in the cardinality set (by extension), it is possible to enumerate them based on predicates and variables (by comprehension). For instance,

$$0 \, \{f(A, B) : g(A)\} \, 2 \leftarrow h(B)$$

means that for each value of B so that $h(B)$ is true, all values of A so that $g(A)$ is true are enumerated in order to create predicates $f(A, B)$, and at most two are kept. Typically, this might create a number of answer sets that depend on the existing predicates g and h.

The grounding and solving steps are usually tackled by specialized software. For the present chapter, we use Clingo[1] (Gebser et al., 2016), which is a combination of a grounder and a solver. In the rest of this chapter, we use ASP to tackle the problems of enumerating all stable states and attractors of a given AAN model.

7.3.3 Scripting

Since its version 5.0, Clingo offers scripting capabilities using either Lua or Python. In both cases, this scripting allows us to enrich the usual solving of an ASP program with

- grounding and solving control,
- external predicates (provided by external data or software),
- on-the-fly modification of programs,
- post-processing of solutions.

For instance, in Python, this scripting allows us to programmatically start the solving and interrupt it for each solution found to execute some processing on the result with Python. The introduction of a more classical imperative language into the solving has many advantages, such as allowing to easily compute some parts of the program that are trivially done in imperative scripting but difficult in ASP.

The solution that we present in this work comes in three versions that make a progressive usage of this scripting:

1. The first version uses no scripting and thus outputs duplicate solutions that are detailed later.
2. In the second version, each answer set found by Clingo is provided to a Python script that checks if it is a duplicate or not and outputs it only if it is a genuine solution; this solution uses scripting as a mere post-processing tool.

---

[1] We used Clingo version 5.4: https://potassco.org/

3. In the third version, for each answer set found by Clingo, a constraint is added to the ASP program in order to avoid enumerating a duplicate solution; this solution uses on-the-fly modification of the ASP program for the same result.

7.4 Dynamic Modelling Using Asynchronous Automata Networks

This section introduces the formal definitions required to model and analyze AANs.

7.4.1 Motivation: Using ASP to Analyze the Dynamics

In order to study the dynamics of a system, one of the main approaches is to model it in a formal way. This allows us to understand or control the behavior of this system by checking properties on its model.

The modelling process can be considered as a particular form of operationalization of the knowledge representation. ASP logic programs follow the "generate and test" strategy. This strategy includes four steps:

- enumerate with facts,
- explain and extend with rules,
- generate all possibilities with cardinalities, and finally
- filter with constraints.

In the rest of this chapter, we propose ASP programs whose semantics and rule syntax are based on those presented here. The idea of using ASP to analyze the dynamics of a discrete model has its origin in the combinatorial explosion of the number of states. As ASP was designed to demonstrate a certain efficiency in enumerating sets satisfying a set of rules, this framework seems appropriate to study some dynamic properties of interest. In the following, we will focus on one type of dynamic model, AANs, that we introduce formally in this section. The conversion of such an AAN model into ASP is the subject of Section 7.5.

7.4.2 Definition of Asynchronous Automata Networks

AANs are a recently introduced formalism allowing to model discrete dynamical systems. Although mostly used to model BRNs, AANs are general purpose and could be applied to other fields. AANs are conveniently encoded into ASP because of their simple form that can be straightforwardly described in terms of a few simple rules. Its two most used update semantics are also easily represented in ASP (as explained later in Section 7.5.3). Moreover, AANs have been chosen for this chapter as they allow us to encompass more widespread frameworks such as René Thomas modelling (Thomas, 1973) and Boolean networks (Kauffman, 1969). The results of this chapter are thus applicable to these frameworks too.

Definition 7.1 introduces the formalism of AANs (Folschette et al., 2015), which allows us to model a finite number of discrete levels, called *local states*, into several *automata*. A local state is denoted a_i, where a is the name of the automaton, corresponding usually to a biological component, and i is a level identifier within a. At any time, exactly one local state of each automaton is *active*, modelling the current level of activity of this automaton or, equivalently, its internal state. The set of all active local states at a given time is called the *global state* of the network. Figure 7.1 depicts the structure of a simple AAN, which is more thoroughly detailed later.

The possible local evolutions inside an automaton are defined by *local transitions*. A local transition is a triple noted $a_i \xrightarrow{\ell} a_j$ and is responsible, inside a given automaton a, for the change of the active local state (a_i) to another local state (a_j), conditioned by a set ℓ of local states belonging to other automata. Such a local transition is *playable* if and only if a_i and all local states in the set ℓ are active. Thus, it can be read as "all the local states in ℓ can cooperate to change the active local state of a by making it switch from a_i to a_j." It is required that a_i and a_j are two different local states in automaton a and that ℓ contains no local state of a. We also note that ℓ should contain at most one local state per automaton; otherwise, the local transition is unplayable. Conversely, if ℓ is empty, the transition is playable if and only if a_i is active.

Definition 7.1 *(Asynchronous Automata Network)* An *AAN* is a triple $(\Sigma, S, \mathcal{T})$ where

- $\Sigma = \{a, b, \dots\}$ is the finite set of *automata* identifiers;
- For each $a \in \Sigma$, $b(a) \in \mathbb{N} \setminus \{0\}$ denotes the upper bound of automaton a and $S_a = \{a_0, \dots, a_{b(a)}\}$ is the finite set of *local states* of automaton a; $S = \prod_{a \in \Sigma} S_a$ is the finite set of *global states*; $\mathcal{LS} = \cup_{a \in \Sigma} S_a$ denotes the set of all the local states.
- For each $a \in \Sigma$, $\mathcal{T}_a \subset \{a_i \xrightarrow{\ell} a_j \in S_a \times \wp(\mathcal{LS} \setminus S_a) \times S_a | a_i \neq a_j\}$ is the set of *local transitions* on automaton a, where $\wp$ denotes the power set; $\mathcal{T} = \cup_{a \in \Sigma} \mathcal{T}_a$ is the set of all local transitions in the model.

266 | 7 Analyzing Long-Term Dynamics of Biological Networks With Answer Set Programming

For a given local transition $\tau = a_i \xrightarrow{\ell} a_j$, a_i is called the *origin* or τ, ℓ the *condition*, and a_j the *destination*, and they are respectively noted $\text{ori}(\tau)$, $\text{cond}(\tau)$, and $\text{dest}(\tau)$.

Example 7.4 Figure 7.1 represents an AAN $(\Sigma, S, \mathcal{T})$ with 4 automata (among which 2 contain 2 local states and the 2 others contain 3 local states) and 12 local transitions:

- $\Sigma = \{a, b, c, d\}$,
- $b(a) = 1, b(b) = 2, b(c) = 1, b(d) = 2$,
- $S_a = \{a_0, a_1\}, S_b = \{b_0, b_1, b_2\}, S_c = \{c_0, c_1\}, S_d = \{d_0, d_1, d_2\}$,
- $\mathcal{T} = \{ a_0 \xrightarrow{\{c_1\}} a_1, a_1 \xrightarrow{\{b_2\}} a_0, b_0 \xrightarrow{\{d_0\}} b_1, b_0 \xrightarrow{\{a_1,c_1\}} b_2, b_1 \xrightarrow{\{d_1\}} b_2, b_2 \xrightarrow{\{c_0\}} b_0,$
  $c_0 \xrightarrow{\{a_1,b_0\}} c_1, c_1 \xrightarrow{\{d_2\}} c_0, d_0 \xrightarrow{\{b_2\}} d_1, d_0 \xrightarrow{\{a_0,b_1\}} d_2, d_1 \xrightarrow{\{a_1\}} d_0, d_2 \xrightarrow{\{c_0\}} d_0 \}$.

The local transitions given in Definition 7.1 define concurrent interactions between automata. They have by nature only local consequences, but they can be combined in different ways to create a global dynamics for the model. When such a way of combining local transitions, also called *semantics*, is decided, it becomes possible to compute *global transitions* between global states. In the following, we will only focus on the (purely) asynchronous and (purely) synchronous semantics, which are the most widespread in the literature. The choice of such a semantics mainly depends on the considered biological phenomena modeled and the mathematical abstractions chosen by the modeler.

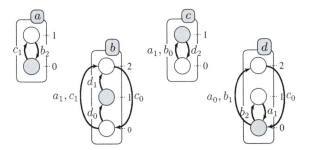

Figure 7.1 An example of an AAN model with four automata: *a*, *b*, *c*, and *d*. Each box represents an automaton (modelling a biological component), circles represent their local states (corresponding to their discrete expression levels) and the local transitions are represented by arrows labeled by their necessary conditions (consisting of a set of local states from other automata). The automata *a* and *c* are either at level 0 or 1, and *b* and *d* have 3 levels (0, 1, and 2). The grayed local states stand for the global state $\langle a_0, b_1, c_1, d_0 \rangle$.

7.4.3 Semantics and Dynamics of Asynchronous Automata Networks

As explained in the previous section, a global state of an AAN is a set of local states of automata containing exactly one local state for each automaton. In the following, we give some notations related to global states, then we define the global dynamics of an AAN.

The active local state of a given automaton $a \in \Sigma$ in a global state $\zeta \in S$ is noted $\zeta[a]$. For any given local state $a_i \in \mathcal{LS}$, we also note $a_i \in \zeta$ if and only if $\zeta[a] = a_i$. For a given set of local states $X \subseteq \mathcal{LS}$, we extend this notation to $X \subseteq \zeta$ if and only if $\forall a_i \in X, a_i \in \zeta$, meaning that all local states of X are active in ζ.

Furthermore, for any given local state $a_i \in \mathcal{LS}$, $\zeta \pitchfork a_i$ represents the global state that is identical to ζ, except for the local state of a, which is substituted with a_i: $(\zeta \pitchfork a_i)[a] = a_i \wedge \forall b \in \Sigma \setminus \{a\}, (\zeta \pitchfork a_i)[b] = \zeta[b]$. We generalize this notation to a set of local states $X \subseteq \mathcal{LS}$ containing at most one local state per automaton, that is, $\forall a \in \Sigma, |X \cap S_a| \leq 1$ where $|\cdot|$ denotes the cardinality of a set; in this case, $\zeta \pitchfork X$ is the global state ζ where the local state of each automaton has been replaced by the local state of the same automaton in X, if it exists: $\forall a \in \Sigma, (X \cap S_a = \{a_i\}) \Rightarrow (\zeta \pitchfork X)[a] = a_i) \wedge (X \cap S_a = \emptyset \Rightarrow (\zeta \pitchfork X)[a] = \zeta[a])$.

In Definition 7.2, we formalize the notion of playability of a local transition which was informally presented in the previous section. Playable local transitions are not necessarily used as such but combined depending on the chosen semantics, which is the subject of the rest of the section.

Definition 7.2 (*Playable Local Transitions*) Let $\mathcal{AAN} = (\Sigma, S, \mathcal{T})$ be an AAN and $\zeta \in S$ a global state. The set of playable local transitions in ζ is called P_ζ and defined by $P_\zeta = \{a_i \xrightarrow{\ell} a_j \in \mathcal{T} \mid \ell \subseteq \zeta \wedge a_i \in \zeta\}$.

The dynamics of the AAN is a composition of global transitions between global states that consist in selecting and playing a set of local transitions. Such sets are different depending on the chosen semantics. In the following, we give the definition of the asynchronous and synchronous semantics by characterizing the sets of local transitions that can be "played" as global transitions. The sets of the asynchronous semantics (Definition 7.3) are made of exactly one playable local transition; thus, a global asynchronous transition changes the local state of exactly one automaton. On the other hand, the sets of the synchronous semantics (Definition 7.4) consist of exactly one playable local transition for each automaton, except for the automata where no local transition is playable; in other words, a global synchronous transition changes the local state of all automata that can evolve at a time. The empty set is not a valid set of local transitions for both semantics, meaning that they both cannot produce a global transition that changes no automaton (also known as self-transition).

Definition 7.3 *(Asynchronous Semantics)* Let $\mathcal{AAN} = (\Sigma, S, \mathcal{T})$ be an AAN and $\zeta \in S$ a global state. The set of global transitions playable in ζ for the *asynchronous semantics* is given by

$$\mathcal{U}^{\text{asyn}}(\zeta) = \{\{a_i \xrightarrow{\ell} a_j\} \mid a_i \xrightarrow{\ell} a_j \in P_\zeta\}$$

Definition 7.4 *(Synchronous Semantics)* Let $\mathcal{AAN} = (\Sigma, S, \mathcal{T})$ be an AAN and $\zeta \in S$ a global state. The set of global transitions playable in ζ for the *synchronous* semantics is given by

$$\mathcal{U}^{\text{syn}}(\zeta) = \{u \subseteq \mathcal{T} \mid u \neq \emptyset \wedge \forall a \in \Sigma, (P_\zeta \cap \mathcal{T}_a = \emptyset$$
$$\Rightarrow u \cap \mathcal{T}_a = \emptyset) \wedge$$
$$(P_\zeta \cap \mathcal{T}_a \neq \emptyset \Rightarrow |u \cap \mathcal{T}_a| = 1)\}$$

Once a semantics has been chosen, it is possible to compute the corresponding dynamics of a given AAN. Thus, in the following, when it is not ambiguous and when results apply to both of them, we will denote by $\mathcal{U}$ a chosen semantics among $\mathcal{U}^{\text{asyn}}$ and $\mathcal{U}^{\text{syn}}$. Definition 7.5 formalizes the notion of a global transition depending on a chosen semantics $\mathcal{U}$.

Definition 7.5 *(Global Transition)* Let $\mathcal{AAN} = (\Sigma, S, \mathcal{T})$ be an AAN, $\zeta_1, \zeta_2 \in S$ two states and $\mathcal{U} \in \{\mathcal{U}^{\text{asyn}}, \mathcal{U}^{\text{syn}}\}$ a semantics. The *global transition* relation between two states ζ_1 and ζ_2 for the semantics $\mathcal{U}$, noted $\zeta_1 \to_\mathcal{U} \zeta_2$, is defined by

$$\zeta_1 \to_\mathcal{U} \zeta_2 \iff \exists u \in \mathcal{U}(\zeta_1) \wedge \zeta_2 = \zeta_1 \mathbin{\text{\reflectbox{m}}} \{\text{dest}(\tau) \in \mathcal{LS} \mid \tau \in u\}$$

The state ζ_2 is called a *successor* of ζ_1.

Example 7.5 Figures 7.2 and 7.3 illustrate the asynchronous and synchronous semantics, respectively, on the model of Figure 7.1. Each global transition is depicted by an arrow between two global states. Only an interesting subset of the whole dynamics is depicted in both figures.

A semantics is called *deterministic* if it makes each global state to have at most one successor. However, in general, semantics are *non-deterministic*: each global state can have several successors. Indeed, the two semantics

7.4 Dynamic Modelling Using Asynchronous Automata Networks

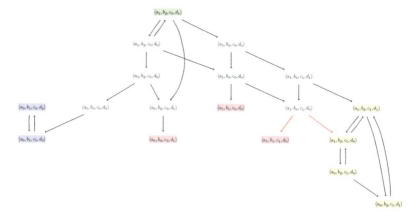

Figure 7.2 A part of the state-transition graph of the AAN given in Figure 7.1 for the **asynchronous** semantics, computed from the initial state: $\langle a_1, b_2, c_0, d_1 \rangle$ until reaching attractors. We can observe three stable states: $\langle a_1, b_1, c_1, d_0 \rangle$, $\langle a_1, b_1, c_0, d_0 \rangle$, and $\langle a_0, b_0, c_0, d_1 \rangle$; an attractor of size 2: $\{\langle a_0, b_1, c_0, d_0 \rangle, \langle a_0, b_1, c_0, d_2 \rangle\}$, and an attractor of size 4: $\{\langle a_1, b_2, c_1, d_1 \rangle, \langle a_0, b_2, c_1, d_1 \rangle, \langle a_0, b_2, c_1, d_0 \rangle, \langle a_1, b_2, c_1, d_0 \rangle\}$.

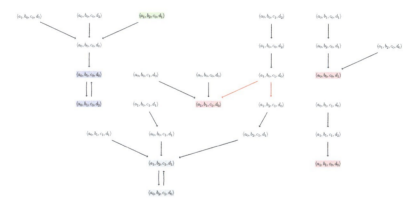

Figure 7.3 A part of the state-transition graph of the AAN given in Figure 7.1 for the **synchronous** semantics, computed from several initial states, such as $\langle a_1, b_2, c_0, d_1 \rangle$, until reaching attractors. It features non-deterministic global transitions, depicted by the two arrows from state $\langle a_1, b_0, c_1, d_0 \rangle$. We can observe the same three stable states than for the asynchronous semantics of Figure 7.2, but instead two attractors of size 2: $\{\langle a_0, b_1, c_0, d_0 \rangle, \langle a_0, b_1, c_0, d_2 \rangle\}$ and $\{\langle a_1, b_2, c_1, d_1 \rangle, \langle a_0, b_2, c_1, d_0 \rangle\}$.

studied here are non-deterministic in general (some particular models may not show the non-determinism).

In the case of the asynchronous semantics, the non-determinism may come from concurrent local transitions, but it actually mainly comes from the fact that exactly one local transition is taken into account for each global transition (see Definition 7.3). Thus, for a given state $\zeta \in S$, as soon as $|P_\zeta| > 1$, several successors may exist. In the model of Figure 7.1, for example, the global state $\langle a_1, b_2, c_0, d_1 \rangle$ (in green on Figure 7.2) has three successors because $\langle a_1, b_2, c_0, d_1 \rangle \to \mathcal{U}^{\text{asyn}} \langle a_0, b_2, c_0, d_1 \rangle$, $\langle a_1, b_2, c_0, d_1 \rangle \to \mathcal{U}^{\text{asyn}} \langle a_1, b_0, c_0, d_1 \rangle$, and $\langle a_1, b_2, c_0, d_1 \rangle \to \mathcal{U}^{\text{asyn}} \langle a_1, b_2, c_0, d_0 \rangle$.

In the case of the synchronous semantics (see Definition 7.4), however, the non-determinism on the global scale is only generated by local transitions that create non-determinism inside an automaton, that is, local transitions that have the same origin, are together playable in at least one global state, but have different destinations. For example, the model of Figure 7.1 features two local transitions $b_0 \xrightarrow{\{d_0\}} b_1$ and $b_0 \xrightarrow{\{a_1, c_1\}} b_2$ that can produce the two following global transitions from the same state (depicted by the two arrows from state $\langle a_1, b_0, c_1, d_0 \rangle$ and $\langle a_1, b_0, c_1, d_0 \rangle \to \mathcal{U}^{\text{syn}} \langle a_1, b_2, c_1, d_0 \rangle$. Note that for this particular case, these transitions also exist for the asynchronous semantics (also depicted by red arrows on Figure 7.2).

Finally, Definition 7.6 introduces the notions of *path* and *trace*, which are used to characterize a set of successive global states with respect to a global transition relation. Paths are useful for the characterization of attractors that are the topic of this work. The trace is the set of all global states traversed by a given path (thus disregarding the order in which they are visited). Thus, a path is a sequence while a trace is a set.

Definition 7.6 (Path and Trace) Let $\mathcal{AAN} = (\Sigma, S, \mathcal{T})$ be an AAN, $\mathcal{U}$ a semantics, and $n \in \mathbb{N} \setminus \{0\}$ a strictly positive integer. A sequence $\mathbf{H} = (\mathbf{H}_i)_{i \in [\![0;n]\!]} \in S^{n+1}$ of global states is a *path of length n* if and only if $\forall i \in [\![0; n-1]\!]$, $\mathbf{H}_i \to \mathcal{U} \mathbf{H}_{i+1}$. $\mathbf{H}$ is also simply called a *path* if its length n is not known or relevant. The *trace* of $\mathbf{H}$ is the set of global states it contains: $\text{trace}(\mathbf{H}) = \{ \mathbf{H}_j \in S | j \in [\![0; n]\!] \}$.

In the following, when we define a path $\mathbf{H}$ of length n, we use the notation $\mathbf{H}_i$ to denote the ith element in the sequence $\mathbf{H}$, with $i \in [\![0; n]\!]$. We also use the notation $|\mathbf{H}| = n$ to denote the length of a path $\mathbf{H}$, allowing to write $\mathbf{H}_{|\mathbf{H}|}$ to refer to its last element. $\mathbf{H}$ is said to *start from* a given global state $\zeta \in S$ if and only if $\mathbf{H}_0 = \zeta$; it is said to *end in* a given global state $\zeta' \in S$ if and only if $\mathbf{H}_n = \zeta'$. Finally, we note that a path of length n models the succession of n global transitions and thus features at most $n+1$ states: $\text{trace}(\mathbf{H}) \leq n+1$. When at least one state is visited more than once in $\mathbf{H}$, the inequality becomes strict.

Example 7.6 The following sequence is a path of length 6 for the asynchronous semantics:

$$\mathbf{H} = (\langle a_1, b_2, c_1, d_1\rangle; \langle a_0, b_2, c_1, d_1\rangle; \langle a_1, b_2, c_1, d_1\rangle; \langle a_1, b_2, c_1, d_0\rangle;$$
$$\langle a_0, b_2, c_1, d_0\rangle; \langle a_0, b_2, c_1, d_1\rangle; \langle a_1, b_2, c_1, d_1\rangle)$$

We have trace($\mathbf{H}$) = $\{\langle a_1, b_2, c_1, d_1\rangle, \langle a_0, b_2, c_1, d_1\rangle, \langle a_1, b_2, c_1, d_0\rangle, \langle a_0, b_2, c_1, d_0\rangle\}$. Although $|\mathbf{H}| = 6$, we note that $|\text{trace}(\mathbf{H})| = 4$ because $\mathbf{H}_0 = \mathbf{H}_2 = \mathbf{H}_6$ and $\mathbf{H}_1 = \mathbf{H}_5$.

7.4.4 Stable States and Attractors in Asynchronous Automata Networks

Studying the dynamics of biological networks was the focus of many works, explaining the diversity of existing frameworks dedicated to modelling and the different methods developed in order to identify some patterns, especially attractors (Skodawessely and Klemm, 2011, Zhang et al., 2007, Mushthofa et al., 2014, Akutsu et al., 2012, Berntenis and Ebeling, 2013). In this chapter, we focus on several sub-problems related to this: we seek to identify the *stable states* and the *attractors* of a given network. The stable states and the attractors are the two long-term structures in which any dynamics eventually falls into. Indeed, they consist in terminal (sets of) global states that cannot be escaped and in which the dynamics always ends. In the following, we formally define these dynamical properties.

A *stable state* (sometimes called *fixed point*) is a global state that has no successor, as given in Definition 7.7. The existence of several of these states is called multistability and implies bifurcations in the dynamics (Wuensche, 1998). At this point, it is important to remind that the empty set never belongs to the semantics defined above: $\forall \zeta \in S$, $\emptyset \notin \mathcal{U}^{\text{asyn}}(\zeta) \land \emptyset \notin \mathcal{U}^{\text{syn}}(\zeta)$. The consequence on the dynamics is that a global state can never be its own successor. In other words, even when no local transition can be played in a given global state (i.e., $P_\zeta = \emptyset$), we do not add a self-transition on this state. Instead, this state has no successor and is thus structurally a sink node in the state-transition graph.

Definition 7.7 *(Stable State)* Let $\mathcal{AAN} = (\Sigma, S, \mathcal{T})$ be an AAN and $\mathcal{U} \in \{\mathcal{U}^{\text{asyn}}, \mathcal{U}^{\text{syn}}\}$ be a semantics. A global state $\zeta \in S$ is called a *stable state* if and only if no global transition can be played in this state:

$$\mathcal{U}(\zeta) = \emptyset.$$

It is interesting to note that the set of stable states of a model is the same in both the asynchronous and the synchronous semantics (Klarner et al., 2015, de Espanés et al., 2016): $\forall \zeta \in S,\ U^{asyn}(\zeta) = \emptyset \iff U^{syn}(\zeta) = \emptyset$. This comes from the fact that both semantics rely on the same definition of a set of playable transitions.

Example 7.7 The state-transition graphs of Figures 7.2 and 7.3 depict three stable states colored in red: $\langle a_1, b_1, c_1, d_0 \rangle$, $\langle a_1, b_1, c_0, d_0 \rangle$, and $\langle a_0, b_0, c_0, d_1 \rangle$. Visually, they can be easily recognized because they have no outgoing arrows (meaning that they have no successors). Although these figures do not represent the whole dynamics, they allow us to check that in both semantics, the stable states are shared. Actually, the model contains no other stable state than the ones depicted in these figures.

Another complementary dynamical pattern consists in the notion of *trap domain* (Definition 7.8), which is a non-empty set of states that the dynamics cannot escape, and thus in which the system indefinitely remains. Relying on this, an *attractor* (Definition 7.9) is a minimal trap domain in terms of set inclusion. In this work, we focus more precisely on *non-singleton attractors*, that is, attractors made of at least two states. Formally speaking, stable states are trap domains of size 1 and could thus be considered attractors. However, in the scope of this chapter and for the sake of clarity, in the following, we call "attractors" only non-singleton attractors. This is justified by the very different approaches developed to enumerate stable states and attractors in the next sections.

Definition 7.8 (*Trap Domain*) Let $\mathcal{AAN} = (\Sigma, S, T)$ be an AAN and U a semantics. A set of global states $T \subseteq S$, with $T \neq \emptyset$, is called a *trap domain* (regarding a semantics U) if and only if the successors of each of its elements are also in T:

$$\forall \zeta_1 \in T \wedge \forall \zeta_2 \in S, \zeta_1 \to U\zeta_2 \Rightarrow \zeta_2 \in T$$

Definition 7.9 (*Attractor*) Let $\mathcal{AAN} = (\Sigma, S, T)$ be an AAN and U a semantics. A set of global states $A \subseteq S$, with $|A| \geq 2$, is called an *attractor* (regarding semantics U) if and only if it is a minimal trap domain in terms of set inclusion.

The previous definition is hard to encode into ASP because of the notion of minimality. In the following, we use the notion of *cycle* (Definition 7.10), which is a looping path, in order to give an alternative definition for an attractor (Lemma 7.3).

7.4 Dynamic Modelling Using Asynchronous Automata Networks

Definition 7.10 *(Cycle)* Let $\mathcal{AAN} = (\Sigma, S, \mathcal{T})$ be an AAN, $\mathcal{U}$ a semantics, and **C** a path for this semantics. **C** is called a *cycle* (regarding the semantics $\mathcal{U}$) if and only if it starts from and ends in the same state:

$$\mathbf{C}_0 = \mathbf{C}_{|\mathbf{C}|}$$

Example 7.8 The path **H** of length 6 given in Example 7.6 is a cycle because $\mathbf{H}_0 = \mathbf{H}_6$.

Definition 7.11 recalls the definition of a *strongly connected subgraph* (SCSG) in the scope of the dynamics of an AAN: it is a subgraph of the global state-transition graph in which there exists a path between any pair of states. Note that because we consider here only the synchronous and asynchronous semantics, a subgraph made of exactly one state ζ cannot be a SCSG because there never exists a self-transition $\zeta \to \zeta$.

The definition of SCSG is then used in two lemmas that later allow us to give an alternative definition of an attractor. Lemma 7.1 states that the set of (traces of) cycles in a model is exactly the set of SCSGs. Indeed, a cycle allows to "loop" between all states that it contains, and conversely, such a cycle can be built from the states of any SCSG. Lemma 7.2 states that any attractor is also a SCSG. This well-known result comes from the minimality of an attractor and implies that an attractor is always made of one or several loops.

Definition 7.11 *(Strongly Connected Subgraph)* Let $\mathcal{AAN} = (\Sigma, S, \mathcal{T})$ be an AAN, $\mathcal{U}$ a semantics, and $G \subseteq S$ a set of states. G is a *SCSG* (regarding the semantics $\mathcal{U}$) or *SCSG* for short if and only if for all pairs of states $(\zeta_1, \zeta_2) \in G^2$ in this set, there exists a path **H** made of states in G that starts from ζ_1 and ends in ζ_2, that is:

$$\mathbf{H}_0 = \zeta_1 \wedge \mathbf{H}_{|\mathbf{H}|} = \zeta_2 \wedge \forall \zeta \in \mathrm{trace}(\mathbf{H}), \zeta \in G$$

Lemma 7.1 *(The Traces of Cycles Are the SCSGs)* The traces of the cycles are exactly the SCSGs.

Proof: ($\Rightarrow$) From any state of a cycle, it is possible to reach all the other states (by possibly cycling). Therefore, the trace of this cycle is a SCSG. ($\Leftarrow$) Let G be a SCSG. Consider, without loss of generality, an arbitrary state $\zeta_0 \in G$. For each other state $\zeta' \in G \setminus \{\zeta_0\}$, by definition of a SCSG, there exists two paths $\mathbf{J}^{\zeta'}$ and $\mathbf{K}^{\zeta'}$ made of states of G so that $\mathbf{J}^{\zeta'}_0 = \zeta_0$, $\mathbf{J}^{\zeta'}_{|\mathbf{J}^{\zeta'}|} = \zeta'$, $\mathbf{K}^{\zeta'}_0 = \zeta'$ and $\mathbf{K}^{\zeta'}_{|\mathbf{K}^{\zeta'}|} = \zeta_0$. We denote $\mathbf{H}^{\zeta'}$ the concatenation of $\mathbf{J}^{\zeta'}$ and $\mathbf{K}^{\zeta'}$, which is possible because $\mathbf{J}_{|\mathbf{J}^{\zeta'}|}^{\zeta'} = \zeta' = \mathbf{K}_0^{\zeta'}$. $\mathbf{H}^{\zeta'}$ is a cycle because $\mathbf{H}_0^{\zeta'} = \mathbf{J}_0^{\zeta'} = \zeta_0 = \mathbf{K}_{|\mathbf{K}^{\zeta'}|}^{\zeta'} = \mathbf{H}_{|\mathbf{H}^{\zeta'}|}^{\zeta'}$. Moreover, $\mathbf{H}^{\zeta'}$ is only made of states of G. By considering, without loss of generality, an arbitrary sequence $A = (\zeta_i')_{i \in [\![1;|G \setminus \{\zeta_0\}|]\!]}$ of all the states in $G \setminus \{\zeta_0\}$, that is: $\forall \zeta' \in G \setminus \{\zeta_0\}, \exists i \in [\![1;|G \setminus \{\zeta_0\}|]\!], A_i = \zeta_0$, then we can build a path $\mathbf{H}$ by concatenating $\mathbf{H}^{A_0}, \mathbf{H}^{A_1}, \ldots, \mathbf{H}^{A_{|A|}}$. $\mathbf{H}$ is a cycle because $\mathbf{H}_0 = \mathbf{H}_{|\mathbf{H}|} = \zeta_0$. Moreover, $\text{trace}(\mathbf{H}) \subseteq G$ because $\mathbf{H}$ is only made of states in G, and $G \subseteq \text{trace}(\mathbf{H})$ because for each state in G, at least one sub-cycle of $\mathbf{H}$ contains it. Therefore, $\mathbf{H}$ is a cycle so that $\text{trace}(\mathbf{H}) = G$. □

Lemma 7.2 *(An Attractor is a SCSG)* An attractor is a SCSG.

Proof: Let A be an attractor. Suppose that A is not a SCSG in order to make a proof by contradiction. Then, there exists $\zeta_1, \zeta_2 \in A$ so that ζ_2 cannot be reached from ζ_1. Let X be the set containing exactly ζ_1 and all states that can be reached from ζ_1. Obviously, $X \neq \emptyset$. By definition of an attractor (Definition 7.9), A is also a trap domain; therefore, the successors of each state in A are also in A, which recursively gives $X \subset A$ because $\zeta_1 \in A$. However, $\zeta_2 \notin X$ by definition of ζ_2 and X, thus $X \subsetneq A$. X is a trap domain by construction (if a state can be reached from ζ_1, then it belongs to X). This contradicts the fact that A is a minimal trap domain in terms of set inclusion (Definition 7.9). To conclude, A is a SCSG. □

In Definition 7.9, attractors are characterized in the classical way, that is, as minimal trap domains. However, we use an alternative characterization of attractors in this chapter because of the specifics of ASP: Lemma 7.3 states that an attractor can alternatively be defined as a trap domain that is also a cycle and conversely. In other words, the minimality requirement is replaced by a cyclicity requirement.

Lemma 7.3 *(The Attractors Are the Cyclic Trap Domains)* The attractors are exactly the traces of cycles that are trap domains.

Proof: ($\Rightarrow$) By Definition 7.9, an attractor is a trap domain. From Lemma 7.2, it is also a SCSG, and thus, from Lemma 7.1, it is the trace of a cycle. ($\Leftarrow$) Let $\mathbf{C}$ be a cycle whose trace is a trap domain. From Lemma 7.1, $\mathbf{C}$ is also a SCSG. Let us prove by contradiction that $\mathbf{C}$ is a minimal trap domain, by assuming that it is not minimal. This means that there exists a smaller trap domain $\mathbf{D} \subsetneq \mathbf{C}$. Let us consider $\zeta_1 \in \mathbf{D}$ and $\zeta_2 \in \mathbf{C} \setminus \mathbf{D}$. Because $\mathbf{D}$ is a trap domain, there exists no path between ζ_1 and ζ_2; this is in contradiction with $\mathbf{C}$ being a SCSG (as both ζ_1 and ζ_2 belong to $\mathbf{C}$). Therefore, $\mathbf{C}$ is a minimal trap domain and thus an attractor. □

As explained before, Lemma 7.3 will be used in Section 7.5.3 to enumerate attractors. Indeed, directly searching for minimal trap domains would be too cumbersome; instead, we enumerate cycles of length n in the dynamics of the model and filter out those that are not trap domains.

Example 7.9 The state-transition graphs of Figures 7.2 and 7.3 feature different attractors:

- $\{\langle a_0, b_1, c_0, d_0\rangle, \langle a_0, b_1, c_0, d_2\rangle\}$ is depicted in blue and appears in both figures. It is a simple attractor because it contains a unique cycle.
- $\{\langle a_0, b_2, c_1, d_0\rangle, \langle a_0, b_2, c_1, d_1\rangle, \langle a_1, b_2, c_1, d_1\rangle, \langle a_1, b_2, c_1, d_0\rangle\}$ is only present for the asynchronous semantics and is depicted in yellow on Figure 7.2. It is a complex attractor, that is, a composition of several cycles.
- $\{\langle a_1, b_2, c_1, d_1\rangle, \langle a_0, b_2, c_1, d_0\rangle\}$ is, on the contrary, only present for the synchronous semantics and is depicted in gray on Figure 7.3. It is also a simple attractor.

For each of these attractors, the reader can check that they can be characterized as cycles that are trap domains. For instance, the second attractor can be found by considering the following cycle:

$$\mathbf{A} = (\langle a_0, b_2, c_1, d_0\rangle; \langle a_0, b_2, c_1, d_1\rangle; \langle a_1, b_2, c_1, d_1\rangle; \langle a_1, b_2, c_1, d_0\rangle;$$
$$\langle a_0, b_2, c_1, d_0\rangle)$$

and checking that its trace is a trap domain (which is visually confirmed in Figure 7.2 by the absence of outgoing arrows from any of the yellow states). On the other hand, the following cycle is not an attractor:

$$\mathbf{C} = (\langle a_1, b_2, c_0, d_1\rangle; \langle a_1, b_2, c_0, d_0\rangle; \langle a_1, b_2, c_0, d_1\rangle)$$

Indeed, although it is a cycle, it features states having outgoing transitions (such as, for instance, transition $\langle a_1, b_2, c_0, d_0\rangle \rightarrow \mathcal{U}^{\text{asyn}}\langle a_0, b_2, c_0, d_0\rangle$) and thus is not a trap domain.

The aim of the rest of this chapter is to tackle the enumeration of stable states (Section 7.5.2) and attractors (Section 7.5.3) in an AAN using ASP.

7.5 Encoding into Answer Set Programming

This section presents the three facets of this work in terms of ASP:

1. How to encode an AAN model to use with the two other facets (Section 7.5.1),

2. How to encode the stable-state enumeration (Section 7.5.2),
3. How to encode the attractor enumeration (Section 7.5.3).

All parts were initially presented in Abdallah et al. (2015) and Abdallah et al. (2017). The first two are identical, while the last one has been significantly improved by fixing mistakes and adding filtering methods to avoid redundant answers in attractor enumeration. Indeed, a straightforward approach for this problem, as in Abdallah et al. (2017), produces a lot of duplicate solutions because of the enumeration method, as explained at the end of Section 7.5.3.2. In this chapter, we propose two additions based on the combination of Python scripting with ASP to solve this issue.

All code presented in this section is available as free software online.[2] The ASP scripts are exactly the lines of the code presented below, where the symbol ← in rules is replaced by the characters colon and dash ":-" for Clingo compatibility. Rules are sometimes written on several lines, as they are ended only by a period "..". Lines beginning with a percent sign "%" are comments.

7.5.1 Translating Asynchronous Automata Networks into Answer Set Programs

Before any analysis of an AAN model, we first need to express it with ASP rules. We developed a dedicated converter named an2asp.py and we detail its principle in the following.

First, the predicate automaton_level(Automaton, Level) is used to define each automaton Automaton along with its local state Level. Each local transition is then represented with two predicates: condition which is used several times to define each element of the condition along with the origin, and target which is used once to define the target of the local transition. Each local transition is labeled by an identifier that is the same in its condition and target predicates.

Example 7.10 *(Representation of an AAN Model in ASP)*
Here is the representation of the AAN model of Figure 7.1 in ASP:

```
1 % Automata and local states
2 automaton_level("a", 0..1). automaton_level("b", 0..2).
3 automaton_level("c", 0..1). automaton_level("d", 0..2).
4 % Local transitions on a
5 condition(t1, "a", 0). target(t1, "a", 1).
  condition(t1, "c", 1). % $a_0 \xrightarrow{\{c_1\}} a_1$
```

---

2 All programs and benchmarks are available as additional files and at https://zenodo.org/record/6534531.

```
 6  condition(t2, "a", 1). target(t2, "a", 0). condition(t2, "b", 2).
    % a₁ ─{b₂}→ a₀
 7  % Local transitions on b
 8  condition(t3, "b", 0). target(3, "b", 1). condition(t3, "d", 0).
 9  condition(t4, "b", 0). target(4, "b", 2). condition(t4, "a", 1).
    condition(t4, "c", 1).
10  condition(t5, "b", 1). target(5, "b", 2). condition(t5, "d", 1).
11  condition(t6, "b", 2). target(6, "b", 0). condition(t6, "c", 0).
12  % Local transitions on c
13  condition(t7, "c", 0). target(7, "c", 1). condition(t7, "a", 1).
    condition(t7, "b", 0).
14  condition(t8, "c", 1). target(8, "c", 0). condition(t8, "d", 2).
15  % Local transitions on d
16  condition(t9, "d", 0). target(9, "d", 1). condition(t9, "b", 2).
17  condition(t10, "d", 0). target(10, "d", 2).
    condition(t10, "a", 0). condition(t10, "b", 1).
18  condition(t11, "d", 1). target(11, "d", 0).
    condition(t11, "a", 1).
19  condition(t12, "d", 2). target(12, "d", 0).
    condition(t12, "c", 0).
```

In lines 2–3, we define the model's automata with their local states. For example, the automaton "a" has two levels numbered 0 and 1. For this, we use the construct m..n in rule automaton_level("a", 0..1). of line 2, which is actually a shortcut for the two following rules:

automaton_level("a", 0). automaton_level("a", 1).

All the local transitions of the network are defined in lines 5–19; for instance, all the predicates in line 5 declare the transition $\tau_1 = a_0 \xrightarrow{\{c_1\}} a_1$, which is internally labeled t1. We declare as many predicates condition as necessary in order to fully define a local transition τ that has potentially several elements in its condition cond(τ). For instance, transition $b_0 \xrightarrow{\{a_1,c_1\}} b_2$ is defined in line 9 with label t4 and requires three of these predicates for b_0, a_1, and c_1.

As the names of the biological components may start with a capital letter or contain non-alphabetic characters, it is preferable to use double quotes (" ") around the automata names in the parameters of predicates to ensure that they are correctly interpreted as constants by the ASP grounder.

Finally, lines lines 21–24 extend the facts presented above and are thus always defined in other scripts to be used in more complex rules. Predicate automaton gathers all the existing automata names in the model. The underscore symbol "_" in the parameters of a predicate is a placeholder for any value that is not used elsewhere; it could be replaced by a variable name occurring only here in the rule, but this practice reduces readability. Predicate `local_transition` comes in two forms: with arity 1 (i.e., one argument), it gathers all transition labels, and with arity 2 (i.e., two arguments), it gathers transition labels and the automaton they target. Note that two predicates with the same name but different arities are completely distinct; nevertheless, using the same name allows us to remind that they have a close meaning.

```
20  % Automata names
21  automaton(Automaton) ← automaton_level(Automaton, Level).
22  % Local transition names
23  local_transition(Transition) ← target(Transition, _, _).
24  local_transition(Transition, Automaton) ←
    target(Transition, Automaton, _).
```

7.5.2 Stable-State Enumeration

The first aspect of this work is the enumeration of a particular type of minimal trap domains: stable states (also called fixed points or steady states) that are composed of only one global state (see Definition 7.7). They can be studied separately from attractors because their enumeration follows a different pattern that is more specific to this problem. The encoding presented here is equivalent to Abdallah et al. (2017), which itself is an extension of (Abdallah et al., 2015) that tackled a more restrictive class of automata.

To summarize roughly, the enumeration of stable states requires to encode the definition of a stable state (given in Definition 7.7) as an ASP program through logic rules. The first step of this process is to browse all the possible states of the network; in other words, all possible combinations of local states are generated by choosing exactly one local level for each automaton. This is performed by lines 26–27, which create several *candidate* solutions, more precisely as many as there are sets respecting the cardinality constructs (around the curly braces). More precisely, for each value of `Automaton` so that `automaton(Automaton)` is true, we enumerate all values of `Level` so that `automaton_level(Automaton, Level)` is true and create as many `local_state(Automaton, Level)` predicates; only one is kept in each answer set because of the lower and upper cardinalities of 1. Therefore, we theoretically obtain as many answer sets as there are possible states in the model. Of course, among these candidate solutions, we want to filter out all those that are not valid answers to our problem of stable-state enumeration.

```
25  % Enumerate all possible states (one local state per automaton)
26  1 { local_state(Automaton, Level) :
    automaton_level(Automaton, Level) } 1 ←
27     automaton(Automaton).
```

7.5 Encoding into Answer Set Programming

For each enumerated state, we want to filter out those featuring at least one *playable* local transition, that is, a local transition for which all conditions are met. It is not possible to express playable transitions directly because ASP naturally expresses existentiality (∃) rather than universality (∀). Nevertheless, universality in ASP can be expressed as a negation of existentiality. Therefore, we first express *unplayable* local transitions, that is, transitions such that there exists a condition that is not met, with predicate unplayable (lines 29–32). This predicate can then be used with a negation by default (not) to find *playable* transitions: indeed, a transition T is playable if unplayable(T) is not part of the answer set, and thus if not unplayable(T) holds. Finally, using this, we can filter out all candidate solutions (i.e., states) that contain at least one playable local transition, with the constraint of line 34.

```
28 % Compute not playable transitions in the current state
29 unplayable(Transition) ←
30    local_state(Automaton, LevelI),
31    condition(Transition, Automaton, LevelJ),
32    LevelI != LevelJ.
33 % Constraint: discard states with a playable transition
34 ← not unplayable(Transition), local_transition(Transition).
```

The following special clause can be added to the Clingo script in order to show only the predicates local_state of arity 2 and hide all the others:

```
35 #show local_state/2.
```

Example 7.11 *(Stable-State Enumeration)*
The AAN model of Figure 7.1 contains four automata: a and c have two local states while b and d have three; therefore, the whole model has $2 * 2 * 3 * 3 = 36$ states (whether they can be reached or not from a given initial state). We can check that this model contains exactly three stable states: $\langle a_0, b_0, c_0, d_1 \rangle$, $\langle a_1, b_1, c_1, d_0 \rangle$ and $\langle a_1, b_1, c_0, d_0 \rangle$. All of them are represented in both Figures 7.2 and 7.3. In this model, no other state verifies this property. We recall that the stable states are identical for the synchronous and asynchronous semantics (Klarner et al., 2015).

To enumerate stable states, one can execute the ASP program detailed above (lines 26–35) alongside with the AAN model given in Example 7.10 (lines 1–19) and the extended facts (lines 21–24). This can also be done with the following command line on the supplementary material:

```
clingo 0 stabe-states.lp ../models/asp/example.lp
```

The output of Clingo is the following, matching the expected result:

```
Answer: 1
local_state("a",0) local_state("b",0) local_state("c",0)
  local_state("d",1)
Answer: 2
local_state("a",1) local_state("b",1) local_state("c",1)
  local_state("d",0)
Answer: 3
local_state("a",1) local_state("b",1) local_state("c",0)
  local_state("d",0)
```

The solving is performed in about a hundredth of a second.

7.5.3 Attractors

The previous section offered a method to enumerate all stable states of a given model. In a sense, a stable state can be considered as an attractor: it cannot be escaped and its size ($n = 1$) makes it trivially minimal. However, attractors in the general case are made of several states. In the rest of this section, we exclude one-state attractors and focus on attractors that are made of several states (following Definition 7.9). We describe how to obtain all the attractors up to a given *length* in a model for the asynchronous semantics, where the length is the number of steps of the minimal path covering the whole attractor (see Definition 7.6). Obtaining all attractors of any length can be theoretically tackled by providing a length that is high enough, but attractors being often of small size, results are usually obtained even for small lengths. For the synchronous semantics, the method still holds but only a part of the attractors are returned.

The computational method to enumerate all attractors of length n in AAN models consists in three steps:

1. Enumerate all paths of length n,
2. Remove all paths that do not contain a cycle,
3. Remove all paths that are not trap domains.

Once all steps are passed, each trace of the remaining paths is an attractor (following Lemma 7.3).

This whole section is largely similar to Abdallah et al. (2017), with several improvements:

- The returned attractors can now be of length inferior or equal to n (and not necessarily exactly n);
- Minor fixes for bugs that occurred in rare cases;
- Python scripting now allows to avoid duplicate answer sets;

- Specific optimizations in the ASP scripts have been removed as they were deprecated.

7.5.3.1 Cycle Enumeration

The approach presented here first consists in enumerating all possible paths of a given length n in the AAN model (Definition 7.6) and attempting to find a cycle (Definition 7.10) among the first states of these paths. In an ASP program, it is possible to instantiate constants, whose values are defined by the user at each execution: this is the role of the lowercase n in step(0..n) (line 37), which represents the number of considered steps. For instance, if n is initialized at value 2, the predicate above becomes step(0..2), which in turns means: step(0). step(1). step(2).; in other words, three successive global states are considered, that is, a path of length 2 because it contains two transitions. Then, predicate main_cycle_length(N) states that we make the assumption that the states 0 and N are identical and thus characterize a cycle; this will be checked afterward. This predicate is instantiated in line 39 with a cardinality construct that means that as many candidate solutions are created as there are possible values for N. In other words, all values for N are tested, up to n (the constant). Finally, we differentiate the steps before and equal to N, which make up the *main cycle*, with predicate cycle_step, and the steps after N, with predicate after_cycle_step (lines 41–43).

```
36 % Steps in the whole path
37 step(0..n).
38 % Length of the main cycle (i.e., a cycling sub-path)
39 1 {main_cycle_length(N) : step(N), N > 0 } 1.
40 % Steps in the main cycle
41 cycle_step(0..N) ← main_cycle_length(N).
42 % Steps after the main cycle
43 after_cycle_step(N+1..n) ← main_cycle_length(N).
```

In order to enumerate all the possible paths, step 0 should take the value of all the possible (global) initial states, in a similar way to the stable-state enumeration. For this, the cardinality construct in lines 45–46 allows to create as many candidate solutions as there are possible initial states by activating exactly one local state for each automaton, and thus, all combinations are tested. Here and in the following, the fact that local state L of automaton A is active in a given step S is denoted by a predicate: active(level(A, L), S).

```
44 % Select randomly one initial state (step 0)
45 1 { active(level(Automaton, Level), 0) :
       automaton_level(Automaton, Level) } 1 ←
46    automaton(Automaton).
```

Then, identifying the successors of a given global state requires to identify the set of its playable local transitions. We recall that a local transition is playable in a global state when its origin and all its conditions are active in that global state (see Definition 7.2). Similar to Section 7.5.2, we define an ASP predicate `unplayable(T,S)` in lines 48–52 stating that transition T is *not* playable in step S because at least one of its conditions is not satisfied. Obviously, each local transition that is not flagged as unplayable is playable.

```
47 % Compute not playable transitions for each step
48 unplayable(Transition, Step) ←
49   active(level(Automaton, LevelI), Step),
50   condition(Transition, Automaton, LevelJ),
51   LevelI != LevelJ,
52   step(Step).
```

At this point, one of the two semantics, asynchronous or synchronous, can be applied in order to compute the possible paths from each of the initial states. Each semantics comes in a different piece of ASP script, so that the chosen semantics can be executed alongside the main script solving the attractor enumeration. We show in the following how to compute the evolution of the model through the asynchronous or synchronous semantics, as presented in Section 7.4.3. The piece of program that computes the attractors, given afterward, is common to both semantics.

The possible evolutions of a model from a given state, that is, the different resulting paths after playing a set of global transitions, can be enumerated with cardinality rules given below. Thus, the rules below reproduce all possible paths in the dynamics of the model by representing each possible successor of a considered state as an answer set and so on. This enumeration encompasses the non-deterministic behavior (in both semantics).

To enforce the strictly asynchronous dynamics, which requires that exactly one automaton changes during a global transition (see Definition 7.3), we use the constraint of lines 54–59 to choose exactly one local transition to play in each step. This local transition plays the role of a global transition by itself and will be used later to compute the contents of the following global state. From this new state, the same constraint will be applied to compute the following state, and so on, up to n steps.

```
53 % Asynchronous semantics: select exactly 1 local transition
   % to play
54 1 {
55     played(Transition, Step) :
56       local_transition(Transition),
57       not unplayable(Transition, Step)
58 } 1 ←
59   step(Step).
```

The second semantics corresponds to the synchronous dynamics in which a maximal set of playable transitions, with at most one local transition per automaton, has to be played (see Definition 7.4). For this, in lines 61–64, we first search the automata that have at least one playable local transition in the current step with predicate has_playable. Then, the constraint of lines 66–71 selects exactly one transition for each automaton in this case. Finally, we forbid "empty" global transitions, even when no transition is playable (line 73) as this would create a self-transition, which we deliberately exclude here. For this, we use a cardinality construct in the body of a rule, which does not create multiple candidate answer sets but only acts as an atom that "verifies" that the cardinalities are verified. Here, if this cardinality construct is true, then the constraint applies and removes the current candidate solution.

```
60 % Automata that have at least one playable local transition
61 has_playable(Automaton, Step) ←
62    not unplayable(Transition, Step),
63    local_transition(Transition, Automaton),
64    step(Step).
65 % Synchronous semantics: select 1 transition to play for each
   % automaton, if possible
66 1 {
67    played(Transition, Step) :
68       not unplayable(Transition, Step),
69       local_transition(Transition, Automaton)
70 } 1 ←
71    has_playable(Automaton, Step).
72 % Constraint: play at least one local transition
73 ← 0 { played(_, Step) } 0, step(Step).
```

In a nutshell, one should choose one of both pieces of program presented above, that is, either lines 54–59 for the asynchronous semantics or lines 61–73 for the synchronous one. The point of either of these pieces of programs is to produce a collection of answer sets, where each one is a possible path of length n (that is, up to step n) and starting from any initial state (at step 0).

However, to actually produce each step, we first need to compute the resulting global step following of each global transition proposed by the semantics. For this, the predicate change in lines 75–78 allows us to witness the fact that in a given step, a local transition is played (because it has been chosen by the semantics) and thus the local active state of the related automaton must be updated. If a change is witnessed, then line 80 defines the contents of predicate active in the next step based on the target of the local transition. If there is no such change, then the local level of this automaton stays the same, as stated by line 84.

```
74 % In Step, Automaton uses Transition to change from LevelI
   % to LevelJ
75 change(Transition, Automaton, LevelI, LevelJ, Step) ←
76   played(Transition, Step),
77   target(Transition, Automaton, LevelJ),
78   condition(Transition, Automaton, LevelI).
79 % Change for the new active level if there is a change on
   % Automaton
80 active(level(Automaton, LevelK), Step + 1) ←
81   change(_, Automaton, _, LevelK, Step),
82   Step < n.
83 % Keep the same active level if no change on Automaton
84 active(level(Automaton, LevelK), Step + 1) ←
85   not change(_, Automaton, _, _, Step),
86   active(level(Automaton, LevelK), Step),
87   step(Step),
88   Step < n.
```

Now that the path of length n (constant) is fully defined, it is time to check if it actually contains a cycle of length N (variable), as this is a necessary and sufficient condition to be a SCSG (see Lemma 7.1) and thus a necessary condition to be an attractor. For this, we simply need to check if the states corresponding to steps 0 and N are identical. As this cannot be done directly (because it would require expressing universality), we first define a predicate different_states_on that checks whether a given automaton has the same local state in two different steps (lines 90–97). We perform this computation for each automaton and each pair of steps because this result will be used again later. Based on this, predicate different_states (line 99) generalizes this to any global pair of steps, thus witnessing globally different states. Obviously, states that are not different are identical, which is expressed by predicate same_state (lines 101–105). As a consequence, this means that there exists a cycle between both steps (starting from the earliest to the latest). When it is not the case between steps 0 and N, it means that the assumption main_cycle_length(N) is not true, and the current answer set must be rejected, which is done by the constraint of line 107.

```
89 % States of Step1 and Step2 are different on Automaton,
   % with Step1 < Step2
90 different_states_on(Step1, Step2, Automaton) ←
91   active(level(Automaton, LevelI), Step1),
92   active(level(Automaton, LevelJ), Step2),
93   LevelI != LevelJ,
94   step(Step1),
95   step(Step2),
96   automaton(Automaton),
97   Step1 != Step2.
98 % States of Step1 and Step2 are different on at least one
   % automaton
```

```
 99 different_states(Step1, Step2) ← different_states_on(Step1,
      Step2, _).
100 % States of Step1 and Step2 are identical (thus there is a
    % cycle)
101 same_state(Step1, Step2) ←
102    not different_states(Step1, Step2),
103    step(Step1),
104    step(Step2),
105    Step1 != Step2.
106 % Constraint: remove answer sets that are not cyclic on the
    % main cycle (steps 0 and N)
107 ← not same_state(0, N), main_cycle_length(N).
```

We now have the certainty that the steps in the main cycle, that is, from 0 to N, form a SCSG. However, we must still check that the states after this main cycle, that is, steps N+1 to n, are part of the same SCSG; otherwise, the rest is not applicable. For this, we simply check that each state after the main cycle is equal to at least one state in the main cycle, with predicate valid_state_after_main_cycle (lines 109–112). Paths containing at least one state that does not respect this are removed by the constraint of line 114.

```
108 % Check that the states after the main cycle are already
    % visited in the main cycle
109 valid_state_after_main_cycle(Step2) ←
110    same_state(Step1, Step2),
111    cycle_step(Step1),
112    after_cycle_step(Step2).
113 % Constraint: remove answer sets that visit new states after
    % the main cycle
114 ← not valid_state_after_main_cycle(Step1),
      after_cycle_step(Step1).
```

As stated in Lemma 7.1, all the remaining paths are now SCSGs. We finally need to verify that they are trap domains (Lemma 7.3) in order to discriminate attractors.

7.5.3.2 Attractor Enumeration

Because of the non-deterministic nature of the dynamics handled in this work, each state in the state-transition graph of a given AAN might have several successors. Therefore, a cyclic path is not necessarily an attractor. Theoretically, the only exception is the case of Boolean models under the synchronous semantics, which is always deterministic; in this case, the computation could be stopped at this point because a cycle is necessarily an attractor (Dubrova and Teslenko, 2009, Qu et al., 2015, Hayashida et al., 2008). In the following, we consider the general and non-deterministic case.

At this point, it is thus necessary to express the fact of *not* being a trap domain (see Lemma 7.3) in order to filter out these cases. For instance, in the partial state-transition graph of Figure 7.2, we can spot many cycles of various lengths but not all of them are attractors. In particular, the initial global state is part of a cycle of length 2, which is not an attractor, and whose trace is $\{\langle a_1, b_2, c_0, d_1 \rangle, \langle a_1, b_2, c_0, d_0 \rangle\}$. In the following, we will only consider the main cycle (steps 0 to N) because we have constrained above that the part of the path after the main cycle is only made of duplicate states of the main cycle (see lines 109–114).

The filtering that we present in the following applies fully to the asynchronous semantics only. In the case of the synchronous semantics, however, it filters out some legitimate attractors. More precisely:

- In the Boolean case (if all automata have only two possible levels), all attractors are correctly returned;
- More generally, *simple attractors*, which are made of a simple loop and not of a composition of several loops, are correctly returned;
- A *complex attractor*, which is made of a composition of loops, is correctly returned only if there exists a covering path so that all the other global transitions in this attractor (that are not part of this path) are composed of only one local transition (similarly to the asynchronous semantics);
- All other complex attractors are erroneously filtered out.

This filtering is performed with a constraint, which, once again, is the most suitable solution. In order to define such a constraint, we need to describe the behavior that we must not observe, namely: escaping the considered main cycle. For this, it is necessary to distinguish, in a given step, the local transitions that were actually chosen to build the main cycle (predicate `played`) and the local transitions that are also playable but have not been chosen for this candidate solution. Such local transitions are gathered with the predicate `also_playable` given in lines 116–120. Then, predicate `evolves_in_main_cycle` of lines 122–128 checks when such a transition makes the dynamics evolve from a state of the main cycle to another state still in the main cycle. If this predicate does not exist for any local transition in any step, it means that this transition makes the dynamics evolve outside of the main cycle, which is thus not a trap domain. Such a case is filtered out by the constraint of line 131.

```
115 % Enumerate transitions also playable in a state (but not
    % chosen to build the path)
116 also_playable(Transition, Step) ←
117     not unplayable(Transition, Step),
118     not played(Transition, Step),
119     local_transition(Transition),
120     step(Step).
```

```
121 % Transition allows to go from Step1 to Step2 which is also
    % in the path
122 evolves_in_main_cycle(Transition, Step1, Step2) ←
123   also_playable(Transition, Step1),
124   target(Transition, Automaton, LevelK),
125   active(level(Automaton, LevelK), Step2),
126   cycle_step(Step2),
127   Step2 != Step1 + 1,
128   1 = { different_states_on(Step1, Step2, _) }.
129 % Constraint: remove answer sets where a local transition is
    % playable and
130 % allows to escape the main cycle (not an attractor)
131 ← also_playable(Transition, Step), not
      evolves_in_main_cycle(Transition, Step, _).
```

We stress out the limitation of the piece of code above: although it is correct for the asynchronous case, line 128 is not completely correct for the synchronous case. Indeed, this line makes the assumption that all global transitions are made of exactly one local transition, which is not the case in general for the synchronous semantics. However, as explained above, it works in several cases, some of which are predictable, and, if it can miss legitimate solutions, it never returns erroneous ones.

Finally, the following line tells Clingo to only show the instances of the active predicate that actually encompasses the states of the final attractors:

```
132 #show active/2.
```

To conclude, we note that at this point, an attractor will always be output in a duplicated manner. Indeed, the pieces of code above enumerate all initial states, all possible main cycle lengths, and all possible dynamical branchings. Any path in this enumeration that covers an attractor is returned by the solver, and therefore, a given attractor will be returned under different forms as it can be covered in different manners. This is due to the fact that each answer set is oblivious of the other ones, and there is no possibility to stop the computation *at the ASP scripting level* when an attractor has been found under at least one form. In the next section, we will provide two ways to filter out these spurious solutions in order to output each attractor only once.

Example 7.12 In the dynamics of the networks presented in Figure 7.1 with the asynchronous semantics, let us consider the following cycle of length 2, which can be seen in Figure 7.2: $\langle a_0, b_1, c_0, d_0 \rangle \rightarrow \mathcal{U}^{\text{asyn}} \langle a_0, b_1, c_0, d_2 \rangle \rightarrow \mathcal{U}^{\text{asyn}} \langle a_0, b_1, c_0, d_0 \rangle$. As this cycle is an attractor, following the pieces of program given above, we can use the following command line and try to enumerate it:

```
clingo 0 --const n=2 asynch.lp attractors.lp../
models/asp/example.lp
```

The two answer sets returned are given below, and the predicates have been reordered for readability:

```
Answer: 1
active(level("a",0),0)    active(level("b",1),0)
active(level("c",0),0)    active(level("d",0),0)
active(level("a",0),1)    active(level("b",1),1)
active(level("c",0),1)    active(level("d",2),1)
active(level("a",0),2)    active(level("b",1),2)
active(level("c",0),2)    active(level("d",0),2)
main_cycle_length(2) played(t10,0) played(t12,1) played(t10,2)

Answer: 2
active(level("a",0),0)    active(level("b",1),0)
active(level("c",0),0)    active(level("d",2),0)
active(level("a",0),1)    active(level("b",1),1)
active(level("c",0),1)    active(level("d",0),1)
active(level("a",0),2)    active(level("b",1),2)
active(level("c",0),2)    active(level("d",2),2)
main_cycle_length(2) played(t12,0) played(t10,1) played(t12,2)
```

Consider the first answer set (`Answer: 1`). The three states in the cycle are labeled 0, 1, and 2, and the active local states they contain are described by the predicate `active` (see lines 45–46, lines 80–82, and line 84). We note that states 0 and 2 are actually identical. Two additional predicates are shown to give supplementary information: predicate `played` shows the transitions (labeled `t10` and `t12`, see lines 17 and 19) allowing us to run through all the states of the cycle, while predicate `main_cycle_length` (see line 39) gives the size of the cycle analyzed (here equal to n, but it could be lesser).

Consider now the second answer set (`Answer: 2`): it actually describes exactly the same attractor but covered starting from the initial state $\langle a_0, b_1, c_0, d_2 \rangle$ instead of $\langle a_0, b_1, c_0, d_0 \rangle$. This second answer set, although correct, can be considered spurious. This kind of duplicated solutions is of course more present when considering the answer sets of bigger size (having more possible initial states) or with internal dynamical branchings (i.e., complex attractors). Indeed, increasing the value of n to 4 allows us to obtain 11 answer sets that in fact represent only the two attractors of the model: four answer sets for the attractor of size 2 (two similar to above and two using a cycle of size N = 4) and seven for the attractor of size 4.

7.5.3.3 Python Scripting

The ASP code presented up to this point consists in the complete ASP program of our solution. However, as shown in Example 7.12, this code produces a lot of duplicated solutions: one for each possible initial state and possible traversal. Filtering out these duplicates is very troublesome in "pure" ASP. Given the implementation choices, it is even impossible, as the different answer sets do not "communicate" one with each other. Python scripting here comes in handy. In the following, we propose a Python script that, used conjointly with the previous ASP code, offers two different approaches to filter out unwanted solutions. In both cases, the scripting simply takes the form of a Python `for` loop that awaits the production of the next answer set. Once such an answer set is raised by Clingo, it is stored in a variable, ready to be processed by Python's classical imperative scripting.

7.5.3.3.1 Post-filtering

With the first approach, each answer set obtained by the Python script is processed in order to represent the contained attractor under a normalized form using a dictionary. This attractor is then compared to the current set of attractors that have already been found. If it is new, then it is output to the terminal and added to the set of found attractors; otherwise, it is simply discarded. This is a mere post-filtering because all answer sets are still enumerated by Clingo, but only a part of them are considered "legitimate" solutions and are actually output.

7.5.3.3.2 Pre-filtering

With the second approach, every answer set obtained by the Python script is processed: each state contained in the attractor is translated into an ASP constraint on step 0 and added to the ASP program being solved. For instance, if an attractor $\{\langle a_1, b_1\rangle\langle a_1, b_2\rangle\}$ is found, the following constraints are added:

```
← active(level("a", 1)), 0), active(level("b", 1)), 0).
← active(level("a", 1)), 0), active(level("b", 2)), 0).
```

Each constraint allows us to never visit the related states of this attractor again, thus ensuring that no duplicate of this attractor will be returned by Clingo. These constraints are arbitrarily applied to step 0 of the traversals, but any other step would lead to the same result. This can be considered pre-filtering as the next answer sets enumerated by Clingo will never be duplicates of already found attractors because all their states are now "forbidden." We note that this approach is possible because attractors are all disjoint (which can be easily proven knowing that an attractor is both a trap domain and a SCSG). Moreover, with this pre-filtering, no more post-filtering is needed.

To call one of these scripting enhancements, simply add to the Clingo command line the `filtering-attractors.lp` script, along with a special constant to define which filtering to apply:

```
clingo 0 --const n=<length> --const filtering=
{pre|post} <model.lp> {asynch|synch}.lp attrac-
tors.lp filtering-attractors.lp
```

7.6 Case Studies

In this section, we exhibit several experiments conducted on biological networks. We first detail the results of our programs on the AAN toy model of Figure 7.1, and on another four-component model of a real system, the bacteriophage lambda. Finally, we sum up the results of benchmarks performed on other models up to 101 components. In general, the time performances are good and the overall results confirm the applicability of ASP for the verification of formal properties or the enumeration of some dynamical patterns in biological systems, although the biggest model of 101 components highlights the limits of this approach.

All experiments were performed on a desktop PC running Ubuntu 18.04 with an Intel Core i7-8565U processor (8 cores at 1.80 GHz) and 8 GB memory. The default settings of Clingo were used, including multithreading options.

7.6.1 Toy Example

We first conducted detailed experiments on the four-component model of Figure 7.1 for the asynchronous semantics only. As detailed in Section 7.4, this network contains 4 automata and 12 local transitions. Its asynchronous state-transition graph comprises 36 different global states and some of them are detailed in the partial state-transition graph of Figure 7.2.

The analytic study of the minimal trap domains on this small network allows us to find the following attractors and stable states for the asynchronous semantics:

- stable states: $\langle a_1, b_1, c_1, d_0 \rangle, \langle a_1, b_1, c_0, d_0 \rangle$, and $\langle a_0, b_0, c_0, d_1 \rangle$;
- attractor of length 2: $\{\langle a_0, b_1, c_0, d_0 \rangle and \langle a_0, b_1, c_0, d_2 \rangle\}$;
- attractor of length 4: $\{\langle a_1, b_2, c_1, d_1 \rangle, \langle a_0, b_2, c_1, d_1 \rangle, \langle a_0, b_2, c_1, d_0 \rangle, and \langle a_1, b_2, c_1, d_0 \rangle\}$.

When given to Clingo, the ASP programs given in the previous sections output the expected solutions. The output for the stable-state enumeration (following the scripts of Section 7.5.2) was given in Example 7.11.

An attractor enumeration (following the scripts of Section 7.5.3) with additional information (printing of the solutions as Python structures and computation of the number of attractors of each size) can be performed by executing the following command line:

```
clingo 0 --const n=4 --const print_solutions=1
--const print_solution_sizes=1 asynch.lp
attractors.lp filtering-attractors.lp../models/asp/
example.lp
```

Pre-filtering is used by default. The computation ends in about 100 ms and the output provided by the Python filtering script is given below, slightly refactored for readability:

```
*** Solutions found: 2; skipped: 0
Automata names in order:
['a', 'b', 'c', 'd']
Unique attractors:
{frozenset({(0, 1, 0, 2), (0, 1, 0, 0)}),
 frozenset({(1, 2, 1, 1), (1, 2, 1, 0), (0, 2, 1, 0),
 (0, 2, 1, 1)})}
Frequencies of attractor sizes:
{2: 1, 4: 1}
```

Each information is given under a Python-compatible format for easy re-use. The information given is as follows:

- the name of each automaton in the order used after;
- the set of attractors under the form of frozen sets (immuable Python sets) where each tuple gives the local state of each automaton in order;
- a dictionary giving the sizes of the attractors and the number of attractors of this size (here: one attractor of size 2 and one of size 4).

Clingo also prints these solutions under the form of predicates (not reproduced here). Note that when the filtering script is not used, the same results are found, but each attractor is duplicated several times.

The computation above requires the knowledge of the length of the biggest attractor (here, 4) in order to give an adequate value to n. When this value is unknown and the model is small enough, it is possible to use bigger values of n to search for bigger attractors. For instance, when using n = 40, the computation ends in about six seconds and yields the same results. The total number of global states in the model can be considered a possible higher value. However, although it is very unlikely, attractors of bigger length might still exist. Moreover, for big models, this number of global states is too big to give good performances.

The same analysis can be performed for the synchronous semantics by replacing `asynch.lp` by `synch.lp` in the command line above, and the results are also compatible with Figure 7.3; thus, no attractors are missed in this case.

7.6.2 Bacteriophage Lambda

This section focuses on applying our methods in detail to a real model of small size that has been studied in existing works, namely, the *bacteriophage lambda*, also known as *lambda phage*.

The bacteriophage lambda is a virus of particular biological interest as it can show two very different responses when infecting host bacteria: either a usual "lytic" response in which the host cell is destroyed to help the virus reproduction or a "lysogenic" response, where the virus DNA is simply merged into the host's without fatal outcome. Four genes, named cI, cII, cro, and N, have been known to be of particular importance in this process. Among them, the switch of gene cI is decisive in the choice between the lytic and lysogenic responses.

To model this behavior, a small model containing exactly these four genes of interest was proposed by Thieffry and Thomas (1995), where the complete dynamics is given as a René Thomas model with discrete parameters. In this model, one of these genes is supposed to have four discrete expression levels, another to have three, and the remaining two are Boolean (two expression levels). Therefore, this model contains 48 different states. This state graph is explored in Thieffry and Thomas (1995) using the classical asynchronous semantics, and the authors show the presence of a stable state and an attractor of size 2, corresponding, respectively, to the lysogenic and lytic responses.

To test this model with our implementation, we fetch it from the GINsim repository[3] and use the following tools to translate it to a logic program representation:

- The existing GINsim[4] tool allows us to export GINML models into the *SBML-qual*[5] formalism;
- The existing bioLQM library[6] (Paulevé, 2016b, Naldi et al., 2015) can convert *SBML-qual* models into AAN models;
- Finally, our script an2asp.py, provided with the supplementary material, converts AAN models into ASP programs, following the principles detailed in Section 7.5.1.

It is noteworthy that each step fully preserves the dynamics between models regarding the asynchronous semantics (Chatain et al., 2014); thus, the final ASP program is bisimilar, under the asynchronous semantics, to the original GINML model.

---
3 http://ginsim.org/node/47
4 http://ginsim.org/
5 http://colomoto.org/formats/sbml-qual.html
6 https://github.com/colomoto/bioLQM

At this point, we can apply our stable-state and attractor enumerations to the final file. The stable-state enumeration outputs a unique stable state, which is valid for both the asynchronous and synchronous semantics. This stable state is coherent with the literature and is known to characterize the lysogenic response:

$$\langle cI = 2; cII = 0; cro = 0; N = 0 \rangle.$$

The attractor enumeration with the asynchronous semantics has been performed for n = 48 (the size of the state space) as it can be considered to enumerate all attractors almost certainly. This is of course only possible because the models are small. The computation finishes in 21 seconds and is also coherent with the literature as it outputs the unique following attractor, corresponding to the lytic response:

$$\{\langle cI = 0; cII = 0; cro = 2; N = 0 \rangle, \langle cI = 0; cII = 0; cro = 3; N = 0 \rangle\}$$

Finally, the attractor enumeration with the synchronous semantics can also be applied, although the original model was not originally created to be used with this semantics. This enumeration, once again performed for n = 48, produces the two following attractors:

$$\{\langle cI = 0; cII = 0; cro = 2; N = 0 \rangle, \langle cI = 0; cII = 0; cro = 3; N = 0 \rangle\}$$

$$\{\langle cI = 2; cII = 0; cro = 1; N = 0 \rangle, \langle cI = 1; cII = 0; cro = 0; N = 0 \rangle\}$$

7.6.3 Benchmarks on Models Coming from the Literature

The problem of finding attractors in a discrete network is NP-hard; therefore, the implementation that we give in this work also faces such a complexity. However, ASP solvers (namely, Clingo in our case) are specialized in tackling such complex problems. This section is dedicated to the results of several computational experiments that we performed on biological networks. We show that our ASP implementation can notably return results in only a few seconds for attractors of small size even on models with 100 components, which is considered large.

We do not give the details of the results of these experiments but rather focus on the computation times and the number of attractors found. We used several preexisting Boolean and multivalued networks inspired from real organisms and found in the literature:

- *Lambda phage*: as detailed in Section 7.6.2 (Thieffry and Thomas, 1995);
- *Trp-reg*: a qualitative model of regulated metabolic pathways of the tryptophan biosynthesis in *Escherichia coli* (Sim ao et al., 2005);

- *Fission yeast*: a cell cycle model of Schizosaccharomyces pombe (Davidich and Bornholdt, 2008);
- *Mamm.*: a mammalian cell cycle model (Fauré et al., 2006);
- *Tcrsig*: a signaling and regulatory network of the TCR signaling pathway in the mammalian differentiation (Klamt et al., 2006);
- *T-helper*: a model of the T-helper cell differentiation and plasticity, which accounts for novel cellular subtypes (Abou-Jaoudé et al., 2015).

Two approaches were used to obtain the models studied in this section. The first one consisted in downloading them from the GINsim model repository[7] (Chaouiya et al., 2012), in GINML format, and applying the automated translations detailed in Section 7.6.2 in order to obtain a model in ASP format. The second method consisted to simply manually translate the model from the literature into AAN and finally only use the final step of this process. The characteristics of each model once translated in AAN are given in Table 7.1. The results of our benchmarks[8] are given in Table 7.2 for the stable-state enumeration and Tables 7.3 and 7.4 for the attractor enumeration. For stable-state enumeration (Table 7.2), the following command line has been used:

```
clingo --quiet=1 0 stable-states.lp <model file>
```

in order to search for all solutions while avoiding their output on the terminal, which considerably slows down the execution when a lot of them are found. Regarding attractor enumeration, the script benchmarks.sh has been used, which itself calls command lines of the form:

```
clingo 0 --const n=<length> --const filtering=
{no|pre|post} --const write_nbr_solutions=<output
file> --quiet=2 --time-limit=100 {synch|asynch}.lp
attractors.lp filtering-attractors.lp <model file>
```

Finally, it has to be noted that the scripts can also be used to search for the mere *presence* of an attractor of length at most n, by changing the first argument of Clingo to 1 (i.e., search for one solution) instead of 0 (i.e., search for all solutions). The computation is then much faster as the first solution found triggers the end of the execution.

---

[7] http://ginsim.org/models_repository
[8] All programs and benchmarks are available as additional files and at https://zenodo.org/record/6534531.

Table 7.1 Brief description of the models used in our benchmarks: number of automata ($|\Sigma|$), maximal local level in the automata ($\max(b(\Sigma))$), number of local transitions ($|\mathcal{T}|$) and number of states in the state-transition graph ($|\mathcal{S}|$).

Models	Model description			
	$|\Sigma|$	$\max(b(\Sigma))$	$|\mathcal{T}|$	$|\mathcal{S}|$
Example	4	2	12	36
Lambda phage[a]	4	3	46	48
Trp-reg[b]	4	2	14	36
Fission yeast[c]	9	2	43	$3 \times 2^9 = 1\,536$
Mamm.[d]	10	1	34	$2^{10} = 1\,024$
Tcrsig[e]	40	1	85	$2^{40} \simeq 10^{12}$
T-helper[f]	101	2	316	$2^{102} \simeq 5.7 \times 10^{31}$

References of the models:
a) (Thieffry and Thomas, 1995) – http://ginsim.org/node/47
b) (Sim ao et al., 2005) – manually translated.
c) (Davidich and Bornholdt, 2008) – http://ginsim.org/node/37
d) (Fauré et al., 2006) – manually translated.
e) (Klamt et al., 2006) – http://ginsim.org/node/78
f) (Abou-Jaoudé et al., 2015) – manually translated.

Table 7.2 Results of our stable-state enumeration implementation.

Models	Stable-state enumeration for both semantics	
	Time (ms)	Number
Example	<5	3
Lambda phage	<5	1
Trp-reg	<5	2
Fission yeast	<5	1
Mamm.	<5	1
Tcrsig	6	7
T-helper	6 774	5 875 504

The successive lines sum up the information regarding models detailed in Table 7.1. For each model, the table shows the computation time for the enumeration of all results and the total number of returned answer sets.

Table 7.3 Results of the attractor enumeration implementation.

		Asynchronous				Synchronous			
		Computation time (in ms) by filtering type			$\|\mathcal{A}\|$	Computation time (in ms) by filtering type			inf
Model	n	None	Post	Pre	Pre	None	Post	Pre	$\|\mathcal{A}\|$[a] Pre
Example	2	20	14	16	1	21	16	16	2
	5	25	26	22	2	24	21	19	2
	10	311	307	88	2	45	49	35	2
	15	7 587	7 608	224	2	67	59	53	2
Lambda phage	2	17	16	17	1	17	17	17	2
	5	65	65	63	1	35	35	36	2
	10	576	567	585	1	115	262	105	2
	15	773	787	788	1	206	204	171	2
Trp-reg	2	13	13	13	0	13	14	13	0
	5	21	22	21	1	18	19	18	1
	10	38	40	38	1	71	71	67	1
	15	62	63	67	1	55	57	52	1
Fisson yeast	2	16	17	17	0	18	17	18	1
	5	57	55	55	0	34	34	34	1
	10	345	340	336	0	192	105	178	1
	15	1 003	1 004	972	0	185	179	178	1
Mamm.	2	16	16	16	0	16	16	16	0
	5	36	36	37	0	31	31	31	0
	10	237	246	243	0	123	125	114	1
	15	2 945	2 903	3 024	0	221	228	187	1
Tcrsig	2	25	25	25	0	26	26	26	0
	5	96	101	93	0	98	98	99	0
	10	1 043	1 024	1 030	0	501	526	436	1
	15	13 109	12 983	13 076	0	1 276	1 223	1 136	1

Table 7.3 (Continued)

		Asynchronous				Synchronous			
		Computation time (in ms) by filtering type				Computation time (in ms) by filtering type			
					$\|A\|$				inf
Model	n	None	Post	Pre	Pre	None	Post	Pre	$\|A\|$[a] Pre
T-helper	2	154	149	150	0	T.O.	T.O.	T.O.	70 544[+,b]
	5	2 656	2 617	2 569	0	T.O.	T.O.	T.O.	73 428[+,b]
	10	92 384	92 315	92 548	0	T.O.	T.O.	T.O.	41 038[+,b]
	15	T.O.	T.O.	T.O.	0+	T.O.	T.O.	T.O.	21 792[+,b]

The first column gives the model name (cf. Table 7.1), the second column (n) gives the maximum size of the sought attractors, and columns 3–6 (resp. 7 to end) give the results of the computation of the attractors with respect to the asynchronous (resp. synchronous) semantics. For each semantics: • The three first columns (**Computation time**) give the total computation time in milliseconds for each filtering mode: no filtering, post-filtering, and pre-filtering. T.O. means that the timeout of 100s has been reached, thus cutting the computation and outputting the number of attractors already found. • The last column ($|A|$ or **inf** $|A|$) gives the total number of attractors computed, according to the pre-filtering. This value is the same for post-filtering, except when there is a timeout. A trailing "+" means that more attractors could possibly be found with more computation time.
a) As the computation for the synchronous semantics might miss some solutions, this value is only an inferior bound of the actual value.
b) The decrease in the number of solutions is an effect of the timeout only as increasing the value of n can only lead to find more solutions.

7.7 Conclusion

In this chapter, we emphasized the merits of ASP, a powerful declarative programming paradigm, for the analysis of dynamical biological systems. Being able to verify properties on the dynamics of biological systems is crucial in many ways. First, it helps validating the models that are designed, thanks to biological expertise or raw data. Indeed, the confrontation between the knowledge on the dynamics of a system and the actual behavior of the constructed model allows us to discriminate the valid models. Second, it gives new knowledge about the system. It is thus possible to discover previously unknown dynamical properties or to obtain useful information for conducting new biological experiments. In this chapter, we decided to focus on a challenging, yet rewarding, scientific issue consisting in the study of attractors.

Table 7.4 Sizes of enumerated attractors.

Model	Asynchronous	Synchronous[a]
Example	1 attractor of size 2 and 1 attractor of size 4	2 attractors of size 2
Lambda phage	1 attractor of size 2	2 attractors of size 2
Trp-reg	1 attractor of size 4	1 attractor of size 4
Fission yeast	No attractor	1 attractor of size 2
Mamm.	No attractor	1 attractor of size 7
Tcrsig	No attractor	1 attractor of size 7
T-helper	No attractor (T.O.)	20512^+ attractors of size 3 and 1280^+ attractors of size 9 (T.O.)[b],[c]

The enumeration has been performed with pre-filtering only and a maximum size of $n = 15$. The first column gives the model name (cf. Table 7.1) and the second (resp. third) column details the sizes of attractors found for the asynchronous (resp. synchronous) semantics. T.O. means that the timeout of 100s has been reached, thus cutting the computation and outputting the number of attractors already found. The "+" means that more attractors could theoretically be found with more computation time.

a) As the computation for the synchronous semantics might miss some solutions, these values are only inferior bounds of the actual values.
b) Because of the timeout, the sum of these values does not correspond to the total in the corresponding line $n = 15$ of Table 7.3. A possible explanation is that this benchmark included an additional attractor size counting step, which leads to less solutions found before timeout.
c) Surprisingly, no attractor of size 2 is found here, despite being found when searching with $n = 2$ (see Table 7.3). This is probably due to Clingo's internal optimization heuristics that were left as defaults for these benchmarks.

We presented a logical approach to efficiently compute the list of all stable states and attractors in BRNs. We formalized our approach using the AAN framework, which is bisimilar to many logical networks (Chatain et al., 2014). All results given here can thus be applied to the widespread Thomas' modelling (Thomas, 1973) in the asynchronous semantics and to the Kauffman modelling in the synchronous semantics (Kauffman, 1969). In addition, this framework could encompass any update rules, such as the ones represented in Gershenson (2004) and Noual and Sené (2017).

In biological models, the identification of attractors is critical, as it gives an insight into the long-term behavior of biological systems. By combining ASP and Python scripting (an interaction recently introduced in the ASP solver we use, Clingo), we exhibited an efficient method to enumerate stable states, cycles, and attractors of large models. The computational framework is based on the AAN formalism assuming non-deterministic dynamics. The major benefit of such a method is to get an exhaustive enumeration of all potential states while still being tractable for models with a hundred of interacting components. This method covers fully the asynchronous semantics but only partially the synchronous semantics. Yet, even when synchronous semantics is considered, it identifies correctly simple attractors and a subset of complex attractors.

This work could be extended by considering adaptations and optimizations of the approach to address larger models. On the one hand, the method can be improved, for instance, by generalizing it to comprehensively tackle the synchronous semantics, along with others, or by developing optimizations to tackle even larger models. On the other hand, close problems and similar dynamical patterns can also be considered for enumeration, such as basins of attraction, gardens of Eden, or bifurcations (Fippo-Fittime et al., 2016).

Acknowledgments

We would like to thank Olivier Roux for his previous involvement in this work.

References

Emna Ben Abdallah, Maxime Folschette, Olivier Roux, and Morgan Magnin. Exhaustive analysis of dynamical properties of biological regulatory networks with answer set programming. In *2015 IEEE International Conference on Bioinformatics and Biomedicine (BIBM)*, pages 281–285. IEEE, 2015.

Emna Ben Abdallah, Maxime Folschette, Olivier Roux, and Morgan Magnin. ASP-based method for the enumeration of attractors in non-deterministic synchronous and asynchronous multivalued networks. *Algorithms for Molecular Biology*, 12 (1): 1–23, 2017.

Wassim Abou-Jaoudé, Pedro T. Monteiro, Aurélien Naldi, Maximilien Grandclaudon, Vassili Soumelis, Claudine Chaouiya, and Denis Thieffry. Model checking to assess T-helper cell plasticity. *Frontiers in Bioengineering and Biotechnology*, 2, 2015.

Tatsuya Akutsu, Sven Kosub, Avraham A. Melkman, and Takeyuki Tamura. Finding a periodic attractor of a Boolean network. *IEEE/ACM Transactions on Computational Biology and Bioinformatics*, 9 (5): 1410–1421, 2012.

Réka Albert and Hans G. Othmer. The topology of the regulatory interactions predicts the expression pattern of the segment polarity genes in *Drosophila melanogaster*. *Journal of Theoretical Biology*, 223 (1): 1–18, 2003.

Chitta Baral. *Knowledge Representation, Reasoning and Declarative Problem Solving*. Cambridge University Press, 2003. ISBN 0521818028.

Nikolaos Berntenis and Martin Ebeling. Detection of attractors of large Boolean networks via exhaustive enumeration of appropriate subspaces of the state space. *BMC Bioinformatics*, 14 (1): 1, 2013.

Claudine Chaouiya, Aurelien Naldi, and Denis Thieffry. Logical modelling of gene regulatory networks with GINsim. In *Bacterial Molecular Networks: Methods and Protocols* (eds.Jacques van Helden, Ariane Toussaint and Denis Thieffry), pages 463–479. New York: Springer 2012.

Thomas Chatain, Stefan Haar, Loïg Jezequel, Loïc Paulevé, and Stefan Schwoon. Characterization of reachable attractors using Petri net unfoldings. In *International Conference on Computational Methods in Systems Biology*, pages 129–142. Springer, 2014.

Stéphanie Chevalier, Christine Froidevaux, Loïc Paulevé, and Andrei Zinovyev. Synthesis of Boolean networks from biological dynamical constraints using answer-set programming. In *2019 IEEE 31st International Conference on Tools with Artificial Intelligence (ICTAI)*, pages 34–41. IEEE, 2019.

Francis S. Collins, Michael Morgan, and Aristides Patrinos. The human genome project: lessons from large-scale biology. *Science*, 300 (5617): 286–290, 2003.

Maria I. Davidich and Stefan Bornholdt. Boolean network model predicts cell cycle sequence of fission yeast. *PloS ONE*, 3 (2): e1672, 2008.

Pablo Moisset de Espanés, Axel Osses, and Ivan Rapaport. Fixed-points in random Boolean networks: the impact of parallelism in the Barabási–Albert scale-free topology case. *Biosystems*, 150: 167–176, 2016.

Elena Dubrova and Maxim Teslenko. A SAT-based algorithm for computing attractors in synchronous Boolean networks. *arXiv preprint arXiv:0901.4448*, 2009.

Elena Dubrova and Maxim Teslenko. A SAT-based algorithm for finding attractors in synchronous Boolean networks. *IEEE/ACM Transactions on Computational Biology and Bioinformatics*, 8 (5): 1393–1399, 2011.

Adrien Fauré, Aurélien Naldi, Claudine Chaouiya, and Denis Thieffry. Dynamical analysis of a generic Boolean model for the control of the mammalian cell cycle. *Bioinformatics*, 22 (14): e124–e131, 2006.

Louis Fippo-Fittime, Olivier Roux, Carito Guziolowski, and Loïc Paulevé. Identification of bifurcations in biological regulatory networks using answer-set programming. In *Constraint-Based Methods for Bioinformatics Workshop*, 2016.

Maxime Folschette, Loïc Paulevé, Morgan Magnin, and Olivier Roux. Sufficient conditions for reachability in automata networks with priorities. *Theoretical Computer Science*, 608: 66–83, 2015.

Abhishek Garg, Luis Mendoza, Ioannis Xenarios, and Giovanni DeMicheli. Modeling of multiple valued gene regulatory networks. In *2007 29th Annual International Conference of the IEEE Engineering in Medicine and Biology Society*, pages 1398–1404. IEEE, 2007.

Martin Gebser, Roland Kaminski, Benjamin Kaufmann, Max Ostrowski, Torsten Schaub, and Philipp Wanko. Theory solving made easy with Clingo 5, 52, 2016.

Michael Gelfond and Vladimir Lifschitz. The stable model semantics for logic programming. In *ICLP/SLP*, pages 1070–1080, 1988.

Carlos Gershenson. Updating schemes in random Boolean networks: do they really matter. In *Artificial Life 9th Proceedings of the Ninth International Conference on the Simulation and Synthesis of Living Systems*, pages 238–243. MIT Press, 2004.

Aitor González, Claudine Chaouiya, and Denis Thieffry. Logical modelling of the role of the Hh pathway in the patterning of the *Drosophila* wing disc. *Bioinformatics*, 24 (16): i234–i240, 2008.

Morihiro Hayashida, Takeyuki Tamura, Tatsuya Akutsu, Shu-Qin Zhang, and Wai-Ki Ching. Algorithms and complexity analyses for control of singleton attractors in Boolean networks. *EURASIP Journal on Bioinformatics & Systems Biology*, 2008 (1): 1, 2008.

Sui Huang, Gabriel Eichler, Yaneer Bar-Yam, and Donald E. Ingber. Cell fates as high-dimensional attractor states of a complex gene regulatory network. *Physical Review Letters*, 94 (12): 128701, 2005.

David James Irons. Improving the efficiency of attractor cycle identification in Boolean networks. *Physica D: Nonlinear Phenomena*, 217 (1): 7–21, 2006.

Stuart A. Kauffman. Metabolic stability and epigenesis in randomly constructed genetic nets. *Journal of Theoretical Biology*, 22 (3): 437–467, 1969.

Tarek Khaled and Belaid Benhamou. An ASP-based approach for Boolean networks representation and attractor detection. In *LPAR*, pages 317–333, 2020.

Steffen Klamt, Julio Saez-Rodriguez, Jonathan A. Lindquist, Luca Simeoni, and Ernst D. Gilles. A methodology for the structural and functional analysis of signaling and regulatory networks. *BMC Bioinformatics*, 7 (1): 1, 2006.

Hannes Klarner, Alexander Bockmayr, and Heike Siebert. Computing maximal and minimal trap spaces of Boolean networks. *Natural Computing*, 14 (4): 535–544, 2015.

Konstantin Klemm and Stefan Bornholdt. Stable and unstable attractors in Boolean networks. *Physical Review E*, 72 (5): 055101, 2005.

Alexandre Lemos, Inês Lynce, and Pedro T. Monteiro. Repairing Boolean logical models from time-series data using answer set programming. *Algorithms for Molecular Biology*, 14 (1): 1–16, 2019.

Mushthofa Mushthofa, Gustavo Torres, Yves Van de Peer, Kathleen Marchal, and Martine De Cock. ASP-G: An ASP-based method for finding attractors in genetic regulatory networks. *Bioinformatics*, page btu481, 2014.

Aurélien Naldi, Pedro T. Monteiro, Christoph Müssel, Hans A. Kestler, Denis Thieffry, Ioannis Xenarios, Julio Saez-Rodriguez, Tomas Helikar, Claudine Chaouiya, et al. Cooperative development of logical modelling standards and tools with CoLoMoTo. *Bioinformatics*, page 1154–1159, 2015.

Mathilde Noual and Sylvain Sené. Synchronism versus asynchronism in monotonic Boolean automata networks. *Natural Computing*, pages 1–10, 2017. ISSN 1572-9796. doi: http://dx.doi.org/10.1007/s11047-016-9608-8.

Loïc Paulevé. Goal-oriented reduction of automata networks. In *International Conference on Computational Methods in Systems Biology*, volume 9859 of *Lecture Notes in Bioinformatics*, pages 252–272. Springer, 2016a.

Loïc Paulevé. Pint, a static analyzer for dynamics of automata networks. In *14th International Conference on Computational Methods in Systems Biology (CMSB 2016)*, 2016b.

Loïc Paulevé, Morgan Magnin, and Olivier Roux. Refining dynamics of gene regulatory networks in a stochastic π-calculus framework. In *Transactions on Computational Systems Biology XIII* (eds. Corrado Priami, Ralph-Johan Back, Ion Petre and Erik de Vink), pages 171–191. Springer, 2011.

Loïc Paulevé, Courtney Chancellor, Maxime Folschette, Morgan Magnin, and Olivier Roux. Analyzing large network dynamics with process hitting. In *Logical Modelling of Biological Systems*, (eds. Luis Fariñas del Cerro and Katsumi Inoue), pages 125–166. Hoboken: Wiley 2014.

Hongyang Qu, Qixia Yuan, Jun Pang, and Andrzej Mizera. Improving BDD-based attractor detection for synchronous Boolean networks. In *Proceedings of the 7th Asia-Pacific Symposium on Internetware*. ACM, 2015.

E. Sim ao, Elisabeth Remy, Denis Thieffry, and Claudine Chaouiya. Qualitative modelling of regulated metabolic pathways: application to the tryptophan biosynthesis in *E. coli*. *Bioinformatics*, 21 (Suppl. 2): ii190–ii196, 2005.

Thomas Skodawessely and Konstantin Klemm. Finding attractors in asynchronous Boolean dynamics. *Advances in Complex Systems*, 14 (03): 439–449, 2011.

Roland Somogyi and Larry D. Greller. The dynamics of molecular networks: applications to therapeutic discovery. *Drug Discovery Today*, 6 (24): 1267–1277, 2001.

Denis Thieffry and René Thomas. Dynamical behaviour of biological regulatory networks–II. Immunity control in bacteriophage lambda. *Bulletin of Mathematical Biology*, 57 (2): 277–297, 1995.

René Thomas. Boolean formalization of genetic control circuits. *Journal of Theoretical Biology*, 42 (3): 563–585, 1973. ISSN 0022-5193.

René Thomas. Regulatory networks seen as asynchronous automata: a logical description. *Journal of Theoretical Biology*, 153 (1): 1–23, 1991.

Andrew Wuensche. Genomic regulation modeled as a network with basins of attraction. In *Pacific Symposium on Biocomputing*, volume 3, pages 89–102, 1998.

Shu-Qin Zhang, Morihiro Hayashida, Tatsuya Akutsu, Wai-Ki Ching, and Michael K. Ng. Algorithms for finding small attractors in Boolean networks. *EURASIP Journal on Bioinformatics & Systems Biology*, 2007 (1): 1–13, 2007.

Zheng Zhao, Chian-Wei Liu, Chun-Yao Wang, and Weikang Qian. BDD-based synthesis of reconfigurable single-electron transistor arrays. In *Proceedings of the 2014 IEEE/ACM International Conference on Computer-Aided Design*, pages 47–54. IEEE Press, 2014.

8

Hybrid Automata in Systems Biology

Alberto Casagrande[1], Raffaella Gentilini[2], Carla Piazza[3], and Alberto Policriti[3]

[1] Department Mathematics and Geosciences, University of Trieste, Trieste, Italy
[2] Department of Mathematics and Computer Science, University of Perugia, Perugia, Italy
[3] Department of Mathematics, Computer Science, and Physics, University of Udine, Udine, Italy

8.1 Introduction

In this chapter, we will discuss a number of formal tools collectively driving us to *Hybrid Automata*. An hybrid automaton is a formalism introduced and developed with the aim of integrating the discrete and continuous ingredients in a single simulation tool. This integration goal is – no wonder – ubiquitous, and it comes as no surprise that it turned out as an especially valuable instrument for the systems biology arena.

Automata have a long and rich history in computer science, and they have been used in a variety of ways to render (formally) many basic and natural ideas. Some of these ideas lie at the ground of the very notion of computation and every computer scientist gets in touch with (finite state) automata early on in her/his career. We are convinced that an early exposure to the automata formalism can also be instructive for any scientists interested in using flexible tools for simulation, especially in bioinformatics and systems biology. Going back to basic, consider, for example, a situation in which you want to model a machine operating in a given (fixed, fully specified) environment *and* using the minimum possible amount of memory. Such a minimal memory can be "distilled" in the ability of the machine to recognize itself in a finite number of possible *states*. Each of such states can then be characterized functionally by a list of accessible transitions, letting the machine choose among a number of accessible states to be reached, if chosen, at the next instant in time.

Systems Biology Modelling and Analysis: Formal Bioinformatics Methods and Tools,
First Edition. Edited by Elisabetta De Maria.
© 2023 John Wiley & Sons, Inc. Published 2023 by John Wiley & Sons, Inc.

8 Hybrid Automata in Systems Biology

Consider the following toy example:

Example 8.1 *(Escherichia coli – Automaton)*
Escherichia coli, also known as *E. coli*, is a bacterium that has been deeply studied since the 1960s. Among its characteristic behaviors, there is the ability of moving to search nourishment by using some appendages called *flagella*. According to the chemical gradient sensed in its environment, *E. coli* operates its flagella motors to alter its motion: it either "runs" and moves in a straight line by moving its flagella counterclockwise (CCW) or it "tumbles," randomly changing its heading by moving its flagella clockwise. Figure 8.1 reports a finite state automaton representing the *E. coli* motion.

As the reader can easily see, even from this simple example a number of issues arise as soon as we start playing with such a simulation tool. For example: is it reasonable or helpful to let our (finite state) machine change its – internal – state *non-deterministically*? Consider, for instance, the situation in the above example in which the sensed gradient is close to a predefined threshold.

Admitting that we want to use non-determinism, is this advantageous in terms of the computation power demand of our tool?

Turning back to our view of automata as limited memory equipped devices, should we want to be able to add more memory power? How can we realize such an increase in potential? How does Nature realize such an upgrade?

Returning to the previous example, consider the case in which we wish to model an *E. coli* decision mechanism keeping into account also quantitative aspects and thresholds. Suppose, for example, that we want a model that takes into account the viscosity of the environmental fluid and the consequent amount of energy necessary for movement. In such a model, every state must be equipped with an energy consumption function and a threshold that, when reached, forces the transition to a different state.

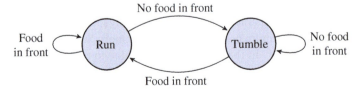

Figure 8.1 A finite state automaton describing the motion of the bacteria *E. coli*.

Notice that this expressive-enrichment of our basic formalism is an indirect way to equip our model with more "memory."

Finally and most importantly, in admitting a *discrete* change of state as in the two examples above, are we not making some strong assumption on the nature of *time* we are using in our model? This last question is a delicate one and leads us directly to the final question discussed on this chapter: transitions and the underlying notion of time are fully defined by the observable events we have been able to explicitly model?

Assume that the environmental conditions triggering *E. coli* reactions, and consequent movements, are so fine-grained that we do not want to explicitly model them in our automaton's states. The situation is such that some of the transitions in our model can only be described by a probability distribution based on a sufficiently rich sampling of different scenarios. Our final section will then be discussing a further *stochastic* dimension that can be added to the automata model. Such dimension needs to be coordinated with the other kinds of knowledge on states and transitions already modelled in the remaining part of the automaton. It is crucial to define such a coordinated formalism in a clear and coherent way.

The chapter is organized in four main sections. In the first section, we define a limited amount of formalism that will be used as the basic language of the subsequent material. All the variations on the notion of automaton presented in subsequent sections start with at least some of the formalism discussed in this section. Then, in the section entitled Events, we discuss possible usages of the simple formalism of finite state automata (FSA). The aspects modelled with automata defined in this section are mainly qualitative. Next, in the next two sections, we introduce, respectively, the two final "ingredients" of our discussion: Time and Uncertainty. This path will lead us from FSA to hybrid automata and their stochastic version.

We conclude with a high-level discussion of the potential of hybrid automata for checking and controlling simulations, thereby proving their ability as tester *in silico* of biological hypotheses.

8.2 Basics

In this section, we introduce first-order languages and theories, and we report on some decidability results over them. A more complete presentation of these topics can be found, for instance, in Enderton (2001) and Mendelson (1997).

8.2.1 Languages and Theories

First-order languages are mathematical frameworks that can describe properties and parts of real systems. Any first-order language $\mathcal{L}$ consists in four components: a set of variables *Var*, a set of constants *Const*, a set of functional operators *Funct*, and a set of relational symbols *Rel*. The variables are symbols, such as X or Z_2, that can assume any possible values in a specified domain, for instance, $\mathbb{R}$. They may be used to express unknown quantities, for the sake of example, the concentration of a substance in a solution or the copies of a specific gene expressed by a cell. The constants assume values in the same domain of the variables, and, among other functions, they can model both quantities and thresholds such as the number of hydrogen atoms per water molecule or the water boiling temperature – e.g. 2 and 100 °C, respectively. The cumulative effects on one gene of multiple interacting proteins – being inhibitors or promoters – can be depicted by a functional operator. The operators "+," "*," and "2·" are more common examples for objects in *Funct*. The relational symbols can constrain variable values, specify laws, or describe dependencies among species. For instance, < and = can be used to discriminate environmental conditions – e.g. $X_t <$ 100 °C, where X_t is a variable representing the temperature – or model the evolution of a quantity – e.g. $X' = X + e^T$, where T and X, X' are three variables representing time, the initial quantity of the investigated species, and the quantity of the investigated specie at time T, respectively.

A *term* is either a variable, a constant, or a functional symbol applied to an appropriate number of terms such as X, Y, 100, or $100 * X + e^Y$. An *atomic formula* φ_a is either one of the two symbols $\top$ or $\bot$ (standing for *true* and *false*, respectively) or a relational operator together with the right number of terms. For instance, $X_t < 100$ and $X' = X + e^T$ are two terms, while $X' = X + e^{T,X}$ is not a valid term because the functional symbol $e^.$ exclusively admits one parameter; thus, neither $e^{T,X}$ nor $X + e^{T,X}$ are valid terms.

A *formula* φ in a first-order language $\mathcal{L}$ has the form

$$\varphi ::= \varphi_a \mid \varphi_1 \vee \varphi_2 \mid \neg \varphi_1 \mid \forall X \, \varphi_1$$

where φ_a is an atomic formula of $\mathcal{L}$, X is a variable in *Var*, and both φ_1 and φ_2 are formulae of $\mathcal{L}$ itself. Parentheses are included in syntax to avoid ambiguities between string representations of formulae and, for instance, to distinguish $\neg(\varphi_1 \vee \varphi_2)$ from $(\neg \varphi_1) \vee \varphi_2$. The symbol $\forall$ is the *universal quantifier*.

It is worth to stress that the formulae of any language are purely syntactic objects and they need an interpretation for constants, functional operators, and relational symbols – or as we will see later on a *model* – to be evaluated. However, from an intuitive point of view, we can anticipate that $\varphi_1 \vee \varphi_2$ – e.g. $X_t < 0 \vee X_t > 100$ – is meant to hold whenever at least one among φ_1 and φ_2 holds, the formula $\neg \varphi_1$ – e.g. $\neg(X_t < 0 \vee X_t > 100)$ – holds if and only if φ_1 does not, and the formula $\forall X \varphi_1$ – e.g. $\forall X X^2 \geq 0$ – holds only if φ_1 holds regardless of the value of X.

Some shortcuts are commonly introduced to ease the language usage: the formulae $\varphi_1 \wedge \varphi_2$, $\varphi_1 \rightarrow \varphi_2$, and $\exists X \varphi_1$ are short hands for $\neg(\neg\varphi_1 \vee \neg\varphi_2)$, $(\neg\varphi_1) \vee \varphi_2$, and $\neg\forall X \neg\varphi_1$ – $\exists$ is the *existential quantifier* –, respectively. These formulae intuitively holds whenever both φ_1 and φ_2 hold, whenever either φ_1 does not hold or both φ_1 and φ_2 hold, and whenever φ_1 holds for some value of X.

The same variable may appear many times in a formula. If an occurrence of variable $X \in Var$ in a formula φ lays in a sub-formula of the kind $\forall X \bar{\varphi}$ or $\exists X \bar{\varphi}$, then that occurrence is said to be *bound* or *quantified*. If it is not bound, it is said to be *free*. For instance, the variable X is free in $\exists T, (X' = X + e^T)$, while T is bound. Despite the fact that, from a formal point of view, a variable may have free and bound occurrences at the same time – e.g. $X < 0 \wedge \forall X X^2 > 0$ –, we can safely assume that the occurrences of any variable are either all free or all bound (e.g. see Casagrande et al. (2008), Dang et al. (2015) for details). The set of free variables occurring in the first-order formula φ is denoted by *Free*(φ). A *sentence* is a formula for which all the variables are bound. For instance, $\forall X X^2 \leq 0$ is a sentence, while $\exists T, (X' = X + e^T)$ is not a sentence because both X and X' are free in it. Later on, we may write $\varphi[X_1, \ldots, X_m]$ to remark that the variables $X_1, \ldots, X_m$ are free in φ, e.g. if φ is $\exists T, (X' = X + e^T)$ we may write $\varphi[X']$.

A *model* of a language $\mathcal{L}$ is a non-empty set M called *support*, on which terms will assume their values, and a set of interpretations Const : *Const* → $C \subseteq M$, Funct, and Rel for constants, functional operators, and relational symbols, respectively. For any functional operator f, Funct(f) is a function mapping the interpretations of f's arguments in values of the support, e.g. in commonly used models, Funct(+) is the addiction that takes two values of the support and returns their sum. Whatever is the relational symbol R, Rel(R) instead associates the interpretations of R's arguments with either ⊤ or ⊥ – to be read as true and false, respectively. For instance, Rel(=) is usually the equality relation that maps pairs of equal values in ⊤ and the remaining pairs in ⊥.

Example 8.2 A possible model for the first-order languages having two constants "O" and "W", an n-ary functional operator "+", and a binary relational symbol "=" is the one in which the support is the set of natural numbers, Const(O) and Const(W) are the numbers 0 and 7, respectively, Funct(+) is the standard addition between natural numbers, and Rel(=) is the usual equality relation. In this model, the formula $X + O + O = Y + W + W$[1] means that the value assigned to the variable X must be that of Y, whatever it is, plus 14.

A less common, but still valid, model for the very same language is the one in which the support is the set of all the molecule kinds, Const(O) and Const(W) are the dioxygen – i.e. O_2 – and water – i.e. H_2O – molecules, respectively, Funct(+) denotes one side of a chemical reaction, and Rel(=) is an asymmetric operator that relates the reaction reactants (on the left of "=") to the products (on the right or "="). In this model, the formula $X + O + O = Y + W + W$ states that there exists a chemical reaction having three reactants, among which two molecules of dioxygen, that produce two water molecules and one unknown compound.

Let $\mathcal{M}$ be a model of $\mathcal{L}$ with support M, $\varphi[X_1, \ldots, X_i, \ldots, X_m]$ be a formula of $\mathcal{L}$, and $c \in$ Const. The expression obtained by replacing X_i by c is denoted by $\varphi[X_1, \ldots, X_{i-1}, c, X_{i+1}, \ldots, X_n]$.

The semantics of $\mathcal{L}$-formulae with respect to a model $\mathcal{M}$ is defined in the standard way (e.g. see Enderton (2001), Mendelson (1997)). In particular, if φ_a is an atomic formula, we say that a formula $\varphi_a[c_1, \ldots, c_m]$ holds in $\mathcal{M}$ if applying the interpretations of the constant, functional, and relational operators we obtain the truth value T. The formula $\varphi_1[c_1, \ldots, c_m] \lor \varphi_2[c_1, \ldots, c_m]$ holds in $\mathcal{M}$ if at least one among $\varphi_1[c_1, \ldots, c_m]$ and $\varphi_2[c_1, \ldots, c_m]$ holds in $\mathcal{M}$. The formula $\neg \varphi_1[c_1, \ldots, c_m]$ holds in $\mathcal{M}$ if $\varphi_1[c_1, \ldots, c_m]$ does not. Finally, $\forall X \, \varphi_1[X, c_1, \ldots, c_m]$ holds in $\mathcal{M}$ if the formula $\varphi_1[c, c_1, \ldots, c_m]$ holds for all $c \in M$.

A formula $\varphi[X_1, \ldots, X_m]$ in $\mathcal{L}$ is *satisfiable* in $\mathcal{M}$ if there exists $c_1, \ldots, c_m \in M$ such that $\varphi[c_1, \ldots, c_m]$ holds in $\mathcal{M}$, while it is *valid* if $\varphi[c_1, \ldots, c_m]$ holds in $\mathcal{M}$ for all $c_1, \ldots, c_m \in M$. When the model $\mathcal{M}$ is clear from the context, we will simply say that a formula holds (is satisfiable or is valid, respectively).

Example 8.3 The formula $X + O + O = Y + W + W$ is satisfiable in both the models depicted in Example 8.2. As far as the numerical-based model may concern, the formula becomes valid by replacing 15 and 1 to X and Y, respectively. In the case of the chemical-based model, the formula represents methane combustion for X and Y replaced by methane – i.e. CH_4 – and carbon dioxide – i.e. CO_2.

---

1 Here, the syntax $a + b + \ldots$ has been used in place of $+(a, b, \ldots)$ to better resemble the usual notation.

We refer to those models having M as support, where M is a nonempty set, as *models over M*. When Const : $Const \to C$ is clear from the context, we write $\langle M, C, \text{Funct}, \text{Rel}\rangle$ in place of $\langle M, \text{Const}, \text{Funct}, \text{Rel}\rangle$.

Example 8.4 Consider the language $\mathcal{L}_R \stackrel{\text{def}}{=} \langle \text{Var}, \text{Const}_{\mathbb{Z}}, \{+, *\}, \{\geq\}, \text{Ar}\rangle$ where $\text{Const}_{\mathbb{Z}} = \{\ldots, \bar{-1}, \bar{0}, \bar{1}, \ldots\}$ is a set of symbols used to represent the integer numbers (e.g. the standard Arabic number system or the Latin system).

The tuple $\mathcal{M}_R \stackrel{\text{def}}{=} \langle \mathbb{R}, \mathbb{Z}, \text{Funct}, \text{Rel}\rangle$, where Funct and Rel are the usual interpretations for $\{+, *\}$ and $\{\geq\}$, respectively, and any $\bar{i} \in \text{Const}_{\mathbb{Z}}$ is interpreted as $i \in \mathbb{Z}$, is a model for $\mathcal{L}_R$.

It is easy to see that for any formula φ_R in the language $\mathcal{L}_R$, there exists a formula φ_0 in the language $\mathcal{L}_0 \stackrel{\text{def}}{=} \langle \text{Var}, \{\bar{0}, \bar{1}\}, \{+, *\}, \{\geq\}, \text{Ar}\rangle$ such that φ_R is satisfiable in $\mathcal{M}_R$ if and only if φ_0 is satisfiable in $\mathcal{M}_0 \stackrel{\text{def}}{=} \langle \mathbb{R}, \{0,1\}, \text{Funct}, \text{Rel}\rangle$. In this sense, $\mathcal{M}_0$ is equivalent to $\mathcal{M}_R$ and, despite providing symbols and constant values limited in number, it can be used in place of the latter without any loss in generality.

A sentence φ is a *logical consequence* of a set Γ of sentences (denoted, $\Gamma \vDash \varphi$) if for each model $\mathcal{M}$, it holds that if each formula of Γ is valid in $\mathcal{M}$ ($\mathcal{M} \vDash \Gamma$), then φ is valid in $\mathcal{M}$ too. Because of the completeness of first-order logic, we may equivalently say that φ is provable from Γ (see Enderton (2001), Mendelson (1997)). A *theory* $\mathcal{T}$ is a set of sentences such that if $\mathcal{T} \vDash \varphi$, then $\varphi \in \mathcal{T}$. Given a language $\mathcal{L}$ and a model $\mathcal{M}$, the *complete theory* $\mathcal{T}(\mathcal{M})$ of $\mathcal{M}$ is the set of all the sentences of $\mathcal{L}$ that are valid in $\mathcal{M}$. We indicate the complete theory of a model $\mathcal{M} \stackrel{\text{def}}{=} \langle M, C, \text{Funct}, \text{Rel}\rangle$, where $\text{Funct} \stackrel{\text{def}}{=} \{f_0, \ldots, f_n\}$ and $\text{Rel} \stackrel{\text{def}}{=} \{r_0, \ldots, r_m\}$, by either $\mathcal{M}$ itself or $\langle M, C, f_0, \ldots, f_n, r_0, \ldots r_m\rangle$. For each model $\mathcal{M}$ and for each sentence φ, either $\varphi \in \mathcal{T}(\mathcal{M})$ or $\neg\varphi \in \mathcal{T}(\mathcal{M})$. Two formulae $\varphi_1[X]$ and $\varphi_2[Y]$, where X and Y are two vectors of variables, are equivalent with respect to a theory $\mathcal{T}$ if it holds that $\mathcal{T} \vDash \forall X, Y(\varphi_1[X] \vDash \varphi_2[Y])$. We say that a theory $\mathcal{T}$ admits the so-called *elimination of quantifiers*, if, for any formula φ, there exists a quantifier free formula ϱ such that φ is equivalent to ϱ with respect to $\mathcal{T}$. If there exists an algorithm for deciding whether a sentence φ belongs to $\mathcal{T}$ or not, we say that $\mathcal{T}$ is *decidable*. The complete theory $\mathcal{T}(\mathcal{M})$ is decidable if and only if both the satisfiability and the validity of formulae in $\mathcal{M}$ are decidable.

Example 8.5 Let us consider the formula $\varphi \stackrel{\text{def}}{=} \exists X (aX^2 + bX + C = 0)$. It is well known that, in the theory $\langle \mathbb{R}, \{0,1\}, +, *, \geq \rangle$, φ holds if and only if the non-quantified formula $b^2 - 4ac \geq 0$ does the same.

Given a language $\mathcal{L}$ and a model $\mathcal{M}$ of $\mathcal{L}$ with support M, we say that a set $S \subseteq M^k$ is *definable* if and only if there exists a formula $\varphi[X_1, \ldots, X_k]$ such that $\varphi[p_1, \ldots, p_k]$ holds in $\mathcal{M}$ if and only if $(p_1, \ldots, p_k) \in S$.

Definition 8.1 (*O-Minimal Theory – van den Dries and Miller (1996)*)
Let $\mathcal{L}$ be a first-order language whose set of relational symbols includes a binary symbol $\geq$ and let $\mathcal{M}$ be a model of $\mathcal{L}$ in which $\geq$ is interpreted as a linear order. The theory $\mathcal{T}(\mathcal{M})$ is *order minimal*, or simply *o-minimal*, if every set definable in $\mathcal{T}(\mathcal{M})$ is a finite union of points and intervals (with respect to $\geq$).

The class of o-minimal theories includes many interesting theories over $\mathbb{R}$. Below we recall a few of them.

The *semi-algebraic theory* $\mathbb{R} = \langle \mathbb{R}, 0, 1, +, *, \geq \rangle$ admits elimination of quantifiers and it is decidable (Tarski (1951)). However, so far, the most time-efficient algorithm to decide whether a formula φ belongs to $\langle \mathbb{R}, 0, 1, +, *, \geq \rangle$ takes exponential time on $O(\Pi_i^l O(n_i))$ where $l - 1$ is the number of quantifier alternations in φ and n_i is the number of variables quantified by the same quantifier in the ith alternation block (see Basu (1997), Basu et al. (1996)).

The theory $\mathbb{R}_{an} = \langle \mathbb{R}, 0, 1, +, *, (f)_{f \in an}, \geq \rangle$, which adds to $\langle \mathbb{R}, 0, 1, +, *, \geq \rangle$ the set an of all the real-analytic functions from $[-1,1]^n$ to $\mathbb{R}$, can describe the behavior of some periodic trajectories such as sine and cosine functions in a bound interval. This theory is model complete (van den Dries (1986)) and, by Khovanskiĭ's finiteness theorem (see Khovanskiĭ (1980)), it is also o-minimal. The theory $\mathbb{R}_{an}$ with the function $1/x$, where $1/0 = 0$ by convention, admits the elimination of quantifiers (van den Dries et al. (1994)).

The theory $\mathbb{R}_{\exp} = \langle \mathbb{R}, 0, 1, +, *, e^x, \geq \rangle$ extends $\langle \mathbb{R}, 0, 1, +, *, \geq \rangle$ with the the exponential function e^x. This theory is model complete (Wilkie (1996)) and, as a direct consequence of Khovanskiĭ's results (Khovanskiĭ (1980)), it is also o-minimal. Van den Dries proved that any extension of $\langle \mathbb{R}, 0, 1, +, *, \geq \rangle$ that adds a family of total real analytic functions admits the elimination of quantifiers if and only if such functions are semi-algebraic (van den Dries (1998)). If Schanuel's conjecture (see Chudnovsky (1984), Macintyre (1991)) holds, then $\mathbb{R}_{\exp}$ is decidable (Macintyre and Wilkie (1996)).

The semi-algebraic theory extended by exponential operator and analytic functions, $\mathbb{R}_{an,\exp} = \langle \mathbb{R}, 0, 1, +, *, (f)_{f \in an}, e^x, \geq \rangle$, is model complete and o-minimal (see, e.g. van den Dries and Miller (1994)). Moreover, the theory $\mathbb{R}_{an,exp,log} = \langle \mathbb{R}, 0, 1, +, *, (f)_{f \in an}, e^x, \log x, \geq \rangle$ admits the elimination of quantifiers (van den Dries et al. (1994)). Finally, semi-algebraic theory extended by total C^∞ functions is o-minimal only for Pfaffian functions (Wilkie (1999)).

8.3 Events

A possible way to model real-world systems is to abstract all the quantitative part of their evolution and exclusively focus on the sequences of discrete events that they produce and that led to some outcome. The *E. coli* model depicted in Figure 8.1 adopts this approach: as described by the model itself, no matter the concentration of food in front of the bacterium or how far it is, if there is food in front of an *E. coli*, then it runs forward; it there is not, it tumbles.

Labeled transition systems describe the evolution of the investigated systems in terms of their transitions from one of its states to another one taking into account the conditions that must be satisfied to achieve this evolution.

Definition 8.2 *(Labeled transition systems – Keller (1976))* A *labeled transition system* is a triple $(S, \Lambda, \rightarrow)$ where

- S is the set of states of the system;
- Λ is the set of labels;
- $\rightarrow \subseteq S \times \Lambda \times S$ is the set of labeled transitions.

Labeled transition systems move from one state to another one according to the relation $\rightarrow$: if $s \in S$ has been reached, then the system can proceed to any state s' such that $(s, \alpha, s') \in \rightarrow$ (also written as $s \xrightarrow{\alpha} s'$); s and s' are the source and the destination of the transition $s \xrightarrow{\alpha} s'$. The transition label refers either to the condition triggering the transition itself or the event produced by it. A *trajectory* is a sequence of transitions $s_0 \xrightarrow{\alpha_0} \ldots s_i \xrightarrow{\alpha_i} \ldots$, being either finite or infinite in number. If $s_0 \xrightarrow{\alpha_0} \ldots$ is a trajectory, then the sequence of states $s_0, s_1, \ldots$ (also written as $(s_i)_{i \in \mathbb{N}}$) is a *path*.

FSA (also known as *Finite State Machines* (FSM)) are finite – in terms of both states and labels – labeled transition systems having an *initial state* and some *final states*. A *run* for any of these automata is a path that begins in the initial state and ends in one among the final states.

Once the investigated system has been described by using a labeling transition system, one might want to formally verify some of its properties or forecast its behaviors under particular conditions. Kripke's structures are a specific kind of (un)labeled transition systems, proposed in the 1960s (Kripke (1963)) and widely embraced in the 1980s (e.g. see Lichtenstein and Pnueli (1985), Emerson and Lei (1985), Clarke et al. (1986)), which have been deeply investigated from this point of view. They are specifically meant to represent reactive systems, i.e. transition systems that moves from one state to another and whose evolution lasts indefinitely long, potentially, forever.

Definition 8.3 *(Kripke's structure – Clarke et al. (1999))* Let AP be a set of atomic propositions. A *Kripke's structure* is a tuple (S, S_0, R, L), where

- S is a finite set of states;
- $S_0 \subseteq S$ is the set of initial states;
- $R \subseteq S \times S$ is a transition relation such that for any $s \in S$, there exists a s' with $(s, s') \in R$;
- $L : S \rightarrow 2^{AP}$ is a labeling function that associates each state with the set of atomic propositions that hold in it.

Kripke's structures (S, S_0, R, L) start their evolution from any of the states in S_0 and, as in general labeled transition systems, they move from one state to another according the relation R: if the state $s \in S$ has been reached, then the Kripke's structure can proceed to any state s' such that $(s, s') \in R$ (also written $s \rightarrow s'$).

The condition imposed on R by definition guarantees the existence of infinite trajectories from any state of the model and, thus, it enables Kripke's structures to represent never-ending evolutions in the portrayed systems.

Any labeled transition system having a finite number of both states and labels can be mapped in a Kripke's structure, which exhibits equivalent behaviors by adding Λ to the set of atomic propositions, by including to the model a sink state s_Ω such that $s_\Omega \rightarrow s_\Omega$ and which should be reachable by any states without outgoing transitions in the original model, and by splitting every transition $e = (s, \alpha, s')$ in two transitions $s \rightarrow s_e$ and $s_e \rightarrow s'$ such that s_e is not included among the set of original states and $L(s_e) = \alpha$.

Example 8.6 *(E. coli – Kripke's structure)*
In Example 8.1, we described a simple automaton representing the *E. coli* motion. In Figure 8.2, we map that automaton into a Kripke's structure.

Kripke's structures cannot natively represent triggering transition labels. Nevertheless, they can be easily be simulated by adding the transition

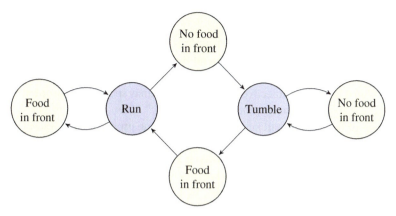

Figure 8.2 A Kripke's structure representing the *E. coli* motion model depicted in Figure 8.1.

labels to the set of atomic propositions and by splitting every transition $e = (s, \alpha, s') \in \rightarrow$ in two new transitions, $s \rightarrow s_e$ and $s_e \rightarrow s'$, such that s_e is a new state and $L(s_e) = \alpha$. In the depicted Kripke's structure, the light-gray-colored nodes encode transition labels; the dark-gray ones are those which were already present in the original transition system.

More formally, for any labeled transition system $T \stackrel{\text{def}}{=} (S, \Lambda, \rightarrow)$ such that both S and Λ are finite in cardinally, there exists a Kripke's structure $K_T \stackrel{\text{def}}{=} (S \cup S_E \cup \{s_\Omega\}, S, \rightarrow_E \cup \rightarrow_\Omega, L_T)$, where

- $S_E \stackrel{\text{def}}{=} \{s_e \mid e \in \rightarrow \text{ and } s_e \notin S\}$;
- $s_\Omega \notin S \cup S_\Lambda$
- $\rightarrow_E \stackrel{\text{def}}{=} \{(s, s_e) \mid e = (s, \alpha, s')\} \cup \{(s_e, s') \mid e = (s, s_\alpha, s')\}$;
- $\rightarrow_\Omega \stackrel{\text{def}}{=} \{(s, s_\Omega) \mid s \in S \text{ and } \forall s' \in S \ \forall \alpha \in \Lambda \ (s, \alpha, s') \notin \rightarrow\}$;
- L_T is such that $L_T(s_e) = \{\alpha\}$ for any $e = (s, \alpha, s') \in \rightarrow$.

It is easy to see that K_T has a trajectory $s_0 \rightarrow s_{e_0} \rightarrow \ldots s_i \rightarrow s_{e_i} \ldots$ where $s_i \neq s_\Omega$ and $e_i = (s_i, \alpha_i, s_{i+1})$ for all $i \in \mathbb{N}$ if and only if T has a trajectory $s_0 \stackrel{\alpha_0}{\rightarrow} \ldots s_i \stackrel{\alpha_i}{\rightarrow} \ldots$. Moreover, $s_0 \rightarrow s_{\alpha_0} \rightarrow \ldots s_i \rightarrow s_{\alpha_i} \ldots$ is a trajectory for K_T such that $s_i \neq s_\Omega$ and $s_j = s_{e_j} = s_\Omega$ for all $j > i$ if and only if $s_0 \stackrel{e_0}{\rightarrow} \ldots s_i$ is a trajectory in T and s_i is not the source of any transition. In this sense, K_T and T exhibit the very same behaviors and they are equivalent. Even though this encoding does not preserve the paths of the investigated model, we can deduce the paths of T from those of K_T.

8.3.1 Temporal Logics

Many of the properties of Kripke's structures related to their evolution can be expressed by *temporal logics* that extend the propositional logic and express properties of transition sequences. Besides using atomic propositions and traditional Boolean connectives, temporal logics may specify time properties such as either "*eventually* property p will hold," "*from now on* property q holds," or "property r will *never* hold," These time properties are described using specific *temporal operators* and *path quantifiers*.

CTL* (Emerson and Halpern (1983)) is a specific temporal logic that distinguishes between two kinds of properties: *state properties*, which may hold on single states, and *path properties*, which instead are features of paths rather than single states. This distinction is preserved in formulas.

Behind the classical logical connectives ¬ (logical negation) and ∨ (non-exclusive disjunction), CTL* also makes use of the path quantifier A – to be read as "on all paths" – and of the temporal operators ° and U – to be read as "next" and "until," respectively. Intuitively, the formula Aπ, where π is a formula describing a path property, holds on a given state s if and only if π holds for all the infinite paths starting from s itself. The formulae °π and $\pi U \pi'$ instead hold on a path p when the first node in p satisfies π and when π is satisfied along the path until π' holds, respectively.

Definition 8.4 *(CTL* - Syntax – Emerson and Halpern (1983))* CTL* *state formulæ* have the form

$$\varphi ::= \top \mid p \mid \neg\varphi \mid \varphi \vee \varphi \mid A\pi$$

where p is an atomic proposition and π is a CTL* path formula. CTL* *path formulæ* have the form

$$\pi ::= \varphi \mid \neg\pi \mid \pi \vee \pi \mid °\pi \mid \pi U \pi$$

where φ is a state formula.

Other connectives and quantifiers, such as ∧, E – to be read "on some paths" –, and ◊ – "eventually" –, are normally used for convenience; however, they can be directly derived from those introduced in Definition 8.4. As a matter of facts, the formulae ⊥, $\varphi_1 \wedge \varphi_2$, $\varphi_1 \rightarrow \varphi_2$, $\varphi_1 \vDash \varphi_2$, $\varphi_1 R \varphi_2$, ◊φ, □φ, and Eφ are used as short-hands for ¬⊤, ¬(¬φ_1 ∨ ¬φ_2), ¬φ_1 ∨ φ_2, ($\varphi_1 \rightarrow \varphi_2$) ∧ ($\varphi_2 \rightarrow \varphi_1$), ¬(¬$\varphi_1 U \neg\varphi_2$), $\top U \varphi$, ¬($\top U \neg\varphi$), and ¬(A¬φ), respectively.

Any CTL* formula may or may not hold depending on the context over which it is evaluated: a state for state formulae or a path for path formulæ.

Definition 8.5 *(CTL* - Semantics – Emerson and Halpern (1983))*
Let $K = (S, S_0, R, L)$ be a Kripke's structure. We say that $s \in S$ *satisfies* the CTL* state formula φ on K, denoted by $K, s \vDash \varphi$, if:

- $K, s \vDash \top$ always holds;
- $K, s \vDash p$ if and only if $p \in L(s)$;
- $K, s \vDash \neg \varphi$ if and only if $K, s \vDash \varphi$ does not hold;
- $K, s \vDash \varphi_1 \vee \varphi_2$ if and only if $K, s \vDash \varphi_1$ or $K, s \vDash \varphi_2$;
- $K, s \vDash \mathsf{A}\pi$ if and only if $K, (s_i)_{i \in \mathbb{N}} \vDash \pi$ for all the paths $(s_i)_{i \in \mathbb{N}}$ in K such that $s_0 = s$.

If $(s_i)_{i \in \mathbb{N}}$ is a path in K, we say that $(s_i)_{i \in \mathbb{N}}$ *satisfies* the CTL* path formula π in K, denoted by $K, (s_i)_{i \in \mathbb{N}} \vDash \pi$ under the following conditions:

- $K, (s_i)_{i \in \mathbb{N}} \vDash \varphi$ if and only if $K, s_0 \vDash \varphi$;
- $K, (s_i)_{i \in \mathbb{N}} \vDash \neg \pi$ if and only if $K, (s_i)_{i \in \mathbb{N}} \vDash \pi$ does not hold;
- $K, (s_i)_{i \in \mathbb{N}} \vDash \pi_1 \vee \pi_2$ if and only if $K, (s_i)_{i \in \mathbb{N}} \vDash \pi_1$ or $K, (s_i)_{i \in \mathbb{N}} \vDash \pi_2$;
- $K, (s_i)_{i \in \mathbb{N}} \vDash {}^\circ\pi$ if and only if $K, (s_i)_{i \in (\mathbb{N} \setminus \{0\})} \vDash \pi$;
- $K, (s_i)_{i \in I} \vDash \pi_1 U \pi_2$ if and only if there exists $k \in \mathbb{N}$ such that $K, s_k \vDash \pi_2$ and $K, s_j \vDash \pi_1$ for each $0 \leq j < k$.

Computation tree logic (CTL) (Emerson and Clarke (1982), Ben-Ari et al. (1983)) and *linear temporal logic* (LTL) (Pnueli (1977, 1979, 1981)) are two useful CTL* sub-logics. The former can express properties over all the paths starting in a given state; the latter specifies features that hold along a single path only.

CTL and CTL* share the same syntax for the state formulæ, while they differ on the path formulae which, on CTL, must have the form

$$\pi ::= \varphi \mid \neg \pi \mid {}^\circ\pi \mid \pi U \pi$$

On the contrary, LTL exclusively admits state formulae of the form $\mathsf{A}\pi$ and its path formulae must be derivable from:

$$\pi ::= \top \mid p \mid \neg \pi \mid \pi \vee \pi \mid {}^\circ\pi \mid \pi U \pi$$

where p is an atomic proposition. The semantics of both CTL and LTL remain unaltered with respect to CTL*.

The property defined by the LTL formula $\mathsf{A}\Diamond\Box p$ cannot be expressed in CTL and, on the other hand, the CTL formula $\mathsf{A}\Diamond\mathsf{A}\Box p$ has no analogous in LTL (Clarke and Draghicescu (1988)). This is enough to conclude that the CTL* formula $\mathsf{A}\Diamond\Box p \vee \mathsf{A}\Diamond\mathsf{A}\Box p$ can be expressed neither by CTL nor by LTL.

8.3.2 Model Checking

Model checking (Clarke et al. (1986, 1999)) is a formal verification technique originally developed to prove alleged Kripke's structure features. This approach, differently from the other ones suggested in the literature, not only verifies whether a state temporal formula holds on a state of a given structure, but, if the formula itself does not hold, it produces a counterexample for it, i.e. a state where a sub-formula of the considered formula does not hold.

Clarke et al. (1986) proposed an algorithm for model checking any CTL formula φ in time $O(|\varphi| * (|S| + |R|))$, where $|\varphi|$ is the number of sub-formulae of φ and S and R are the set of states and transitions, respectively, of the considered Kripke's structure. The CTL model checking problem is P-complete (e.g. see Schnoebelen (2002)).

On the contrary, LTL model checking is PSPACE-complete (Sistla and Clarke (1985)) and there exists a tableau method to solve it in time $O((|S| + |R|) * 2^{O(|\varphi|)})$ (Lichtenstein and Pnueli (1985)). CTL* model checking has the same complexity of LTL model checking (Sistla and Clarke (1985)).

All the cited model checking algorithms rely on two features of Kripke's structures that are a consequence of their finiteness: the ability to decide *reachability* of one state from another one, i.e. the existence of a path connecting them, and the possibility to reduce the infinite paths of a Kripke's structure to finite paths on the underlying graph in which each strongly connected component is contracted to one single node.

8.4 Events and Time

Predicting the behavior of complex living systems involves the consideration of various qualitative, quantitative, and temporal aspects and may rely on different formal methods, depending on the level of detail required and biological questions addressed. The attempt of modelling and predicting the dynamics of interacting genes, for example, has classically been addressed through roughly different approaches (Richard et al. (2006), Barbuti et al. (2020)), where hybrid automata bridge the gap between previous purely discrete (e.g. the well-known Thomas model in Thomas (1978) and Thieffry and Thomas (1995)) or purely continuous model classes. In gene regulatory networks (GRN), genes may reach different expression levels, depending on the states of other genes, called the regulators: thus, activations and inhibitions events are triggered conditionally on the proper expression levels of these regulators. In other words, triggering events – driving the overall discrete time evolution – arise naturally in GRN from some threshold concentrations being reached, while the continuous evolution of trajectories inside the regions confined by such thresholds can be easily formalized and computationally studied.

Overall, GRN stand as a meaningful context to illustrate the usage of events and time within hybrid automata, where the latter formalism provides a powerful tool to study collections of continuous trajectories representing concentrations. Their modelling by means of hybrid automata will therefore be the main subject of the next subsection.

8.4.1 Hybrid Automata and Gene Regulatory Networks

The schematic diagram in Figure 8.3 describes a simple GRN that will be used in this subsection to illustrate the modelling of GRN by means of hybrid automata, following in particular the formalism and notation proposed in Batt et al. (2007a,b).[2] In the GRN in Figure 8.3, the genes a and b code for two repressor proteins A and B. In detail, each protein represses the expression of the other gene above different threshold concentrations. Although simple, this two-gene GRN represent some typical features of regulation, such as inhibition and cross-regulation, and could be found as a module of more complex GRN from a real biological system.

The state of a general n-genes regulatory network is represented by a set of n variables $X = \{x_1, \ldots, x_n\}$, associated with the concentration of the n genes' products of the network. The induced state space $\mathcal{X}$ is therefore an hyper-rectangular subset of $\mathbb{R}^n$: $\mathcal{X} = \Pi_{i=1}^{n}[0, \max_i]$, where $\max_i$ denotes a maximal concentration of the protein encoded by gene i. An analogous hyper-rectangular $\mathcal{P} = \Pi_{j=1}^{m}[\min_{p_j}, \max_{p_j}]$ can also be used to describe a set of possibly unknown quantities or parameters of the networks. X is assumed to evolve, for $i \in \{1, \ldots, n\}$, according to Eq. (8.1), where the functions $\text{Prod}_i(X, P)$ and $\text{Deg}_i(X, P)$ model the production and degradation of each gene. In particular, since the sigmoidal functions involved in gene regulation can be approximated by sign or ramp functions (cfr. Figure 8.4),

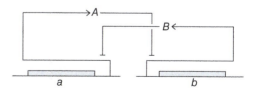

Figure 8.3 Two-gene cross-inibition GRN.

---

[2] The embedding transition systems in Batt et al. (2007a,b) for GRN are easily recast into the hybrid automata formalism, as shown in Bortolussi and Policriti (2008).

8 Hybrid Automata in Systems Biology

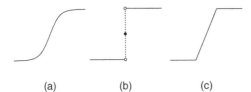

Figure 8.4 (a) Sigmoid curve and (b) sign and (c) ramp function approximations.

$\text{Prod}_i(X, P)$ and $\text{Deg}_i(X, P)$ turn out to be piecewise multi-affine functions, that is multi-variables polynomial of degree at most 1 in each of the variables.

$$\dot{X}_i = f_i(X, P) = \text{Prod}_i(X, P) - \text{Deg}_i(X, P) \quad (8.1)$$

Clearly, the approximation illustrated in Figure 8.4 underlying gene regulation guarantees that the general dynamics in Eq. (8.1) for GRN is amenable to be formalized and computationally studied by means of hybrid automata. As shown in the following example, it also possible to use a temporal logic such as LTL or CTL to formalize a number of expected or unwanted behaviors of the considered GRN. Such properties are fundamental to tune up the parameters involved in the study.

Example 8.7 (*Two- GRN – Batt et al. (2007a)*)
Approximating by ramp functions $r^-(x_a, \theta_a^1, \theta_a^2), r^-(x_b, \theta_b^1, \theta_b^2)$, the sigmoid curves involved in the cross-inhibition of the GRN in Figure 8.3 leads to rewriting Eq. (8.1) as follows:

$$\dot{x}_a = \kappa_a r^-(x_b, \theta_b^1, \theta_b^2) - \gamma_a x_a$$
$$\dot{x}_b = \kappa_b r^-(x_a, \theta_a^1, \theta_a^2) - \gamma_b x_b$$

where κ_a and κ_b (resp. γ_a and γ_b) are synthesis (resp. degradation) parameters and $\theta_a^1, \theta_a^2, \theta_b^1$, and θ_b^2 are threshold concentrations for proteins a and b. Here, the first equation states that the synthesis of protein a is inhibited by protein b (ramp function $r^-(x_b, \theta_b^1, \theta_b^2)$) and that its degradation is not regulated. The property of being *bistable*, which is known for this network, can be expressed in LTL as follows:

$$\psi = (x_a < \theta_a^1 \wedge x_b > \theta_b^2 \rightarrow G(x_a < \theta_a^1 \wedge x_b > \theta_b^2)) \quad (8.2)$$
$$\wedge (x_b < \theta_b^1 \wedge x_a > \theta_a^2 \rightarrow G(x_b < \theta_b^1 \wedge x_a > \theta_a^2)) \quad (8.3)$$

The hybrid automata formalism for a general n-gene GRN naturally arises by considering the $(n-1)$-dimensional threshold hyperplanes $\{x \in \mathcal{X} \mid x_i = \theta_{l_i}\}, 1 \leq l_i \leq p_i$, where the protein encoded by gene i, $1 \leq i \leq n$, is assumed to have p_i threshold concentrations. Such $(n-1)$-dimensional threshold hyperplanes partition the state-space $\mathcal{X}$ into a set $\mathcal{R}$ of (hyper)rectangular regions. In particular, the regulation of gene expression is identical everywhere in each region $r \in \mathcal{R}$. More in detail, the modelling hybrid automaton will have the following components:

- For each hyper-rectangular region $r \in \mathcal{R}$, there is a location v_r.
- The invariant region of the location l_r is exactly the hyper-rectangular region r and can be easily expressed as conjunctions of inequalities. Activation and reset conditions correspond instead to (region) boundaries and identities, respectively.
- The continuous dynamic associated with each location v_r can be expressed simply requiring the existence of a solution for the system of differential equations (8.1) connecting two states in the region r.
- Finally, $\mathcal{E} \subseteq \mathcal{V} \times \mathcal{V}$ is the transition relation defined by $(v_r, u_{r'}) \in \mathcal{E}$ iff
  1. r and r' are either equal or adjacent regions, and
  2. there exist $x \in r, x' \in r', \tau \in \mathbb{R}_{>0}$ and a solution f of the vector field in Eq. (8.1) such that $f(0) = x, f(\tau) = x'$ and for all $0 < \tau' < \tau$, it holds that $f(\tau') \in r \cup r'$.

By way of illustration, consider the two-genes regulatory network in Figure 8.3. As shown in (Batt et al., 2007a, Fig. 3.(a)), the state space $\mathcal{X}$ is partitioned into nine grid regions by the involved threshold concentrations $\theta_a^1, \theta_a^2, \theta_b^1, \theta_b^2$ and maximal values of concentrations $\max_a, \max_b$. The blue lines in that figure denote both the boundaries and the activation regions. (Batt et al., 2007a, Fig. 3.(b)) represents the graph $\langle \mathcal{V}, \mathcal{E} \rangle$ underlying the hybrid automaton of the GRN in Figure 8.3 for some fixed parameter values (as computed in Batt et al. (2007a)). This graph can be properly labeled and used as a Kripke's structure for checking if the GRN satisfies the expected properties, and eventually tune the choice of the parameters.

In the literature, the approach described so far has been used in several papers to model and study by means of hybrid automata roughly complex GRNs. In Farina and Prandini (2007), the authors study an hybrid automaton model of the *lac operon* regulatory network, responsible for the treatment of lactose by the bacterium *E. coli*, identifying two stationary conditions for the system. The *lux regulon* network controlling the phenomenon of bioluminescence production in the marine bacterium *Vibrio fischeri* has been considered in Belta et al. (2005): the authors use the properties of hybrid multi-affine rectangular systems to compute infeasible sets, by evaluating the vector field at the vertices of the rectangular invariants. Batt et al. (2007a,b) focus on parameter tuning in GRNs using the discrete part of the hybrid automaton model as an LTL structure. Ultimately, the hybrid automaton modelling for GRN used in Batt et al. (2007a,b), Belta et al. (2005), and Farina and Prandini (2007), and described in this section relies on the possibility of partially discretizing a well-known continuous model and can therefore also be applied to different biological systems. Example 8.8 concludes this subsection by illustrating the hybrid modelling in Casagrande et al. (2012) of a neural oscillator, which is basically obtained with exactly the same ingredients used so far for GRN.

In Ahmad et al. (2006), the authors propose a complementary approach to model GRN with hybrid automata, which builds up on the discrete model for GRN by Rene' Thomas (cfr. Thomas (1978), Thieffry and Thomas (1995)) adding continuous variables to track time delays for the increase/decrease of expression levels (otherwise abstracted in the underlying Thomas' model). Such a method has been applied to several GRN that includes the mucus production regulatory network in the bacterium *Pseudomonas Aeruginosa* in Ahmad et al. (2006) and the regulatory network of dengue virus pathogenesis in Aslam et al. (2014). A recent proposal on the modelling of metabolic-regulatory networks can be found in Liu and Bockmayr (2020). A large amount of literature in this direction can be found in the proceedings of the *International Workshop on Hybrid Systems Biology*, where the term "hybrid" is also broadened to other formalisms.

Example 8.8 *(Neural oscillator – Casagrande et al. (2014))*
Neural oscillators are rhythmic and repetitive electrical stimuli central for the activities of several brain regions. The following ordinary differential system illustrates the simple continuous model of a single oscillator proposed in Tonnelier et al. (1999), where X_e (resp. X_i) is the output of the excitatory neuron N_e (resp. inhibitory neuron N_i), τ is a characteristic time constant, and λ is the amplification gain.

$$f(\tau, \lambda) = \begin{cases} \dot{X}_e = -\dfrac{X_e}{\tau} + tanh(\lambda \cdot X_e) - tanh(\lambda \cdot X_i) \\ \dot{X}_i = -\dfrac{X_i}{\tau} + tanh(\lambda \cdot X_e) + tanh(\lambda \cdot X_i) \end{cases}$$

Using a sign function to approximate the non-linear components (that ultimately limit the possibility of analyzing a complex system obtained by composing multiple copies of the model) we obtain

$$f(\tau, \lambda) = \begin{cases} \dot{X}_e = -\dfrac{X_e}{\tau} + sgn(X_e) - sgn(X_i) \\ \dot{X}_i = -\dfrac{X_i}{\tau} + sgn(X_e) + sgn(X_i) \end{cases}$$

Figure 8.5 A 4-locations hybrid automaton modelling a neural oscillator.

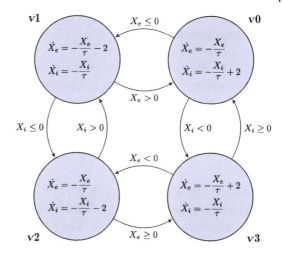

that corresponds to the piecewise affine hybrid model illustrated in Figure 8.5.

Unfortunately, the dynamical behavior of such an hybrid automaton is quite different from the original continuous model. A satisfactory approximation is instead obtained by using the following ramp function to approximate $tan(\lambda \cdot X)$:

$$h_{\lambda,\alpha}(z) = \begin{cases} -1, & \text{if } z < -\frac{\alpha}{\lambda} \\ \frac{\lambda}{\alpha} \cdot z, & \text{if } -\frac{\alpha}{\lambda} \leq z \leq \frac{\alpha}{\lambda} \\ 1, & \text{if } z \geq \frac{\alpha}{\lambda} \end{cases}$$

This leads to a hybrid automaton with nine locations, whose behavior has been computationally studied in Casagrande et al. (2014) using the so-called ϵ-semantics, which roughly takes into account also noise and perturbations (cfr. Section 8.4.2).

8.4.2 Expressibility and Decidability Issues

In order to talk about expressibility and decidability problems, we refer to the following general definition of hybrid automata taken from Casagrande et al. (2014).

Definition 8.6 *(Hybrid automata – syntax)* A *hybrid automaton H of dimension $dim(H) \in \mathbb{N}$ is a tuple $H = \langle \mathbf{X}, \mathbf{X}', T, \mathcal{V}, \mathcal{E}, Inv, Dyn, Act, Res \rangle$ where*

- $X = \langle X_1, \ldots, X_{d(H)} \rangle$ and $X' = \langle X'_1, \ldots, X'_{d(H)} \rangle$ are two tuples of variables ranging over the reals $\mathbb{R}$;
- T is a variable ranging over $\mathbb{R}_{\geq 0}$;
- $\langle \mathcal{V}, \mathcal{E} \rangle$ is a finite directed graph. Each element of $\mathcal{V}$ will be dubbed *location*;
- each location $v \in \mathcal{V}$ is labeled by the two first-order formulae $Inv(v)[X]$ and $Dyn(v)[X, X', T]$ such that if $Inv(v)[\![p]\!]$ is true, then $Dyn(v)[\![p, q, 0]\!]$ is true if and only if $p = q$;
- each edge $e \in \mathcal{E}$ is labeled by the formulae $Act(e)[X]$ and $Res(e)[X, X']$, which are called *activation* and *reset*, respectively.

Intuitively, the formula $Dyn(v)[X, X', T]$ characterizes the dynamics associated with the location v, and it is usually given in terms of a vector field whose solutions are the dynamics. The formula $Inv(v)[X]$ denotes the set of the values admitted during the continuous evolution of the automaton inside v. The formulae $Act(e)[X]$ and $Res(e)[X, X']$ identify the set of continuous values from which the automaton can jump over the edge e and a map that should be applied to the continuous values from which the automaton crosses the edge e.

Formally, the semantics is given as it follows.

Definition 8.7 *(Hybrid automata – semantics)* A *state ℓ of H is a pair $\langle v, r \rangle$, where $v \in \mathcal{V}$ is a location and $s = \langle s_1, \ldots, s_{d(H)} \rangle \in \mathbb{R}^{d(H)}$ is an assignment of values for the variables of X. A state $\langle v, s \rangle$ is admissible if $Inv(v)[\![s]\!]$ holds.* We have two kinds of transitions:

- the *continuous transition relation* $\xrightarrow{t} C$:
  $\langle v, s \rangle \xrightarrow{t} C \langle v, r \rangle \iff$ there exists $f : \mathbb{R}_{\geq 0} \to \mathbb{R}^{d(H)}$ continuous function such that $s = f(0)$, there exists $t \geq 0$ such that $r = f(t)$, and for each $t' \in [0, t]$, both $Inv(v)[\![f(t')]\!]$ and $Dyn(v)[\![s, f(t'), t']\!]$ hold;
- the *discrete transition relation* $\xrightarrow{(v,u)} D$:
  $\langle v, s \rangle \xrightarrow{(v,u)} D \langle u, r \rangle \iff (v, u) \in \mathcal{E}$ and both the formulae $Act((v, u))[\![s]\!]$ and $Res((v, u))[\![s, r]\!]$ holds.

A trace is a sequence of continuous and discrete transitions. A point r is reachable from a point s if there is a trace starting from s and ending in r. We write $\ell \to_C \ell'$ and $\ell \to_D \ell'$ to mean that there exists a $t \in \mathbb{R}_{\geq 0}$ such that $\ell \xrightarrow{t} C\ell'$ and that there exists an $e \in \mathcal{E}$ such that $\ell \xrightarrow{e} D\ell'$, respectively. Moreover, we write $\ell \to \ell'$ to denote either $\ell \to_C \ell'$ or $\ell \to_D \ell'$.

8.4 Events and Time | 325

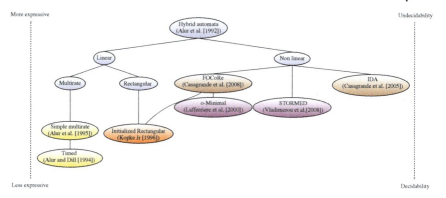

Figure 8.6 Expressibility versus decidability of the reachability problem in hybrid automata. Simple multirate automata, timed automata, o-minimal automata, and STORMED automata have finite bisimulation quotients, while initialized rectangular automata have finite simulation quotient. Neither bisimulation nor simulation quotient is guaranteed to be finite in any of the other classes, including FOCoRe automata and IDA automata. On the other hand, simple multirate automata, timed automata, and initialized automata guarantee the decidability of the problem and the decidability on FOCoRe automata, IDA, o-minimal automata, and STORMED automata depends on the theory used to define the automata themselves. Intriguingly, there exist FOCoRe automata and IDA automata whose bisimulation quotient is infinite and on which the reachability problem is decidable.

The problem of whether a given set of states is reachable from the initial set of states is generally undecidable for hybrid automata (Alur et al. (1993)). Figure 8.6 sketches an overall picture concerning the trade-off between expressibility and decidability (of the reachability problem) in the realm of hybrid automata. Undecidability is not obvious already in the class of hybrid automata where linear constraints over the derivatives define the continuous evolution in each location (the so-called linear hybrid automata in Henzinger (2000)). In this context, rectangular hybrid automata, where the flow of each variable in each location is a rectangular predicate over the derivates, are already on the boundary of the realm of decidability: they are decidable only if *initialized*, i.e. if every variable whose rate constraint is changed upon the traversal of a discrete edge is reset. Lafferriere et al. (2000) proved the existence of a class of decidable hybrid automata not belonging to the broad family of linear hybrid automata, where relevant formulae for continuous and discrete dynamics are formulae of a (decidable) o-minimal theory (cfr. Definition 8.1 in the preliminary section) and each variable is updated to a constant during each discrete jump.[3] *First-order constant reset* (FOCoRe) hybrid automata introduced in Casagrande et al. (2008) generalize (constant reset) o-minimal hybrid automata by considering flows obtained by non-autonomous systems of differential inclusions. While this class does not guarantee finite bisimulation quotient, it preserves the decidability of the reachability problem for systems defined by using a decidable theory. The class of *Independent Dynamics* hybrid automata (IDA), introduced in (Casagrande et al. (2005)), still relies on a (decidable) o-minimal theory to define relevant sets but admits two groups of continuous variables called *dependent* and *independent* variables. Only independent variables are subject to constant reset during discrete jumps. Dependent variables are instead characterized by a mode-independent continuous dynamic. Example 8.9 shows that the framework of IDA hybrid automata is expressive enough to capture the mechanism for movement in *E. coli* also discussed in previous sections. The constant reset limitation is also relaxed in STORMED hybrid systems at the price of semantics conditions on the evolution: if the continuous part is monotonic with respect to some vector and the number of possible discrete transitions is upper-bounded, then the reachability problem on STORMED hybrid systems can be reduced to a formula in the same theory used to define the systems themselves (Vladimerou et al. (2008)).

Example 8.9 *(Escherichia coli – Hybrid automaton – Casagrande et al. (2005))*
E. coli moves in response to the molecular concentration of CheY in a phosphorylated form (Y_P variable in Figure 8.7), which in turn is determined by the bound ligands at the receptors that appear in several forms. The ratio of $y = Y_P/Y_0$ (phosphorylated concentration of CheY w.r.t. its unphosphorylated concentration) determines a bias with an associated probability that flagella will exert a clockwise (CW) rotation. In the model of Figure 8.7, the stochastic effect has been ignored and modelled deterministically. The most important output variable is the angular velocity ω that takes discrete values +1 for CW and −1 for CCW. The more detailed pathway involves other molecules: CheB (either phosphorylated or not, B_p and B_0), CheZ (Z), bound receptors (LT), and unbound receptors (T), while their continuous evolution is determined by a set of differential algebraic equations derived through kinetic mass action formulation. In Figure 8.7, invariants have been omitted and hyper-rectangular regions bounded by the maximum admissible values for the concentrations.

---

3 Without the *constant reset* assumption, o-minimal automata are not decidable anymore.

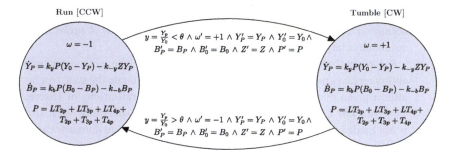

Figure 8.7 A hybrid automaton capturing the run-tumble mechanism of E. coli.

In order to overcome decidability and complexity issues, different abstraction and approximation techniques have been introduced on hybrid automata (see, e.g. Doyen et al. (2018)).

The mentioned results are based on the standard infinite precision semantics of hybrid automata (see Definition 8.7). However, infinite precision does not always exist in nature, where at low concentration levels, one cannot go beyond a single-molecule precision. The ϵ-semantics introduced in Casagrande et al. (2014) aim to obtaining realistic interpretations of hybrid automata behaviors by forbidding to observe the system with infinite precision. As observed in Casagrande et al. (2014), the neural oscillator hybrid automaton of Example 8.8 exhibits an unstable equilibrium in the origin that is never observed in Nature and that can be removed exploiting ϵ-semantics instead of the standard ones.

In this section, we focused our attention on the use of hybrid automata for modelling quite "simple" biological systems. However, being hybrid automata a general and flexible framework, nothing prevent us from using them in more complex scenarios. For instance, *Cancer Hybrid Automata* have been introduced in Loohuis et al. (2014) as a proposal for developing cancer patient-driven therapies. In Cancer Hybrid Automata, driver mutations that lead the progression of the disease provide the discrete components, variables involved in cell functions are the continuous ones, while discrete switches are consequences of either time progression or drug effects.

8.5 Events, Time, and Uncertainty

The notion of *Probabilistic Automaton* dates back to Rabin (1963), where a formal theory of probabilistic automata has been introduced. Probabilistic automata in Rabin's paper generalize the deterministic ones through a probability distribution for each state-symbol pair. Since the acceptance definition depends on a threshold, the expressive power of probabilistic automata is larger than that of classical automata.

Rabin's probabilistic automata are discrete models, where in a sense probability replaces non-determinism when one has to evaluate the state reached after a given event occurs. These are labeled generalizations of Discrete Time Markov Chains (Kemeny and Snell (1976)). In this framework, time is not modelled and only variables ranging within discrete domains can be considered.

On the other hand, different definitions of stochastic automata and timed probabilistic process algebras aim at providing stochastic generalizations of systems involving continuous variables (Hansson (1991), Segala and Lynch (1994)), such as time. In particular, at the low level of description, this can be achieved introducing *Labeled Probabilistic Transition Systems*, which generalize labeled transition systems by partitioning states into *non deterministic* and *probabilistic* (Vardi (1985)). Labeled transitions can only go from the first set of states to the second one and vice versa. Moreover, transitions from probabilistic to non-deterministic states are modelled through probability distributions.

Definition 8.8 *(Labeled probabilistic transition systems)* A *labeled probabilistic transition system is a tuple* $(S, S', \Lambda, \rightarrow, T)$, *where*

- S is the set of *non deterministic* states;
- S' is the set of *probabilistic* states;
- $S \cap S' = \emptyset$;
- Λ is the set of labels;
- $\rightarrow \subseteq S \times \Lambda \times S'$ is the set of non-deterministic transitions;
- $T : S' \rightarrow \mathcal{P}(S)$ is the set of probabilistic transitions, where $\mathcal{P}(S)$ is the set of probability transitions.

As described in D'Argenio and Katoen (2005), time evolution can be modelled in labeled probabilistic transition systems by considering a set of labels of the form $\Lambda = \mathcal{A} \times \mathbb{R}_{>0}$, where $\mathcal{A}$ is the set of event names, while the non-negative real numbers are used to denote the time at which an event occurs. As a consequence, labeled probabilistic transition systems extend generalized semi-Markov processes (Whitt (1980)), which in turn extend the formalism of Continuous Time Markov Chains (Kemeny and Snell (1976)). In Continuous Time Markov Chains, time delays always follow exponential distributions in order to guarantee the memoryless property. In generalized semi-Markov processes, any continuous distribution is allowed for time delays, whereas in labeled probabilistic transition systems also, non-continuous distributions are allowed and non-determinism is introduced.

However, as pointed out by the authors in D'Argenio and Katoen (2005), being labeled probabilistic transition systems a low-level description formalism, soon models become uncountably large and complex to describe. To overcome this limitation, *Stochastic Automata* can be considered. Stochastic Automata have a semantics in terms of labeled probabilistic transition systems, but it is still true that they extend generalized semi-Markov processes. Interestingly Stochastic Automata can also be seen as a generalization of Timed Automata (Alur and Dill (1994)) in which a probability distribution over $\mathbb{R}_{>0}$ is associated with each clock. When a state is reached, some of the clocks are reinitialized according to their probability distributions, then time evolves. As soon as the first clock reaches the value 0, it expires. A transition is enabled when all the clocks in its label are expired. The first enabled transition is taken.

Definition 8.9 **(Stochastic Automata)** A *Stochastic Automaton* is defined as a tuple $(S, \Lambda, C, \rightarrow, \kappa)$, where

- S is the set of states of the system;
- Λ is the set of labels;
- C is the finite set of (random) clocks and each $x \in C$ has an associated distribution F_x;
- $\rightarrow \subseteq S \times (\Lambda \times 2^C) \times S$ is the set of transitions;
- $\kappa : S \rightarrow 2^C$ is the clock setting function.

When the system enters in the state s, the clocks in the set $\kappa(s)$ are randomly initialized each according to its probability distribution. Then, the clocks count down at rate 1 simulating the evolution of time. The first transition involving only clocks that have reached the value 0 is taken and the evolution in a new state starts.

A similar model, named *real-time probabilistic process*, is considered in Alur et al. (1991), where atomic propositions are introduced to label the states of the system and the time spent in a state is bounded. The authors provide a semantics for the logic *TCTL (timed computation tree logic)* over real-time probabilistic processes where an existential quantified path formula is satisfied if the path formula is satisfied with probability greater than 0 and consequently a universal quantified path formula is satisfied if the path formula is satisfied with probability 1. In this framework, the model checking problem is PSPACE-complete.

A probabilistic extension of CTL, named *Probabilistic CTL*, has been first introduced in Hansson and Jonsson (1994) over Discrete Time Markov Chains, whereas the probabilistic extension of TCTL, named *Probabilistic TCTL*, has been considered in Kwiatkowska et al. (1998). A different approach is taken in Ballarini et al. (2011b), where *Hybrid Automaton Stochastic Logic (HASL)* is introduced as a unifying formalism for both model checking and performances evaluation. Such results are at the basis of the development of now well-established verification tools such as PRISM (Kwiatkowska et al. (2006)), UPPAAL (Behrmann et al. (2004)), and COSMOS (Ballarini et al. (2011a)).

While Stochastic Automata and related models mainly focus on stochastic effects on the edge-crossing, in Hu et al. (2000), the notion of Stochastic Hybrid Automata is introduced, adding also the Brownian motion effects on the continuous dynamics. We refer the reader to Agha and Palmskog (2018) and Lavaei et al. (2021) for recent surveys in this context.

In the modelling of biological systems, stochastic hybrid systems are fundamental tools to cope with both unrealistic behaviors exhibited by purely continuous-discrete models (e.g. Zeno and chattering traces) and incomplete knowledge of the underlying parameters. While detailed discussions on the use of stochastic models in systems biology can be found in Zuliani (2015) and Li et al. (2017), here we focus again on some significant examples.

Example 8.10 *(Escherichia coli – Stochastic Hybrid Automaton)*
The model defined in Casagrande et al. (2005) and reported in Figure 8.7 describes *E. coli* movements omitting stochastic effects. Let us assume that we are interested in introducing in the model the following stochastic behavior: the system remains in a given state for a minimum random time period independently of the presence/absence of food. This correctly models the fact that the system cannot keep jumping instantaneously from one state to the other and can be obtained by adding to the system a continuous clock variable C associated with an exponential distribution with parameter λ. The constraint $C = 0$ is added to the activation conditions of both edges, and C is randomly initialized in both states.

Example 8.11 *(Neural oscillator – stochastic effects)*
Let us consider again the neural oscillator of Example 8.8. Beside ϵ-semantics (see Section 8.4.2), another way of removing the unstable equilibrium in the origin would be that of introducing small stochastic perturbations in the equations representing the dynamic of the system, i.e. considering Stochastic Hybrid Automata.

Example 8.12 *(Repressilator – Bortolussi and Policriti (2010))*
Going back to GRN, the repressilator is a well-known system involving three genes/proteins within a feed-back loop. It has been artificially synthesized and studied in Elowitz and Leibler (2000). Being one of the most simple examples exhibiting a periodic behavior, it has been used as a case study in a number of papers on systems biology modelling languages. Purely continuous models based on differential equations show limits in the modelling of the repressilator since only a fine tuning of the parameters ensure the periodicity of the solutions. On the other hand, both hybrid and stochastic hybrid models overcome the problem.

Bortolussi and Policriti (2010) presented a description of the Repressilator in the stochastic Concurrent Constraint Programming (sCCP) process algebra. Semantics of sCCP programs in terms of both hybrid automata and stochastic hybrid automata are defined. As a consequence, translations of the Repressilator in terms of deterministic hybrid automata, non-deterministic hybrid automata, and stochastic hybrid automata are automatically obtained and simulated under different initial conditions. While all models almost always exhibit the correct oscillating traces, thus improving on the purely continuous ones, deterministic hybrid automata still miss the right behavior when at the beginning of the simulation the three proteins have the same concentration. On the other hand, stochasticity in the presence of a small number of molecules makes simulations too noisy, as shown in (Bortolussi and Policriti, 2010, Fig. 13(f)).

These observations lead the authors to conclude that "…the non-deterministic version of the HA is somewhere in the middle between the deterministic and the stochastic one. Non-determinism seems to be a possible way to introduce a small, controlled amount of variability, mimicking the perturbation effects of noise. Further work is required to investigate this issue."

8.6 Conclusions

The path followed in this chapter started with the definition of the basic notion of automaton and evolved in a rich formalism suitable to be used for realistic simulation and hypothesis testing.

The language used in this field is the language of logic, and the notion of (finite state) automaton we started with is a classic in computer science. The material, introduced step by step, begins motivated by a purely qualitative simulation task and then calls into play richer and more ambitious objectives. As a matter of fact, the underlying notion of *time* can be seen as the driving force defining the evolution of the formalism. The initial time instants of simple, finite-state automata models are defined by the occurrence of events. Events can be fully independent and asynchronous; hence, no *absolute* notion of time is necessary. Enriching our models with functions expressing more subtle properties of the simulated field calls into play time as the domain of functions. The automaton formalism must therefore be coherently adapted and full-fledged hybrid automata enter the scene. Hybrid automata's states implicitly represent (through traces) infinitely many states. Moreover, the notion of transition from state to state needs a revision (of the activation mechanism) and a number of more complex computational questions arise. Finally, uncertainty pushes the formalism one last step further and the underlying notion of time undergoes a last modification. The unability/unwillingness to fully specify the dynamic drove us toward more realistic and computationally attractive models. The last notion of underlying time can be described as an absolute time allowing for imprecision/generalization because of the lack of knowledge or computational resources.

Acknowledgement

This work has been partially supported by INdAM group GNCS and PRIN project NiRvAna.

References

Gul Agha and Karl Palmskog. A survey of statistical model checking. *ACM Transactions on Modeling & Computer Simulation (TOMACS)*, 28 (1): 1–39, 2018.

Jamil Ahmad, Gilles Bernot, Jean-Paul Comet, Didier Lime, and Olivier Roux. Hybrid modelling and dynamical analysis of gene regulatory networks with delays. *ComPlexUs*, 3 (4): 231–251, 2006.

Rajeev Alur and David L. Dill. A theory of timed automata. *Theoretical Computer Science*, 126 (2): 183–235, 1994.

Rajeev Alur, Costas Courcoubetis, and David Dill. Model-checking for probabilistic real-time systems. In J. L. Albert, B. Monien, and M. R. Artalejo, editors *International Colloquium on Automata, Languages, and Programming*, pages 115–126. Springer, Berlin Heidelberg, 1991.

Rajeev Alur, Costas Courcoubetis, Thomas A. Henzinger, and Pei-Hsin Ho. Hybrid automata: an algorithmic approach to the specification and verification of hybrid systems. In Robert L. Grossman, A. Nerode, Anders P. Ravn, and H. Rischel, editors, *Hybrid Systems*, pages 209–229. Springer, Berlin Heidelberg, 1993.

Rajeev Alur, Costas Courcoubetis, Nicolas Halbwachs, Thomas A. Henzinger, P.-H. Ho, Xavier Nicollin, Alfredo Olivero, Joseph Sifakis, and Sergio Yovine. The algorithmic analysis of hybrid systems. *Theoretical Computer Science*, 138 (1): 3–34, 1995.

Babar Aslam, Jamil Ahmad, Amjad Ali, Rehan Zafar Paracha, Samar Hayat Khan Tareen, Umar Niazi, and Tariq Saeed. On the modelling and analysis of the regulatory network of dengue virus pathogenesis and clearance. *Computational Biology and Chemistry*, 53: 277–291, 2014.

Paolo Ballarini, Hilal Djafri, Marie Duflot, Serge Haddad, and Nihal Pekergin. COSMOS: A statistical model checker for the hybrid automata stochastic logic. In *2011 8th International Conference on Quantitative Evaluation of SysTems*, pages 143–144. IEEE, 2011a.

Paolo Ballarini, Hilal Djafri, Marie Duflot, Serge Haddad, and Nihal Pekergin. HASL: An expressive language for statistical verification of stochastic models. In *Proceedings of the 5th International ICST Conference on Performance Evaluation Methodologies and Tools*, pages 306–315, 2011b.

Roberto Barbuti, Roberta Gori, Paolo Milazzo, and Lucia Nasti. A survey of gene regulatory networks modelling methods: from differential equations, to Boolean and qualitative bioinspired models. *Journal of Membrane Computing*, 2 (3): 207–226, 2020.

S. Basu. An improved algorithm for quantifier elimination over real closed fields. In *Proceedings of the 38th Annual Symposium on Foundations of Computer Science (FOCS '97)*, pages 56–65. IEEE Computer Society Press, 1997. ISBN 0-8186-8197-7.

S. Basu, R. Pollack, and M.-F. Roy. On the combinatorial and algebraic complexity of quantifier elimination. *Journal of the ACM*, 43 (6): 1002–1045, 1996. ISSN 0004-5411. doi: http://doi.acm.org/10.1145/235809.235813.

Grégory Batt, Calin Belta, and Ron Weiss. Model checking genetic regulatory networks with parameter uncertainty. In *International Workshop on Hybrid Systems: Computation and Control*, pages 61–75. Springer, 2007a.

Grégory Batt, Calin Belta, and Ron Weiss. Model checking liveness properties of genetic regulatory networks. In *International Conference on Tools and Algorithms for the Construction and Analysis of Systems*, pages 323–338. Springer, 2007b.

Gerd Behrmann, Alexandre David, and Kim G. Larsen. A tutorial on UPPAAL. In M. Bernardo and F. Corradini, editors, *Formal Methods for the Design of Real-Time Systems*, vol 3185, pages 200–236. Springer, Berlin Heidelberg, 2004.

Calin Belta, Joel M. Esposito, Jongwoo Kim, and Vijay Kumar. Computational techniques for analysis of genetic network dynamics. *The International Journal of Robotics Research*, 24 (2–3): 219–235, 2005.

Mordechai Ben-Ari, Zohar Manna, and Amir Pnueli. The temporal logic of branching time. *Acta Informatica*, 20: 207–226, 1983.

Luca Bortolussi and Alberto Policriti. Hybrid systems and biology. In M. Bernardo, P. Degano, and G. Zavattaro, editors *Formal Methods for Computational Systems Biology. SFM 2008. Lecture Notes in Computer Science*, vol 5016, pages 424–448. Springer, Berlin Heidelberg, 2008.

Luca Bortolussi and Alberto Policriti. Hybrid dynamics of stochastic programs. *Theoretical Computer Science*, 411 (20): 2052–2077, 2010.

Alberto Casagrande, Venkatesh Mysore, Carla Piazza, and Bud Mishra. Independent dynamics hybrid automata in systems biology. In *Proceedings of the 1st International Conference on Algebraic Biology (AB'05)*, pages 61–73, 2005.

Alberto Casagrande, Carla Piazza, Alberto Policriti, and Bud Mishra. Inclusion dynamics hybrid automata. *Information and Computation*, 206 (12): 1394–1424, 2008.

Alberto Casagrande, Tommaso Dreossi, and Carla Piazza. Hybrid automata and ϵ-analysis on a neural oscillator. In *Proceedings 1st International Workshop on Hybrid Systems and Biology (HSB 2012)*, volume 92, pages 58–72, 2012. doi: https://doi.org/10.4204/EPTCS.92.5.

Alberto Casagrande, Tommaso Dreossi, Jana Fabriková, and Carla Piazza. ϵ-Semantics computations on biological systems. *Information and Computation*, 236: 35–51, 2014.

G. V. Chudnovsky. *Contributions to the Theory of Transcendental Numbers*. Number 19 in Mathematical Surveys and Monographs. American Mathematical Society, 1984.

Edmund Melson Clarke and Ioana Anca Draghicescu. Expressibility results for linear-time and branching-time logics. In J. W. de Bakker, Willem P. de Roever, and Grzegorz Rozenberg, editors, *Linear Time, Branching Time and Partial Order in Logics and Models for Concurrency, School/Workshop, Noordwijkerhout, The Netherlands, May 30–June 3, 1988, Proceedings*, volume 354 of Lecture Notes in Computer Science, pages 428–437. Springer, 1988. doi: https://doi.org/10.1007/BFb0013029.

Edmund Melson Clarke, Ernest Allen Emerson, and Aravinda Prasad Sistla. Automatic verification of finite-state concurrent systems using temporal logic specifications. *ACM Transactions on Programming Languages and Systems (TOPLAS)*, 8 (2): 244–263, 1986. ISSN 0164-0925. doi: http://doi.acm.org/10.1145/5397.5399.

Edmund Melson Clarke, Orna Grumberg, and Doron A. Peled. *Model Checking*. MIT Press, Cambridge, MA, USA, 1999. ISBN 0-262-03270-8.

Thao Dang, Tommaso Dreossi, and Carla Piazza. Parameter synthesis through temporal logic specifications. In *FM 2015: Formal Methods - 20th International Symposium, Oslo, Norway, June 24–26, 2015, Proceedings*, pages 213–230, 2015.

Pedro R. D'Argenio and Joost-Pieter Katoen. A theory of stochastic systems Part I: stochastic automata. *Information and Computation*, 203 (1): 1–38, 2005.

Laurent Doyen, Goran Frehse, George J. Pappas, and André Platzer. Verification of hybrid systems. In E. Clarke, T. Henzinger H. Veith, and R. Bloem, editors *Handbook of Model Checking*, pages 1047–1110. Springer, Cham, 2018.

Michael B. Elowitz and Stanislas Leibler. A synthetic oscillatory network of transcriptional regulators. *Nature*, 403 (6767): 335–338, 2000.

Ernest Allen Emerson and Edmund Melson Clarke. Using branching time temporal logic to synthesize synchronization skeletons. *Science of Computer Programming*, 2 (3): 241–266, 1982. ISSN 0167-6423. doi: https://doi.org/10.1016/0167-6423(83)90017-5.

Ernest Allen Emerson and Joseph Y. Halpern. "Sometimes" and "not never" revisited: on branching versus linear time (preliminary report). In *Proceedings of the 10th ACM SIGACT-SIGPLAN Symposium on Principles of Programming Languages*, POPL '83, pages 127–140, New York, NY, USA, 1983. Association for Computing Machinery. ISBN 0897910907. doi: https://doi.org/10.1145/567067.567081.

Ernest Allen Emerson and Chin-Laung Lei. Modalities for model checking (extended abstract): branching time strikes back. In *Proceedings of the 12th ACM SIGACT-SIGPLAN Symposium on Principles of Programming Languages (POPL '85)*, pages 84–96, New York, NY, USA, 1985. ACM Press. ISBN 0-89791-147-4. doi: http://doi.acm.org/10.1145/318593.318620.

Herbert B. Enderton. *A Mathematical Introduction to Logic*. Harcourt/Academic Press, 2001.

Marcello Farina and Maria Prandini. Hybrid models for gene regulatory networks: the case of lac operon in *E. coli*. In *International Workshop on Hybrid Systems: Computation and Control*, pages 693–697. Springer, 2007.

Hans Hansson. *Time and Probability in Formal Design of Distributed Systems*. Uppsala Univeristy, 1991.

Hans Hansson and Bengt Jonsson. A logic for reasoning about time and reliability. *Formal Aspects of Computing*, 6 (5): 512–535, 1994.

Thomas A. Henzinger. The theory of hybrid automata. In M. K. Inan and R.P. Kurshan, editors *Verification of Digital and Hybrid Systems. NATO ASI Series*, vol 170, pages 265–292. Springer, Berlin, Heidelberg, 2000.

Jianghai Hu, John Lygeros, and Shankar Sastry. Towards a theory of stochastic hybrid systems. In *International Workshop on Hybrid Systems: Computation and Control*, pages 160–173. Springer, 2000.

Robert M. Keller. Formal verification of parallel programs. *Communications of the ACM*, 19 (7): 371–384, July 1976. ISSN 0001-0782. doi: https://doi.org/10.1145/360248.360251.

John G. Kemeny and J. Laurie Snell. *Markov Chains*, volume 6. Springer-Verlag, New York, 1976.

A. G. Khovanskiĭ. On a class of systems of transcendental equations. *Soviet Mathematics Doklady*, 22: 762–765, 1980.

Peter William Kopke Jr. *The Theory of Rectangular Hybrid Automata*. Cornell University, 1996.

Saul Kripke. Semantical considerations on modal logic. *Acta Philosophica Fennica*, 16: 83–94, 1963.

Marta Kwiatkowska, Gethin Norman, Roberto Segala, and Jeremy Sproston. Verifying quantitative properties of continuous probabilistic real-time graphs. *School of Computer Science Research Reports-University of Birmingham CSR*, 1998.

Marta Kwiatkowska, Gethin Norman, and David Parker. Quantitative analysis with the probabilistic model checker prism. *Electronic Notes in Theoretical Computer Science*, 153 (2): 5–31, 2006.

Gerardo Lafferriere, George J. Pappas, and Shankar Sastry. O-minimal hybrid systems. *Mathematics of Control, Signals and Systems*, 13 (1): 1–21, 2000.

Abolfazl Lavaei, Sadegh Soudjani, Alessandro Abate, and Majid Zamani. Automated verification and synthesis of stochastic hybrid systems: a survey. *arXiv preprint arXiv:2101.07491*, 2021.

Xiangfang Li, Oluwaseyi Omotere, Lijun Qian, and Edward R. Dougherty. Review of stochastic hybrid systems with applications in biological systems modeling and analysis. *EURASIP Journal on Bioinformatics & Systems Biology*, 2017 (1): 1–12, 2017.

Orna Lichtenstein and Amir Pnueli. Checking that finite state concurrent programs satisfy their linear specification. In *Proceedings of the 12th ACM SIGACT-SIGPLAN Symposium on Principles of Programming Languages (POPL '85)*, pages 97–107, New York, NY, USA, 1985. ACM Press. ISBN 0-89791-147-4. doi: http://doi.acm.org/10.1145/318593.318622.

Lin Liu and Alexander Bockmayr. Formalizing metabolic-regulatory networks by hybrid automata. *Acta Biotheoretica*, 68 (1): 73–85, 2020.

Loes Olde Loohuis, Andreas Witzel, and Bud Mishra. Cancer hybrid automata: model, beliefs and therapy. *Information and Computation*, 236: 68–86, 2014.

A. Macintyre. Schanuel's conjecture and free exponential rings. *Annals of Pure and Applied Logic*, 51 (3): 241–246, 1991.

A. Macintyre and A. J. Wilkie. On the decidability of the real exponential field. In P. G. Odifreddi, editor, *Kreiseliana, About and Around Georg Kreisel*, pages 441–467. A. K. Peters, Ltd., 1996.

Elliott Mendelson. *Introduction to Mathematical Logic* (Fourth Edition). Chapman & Hall, 1997.

Amir Pnueli. The temporal logic of programs. In *Proceedings of the 18th IEEE Symposium on the Foundations of Computer Science (FOCS-77)*, pages 46–57. IEEE Computer Society Press, Providence, Rhode Island, October 31–November 2 1977. IEEE.

Amir Pnueli. The temporal semantics of concurrent programs. In *Proceedings of the International Sympoisum on Semantics of Concurrent Computation*, pages 1–20. Springer-Verlag, London, UK, 1979. ISBN 3-540-09511-X.

Amir Pnueli. The temporal semantics of concurrent programs. *Theoretical Computer Science*, 13: 45–60, 1981.

Michael O. Rabin. Probabilistic automata. *Information and Control*, 6 (3): 230–245, 1963.

Adrien Richard, Jean-Paul Comet, and Gilles Bernot. Formal methods for modeling biological regulatory networks. In H.A. Gabbar, editor *Modern Formal Methods and Applications*, pages 83–122. Springer, Dordrecht, 2006.

Philippe Schnoebelen. The complexity of temporal logic model checking. In Philippe Balbiani, Nobu-Yuki Suzuki, Frank Wolter, and Michael Zakharyaschev, editors, *Advances in Modal Logic 4, papers from the 4th Conference on "Advances in Modal logic," held in Toulouse, France, 30 September - 2 October 2002*, pages 393–436. King's College Publications, 2002.

Roberto Segala and Nancy Lynch. Probabilistic simulations for probabilistic processes. In *International Conference on Concurrency Theory*, pages 481–496. Springer, 1994.

Aravinda Prasad Sistla and Edmund Melson Clarke. The complexity of propositional linear temporal logics. *Journal of the ACM*, 32 (3): 733–749, July 1985. ISSN 0004-5411. doi: 10.1145/3828.3837.

A. Tarski. *A Decision Method for Elementary Algebra and Geometry*. University California Press, 1951.

Denis Thieffry and René Thomas. Dynamical behaviour of biological regulatory networks—II. Immunity control in bacteriophage lambda. *Bulletin of Mathematical Biology*, 57 (2): 277–297, 1995.

René Thomas. Logical analysis of systems comprising feedback loops. *Journal of Theoretical Biology*, 73 (4): 631–656, 1978.

Arnaud Tonnelier, Sylvain Meignen, Holger Bosch, and Jacques Demongeot. Synchronization and desynchronization of neural oscillators. *Neural Networks*, 12 (9): 1213–1228, 1999.

L. van den Dries. A generalization of the Tarski-Seidenberg theorem and some nondefinability results. *Bulletin of American Mathematical Society (New Series)*, 15 (2): 189–193, 1986.

L. van den Dries. *Tame Topology and O-minimal Structures*, volume 248 of London Mathematical Society Lecture Note Series. Cambridge University Press, 1998.

L. van den Dries and C. Miller. On the real exponential field with restricted analytic functions. *Israel Journal of Mathematics*, 85 (1–3): 19–56, 1994.

L. van den Dries and C. Miller. Geometric categories and o-minimal structures. *Duke Mathematical Journal*, 84: 497–540, 1996.

L. van den Dries, A. Macintyre, and D. Marker. The elementary theory of restricted analytic functions with exponentiation. *Annals of Mathematica*, 140 (1): 183–205, 1994.

Moshe Y. Vardi. Automatic verification of probabilistic concurrent finite state programs. In *26th Annual Symposium on Foundations of Computer Science (SFCS 1985)*, pages 327–338. IEEE, 1985.

Vladimeros Vladimerou, Pavithra Prabhakar, Mahesh Viswanathan, and Geir Dullerud. Stormed hybrid systems. In *International Colloquium on Automata, Languages, and Programming*, pages 136–147. Springer, 2008.

Ward Whitt. Continuity of generalized semi-Markov processes. *Mathematics of Operations Research*, 5 (4): 494–501, 1980.

A. J. Wilkie. Model completeness results for expansions of the ordered field of real numbers by restricted Pfaffian functions and the exponential function. *Journal of the American Mathematical Society*, 9 (4): 1051–1094, 1996.

A. J. Wilkie. A theorem of the complement and some new O-minimal structures. *Selecta Mathematica, New Series*, 5 (4): 397–421, 1999.

Paolo Zuliani. Statistical model checking for biological applications. *International Journal on Software Tools for Technology Transfer*, 17 (4): 527–536, 2015.

9

Kalle Parvinen: Ordinary Differential Equations

Kalle Parvinen[1,2]

[1]*Department of Mathematics and Statistics, University of Turku, Turku, Finland*
[2]*Advancing Systems Analysis Program, International Institute for Applied Systems Analysis, Laxenburg, Austria*

9.1 Introduction

Ordinary differential equations (ODEs) are in general of the following form

$$F\left(t, x, \frac{d}{dt}x, \frac{d^2}{dt^2}x, \ldots\right) = 0 \qquad (9.1)$$

In other words, (9.1) defines relations that a function and its derivatives should satisfy. The derivatives are often denoted by

$$x' = x'(t) = \frac{d}{dt}x, \quad x'' = x''(t) = \frac{d^2}{dt^2}x, \ldots \qquad (9.2)$$

Differential equations arise in many different fields of science. For example, if $x(t)$ describes the position of an object at time t, then $x'(t)$ describes its velocity, and $x''(t)$ describes its acceleration. If $f(t,x)$ is the force affecting the object, we obtain a widely used ODE of classical mechanics, $f(t, x(t)) = mx''(t)$, in which m is the mass of the object. In chemistry, ODEs are widely used to describe how molecule concentrations change in time due to chemical reactions. Systems biology can be viewed as the computational and mathematical analysis and modelling of complex biological systems, which often involves ODEs. Examples of ODE models in systems biology include metabolic networks (Akutsu et al., 2000), cell signaling networks (Angeli et al., 2004), metapopulation models (Hastings, 1983, Hanski, 1999, Metz and Gyllenberg, 2001, Gyllenberg et al., 2002), and eco-evolutionary models (Geritz et al., 1998, Gyllenberg and Parvinen, 2001). ODEs as such are deterministic but can be used in a stochastic context, for example, to describe the dynamics of the probability

Systems Biology Modelling and Analysis: Formal Bioinformatics Methods and Tools,
First Edition. Edited by Elisabetta De Maria.
© 2023 John Wiley & Sons, Inc. Published 2023 by John Wiley & Sons, Inc.

distribution of a continuous-time Markov chain (forward Kolmogorov equations). In contrast, stochastic differential equations (Øksendal, 2003) have a built-in stochastic component and provide an interesting approach to systems biology modelling, which is, however, out of the scope of this chapter.

In this chapter, we will concentrate on explicit autonomic differential equations, which in the space $\mathbb{R}^n$ are of form

$$\frac{d}{dt}x = f(x), \text{ in which } f : \mathbb{R}^n \to \mathbb{R}^n \quad (9.3)$$

We thus consider situations, in which the dynamics of the model variables depend only on each other, through mechanisms that remain the same throughout the investigated time. Throughout this chapter, we assume that the function f is continuous and differentiable.

9.2 Analyzing and Solving Ordinary Differential Equations

9.2.1 Solving Ordinary Differential Equations Analytically

As a solution of an ODE (9.1), we mean a function $x = \varphi(t)$ that satisfies Eq. (9.1) identically. If we have a candidate solution, it is usually rather simple to check whether it really is a solution. This can be done by differentiating the candidate solution and substituting it into the ODE. However, it is often not so easy to find candidate solutions. Although various solution methods for specific types of ODE exist, there are no solution methods that work for all possible ODEs. Therefore, we consider here also other methods using which one can understand the dynamics of an ODE: We analyze the stability of equilibria and use numerical methods to solve the ODE.

Example 9.1 The so-called logistic differential equation is

$$\frac{dx}{dt} = ax\left(1 - \frac{x}{K}\right) \quad (9.4)$$

Theorem 9.1 *The solution of the logistic ODE (9.4) with the initial condition $x(0) = x_0$ is*

$$x(t) = \frac{Kx_0}{(K - x_0)e^{-at} + x_0} \quad (9.5)$$

Proof: Consider the solution at time s. We can write the ODE (9.4) in the form

$$\frac{x'(s)}{x(s)\left(1 - \frac{x(s)}{K}\right)} = r. \quad (9.6)$$

9.2 Analyzing and Solving Ordinary Differential Equations

By integrating both sides, we obtain

$$\int_0^t \frac{x'(s)}{x(s)\left(1 - \frac{x(s)}{K}\right)} ds = \int_0^t a\, ds = at \tag{9.7}$$

By using the partial fraction decomposition $\frac{1}{x\left(1-\frac{x}{K}\right)} = \frac{1}{x} + \frac{1}{K-x}$, we obtain

$$at = \int_0^t \frac{x'(s)}{x(s)} + \frac{x'(s)}{K-x(s)} ds = \int_0^t \ln x(s) - \ln[K - x(s)]$$

$$= \ln \frac{x(t)}{x(0)} - \ln \frac{K - x(t)}{K - x(0)} \tag{9.8}$$

Finally, we apply the exponential function on both sides and denote $x(0) = x_0$, so that we obtain

$$\frac{x(t)}{x_0} \frac{K - x_0}{K - x(t)} = e^{at} \tag{9.9}$$

from which (9.5) follows by solving for $x(t)$. □

Note 9.1 If we are given (9.5) as a candidate solution, it is relatively easy to prove that it really is a solution of (9.4): first, by substituting $t = 0$ into (9.5), we easily see that the initial condition is satisfied.

$$x(0) = \frac{Kx_0}{K - x_0 + x_0} = x_0 \tag{9.10}$$

Then, by differentiating (9.5), we obtain

$$\frac{d}{dt}x(t) = Kx_0 \frac{a(K - x_0)e^{-at}}{((K-x_0)e^{-at} + x_0)^2} = ax(t)\frac{(K-x_0)e^{-at} + x_0 - x_0}{(K-x_0)e^{-at} + x_0}$$

$$= ax(t)\left[1 - \frac{1}{K}\frac{Kx_0}{(K-x_0)e^{-at} + x_0}\right] = ax(t)\left[1 - \frac{x(t)}{K}\right] \tag{9.11}$$

Since $a > 0$, the expression $e^{-at} \to 0$ when $t \to \infty$. Consequently, solution (9.5) of the logistic ODE converges to K from all positive initial conditions, $x_0 > 0$. If $x_0 = 0$, the solution remains at 0. See Figure 9.2b for an illustration.

9.2.2 Equilibria and Their Stability

Definition 9.1 An equilibrium of an ODE $\frac{dx}{dt} = f(x)$ is such value x^* for which $f(x^*) = 0$.

By construction, if at some moment the differential equation has reached a state x^* satisfying $f(x^*) = 0$, we have $\frac{dx}{dt} = 0$ so that the differential equation will remain in state x^* forever. However, if the state is only close to the equilibrium x^*, what happens depends on the model properties. The solution may approach x^*, but it is also possible that the solution goes away from x^*.

Definition 9.2 An equilibrium is (locally asymptotically) stable, if given that $x(0)$ is close to x^* enough, the solution $x(t)$ remains close to x^* for all t and converges to x^*.

Theorem 9.2 *For one-dimensional models, the equilibrium x^* of the ODE $\frac{dx}{dt} = f(x)$ is stable if $f'(x^*) < 0$ and unstable if $f'(x^*) > 0$.*

Proof: By definition, $f(x^*) = 0$. Therefore, when $f'(x^*) < 0$, there exists an $\epsilon > 0$ such that $f(x) > 0$ when $x^* - \epsilon < x < x^*$, and $f(x) < 0$ when $x^* < x < x^* + \epsilon$. (see Figure 9.1a). Therefore, $x(t) \to x^*$ if $|x(0) - x^*| \leq \epsilon$, which means that the equilibrium is stable. Analogously, when $f'(x^*) > 0$, there exists an $\epsilon > 0$ such that $f(x) < 0$ when $x^* - \epsilon < x < x^*$ and $f(x) > 0$ when $x^* < x < x^* + \epsilon$. Therefore, $x(t)$ will go away from x^* at least until $|x(0) - x^*| \geq \epsilon$, which means that the equilibrium is unstable (Figure 9.1b).

We can also investigate the stability of equilibria as follows. Let us denote $y(t) = x(t) - x^*$, so that

$$\frac{d}{dt}x(t) = f(x) = \underbrace{f(x^*)}_{=0} + f'(x^*)(x - x^*) + \underbrace{h(x, x^*)}_{\to 0,\text{ when } x \to x^*}(x - x^*) \quad (9.12)$$

The linear model $\frac{d}{dt}y(t) = f'(x^*)y(t)$ gives an approximation of the original model in the neighborhood of the equilibrium x^*. The linear model has the solution $y(t) = y(0)e^{f'(x^*)t}$. If $f'(x^*) < 0$, the linear approximation $y(t) \to 0$, which means that $x(t) \to x^*$, and the equilibrium is stable. On the other

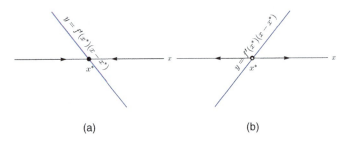

(a)   (b)

Figure 9.1 For one-dimensional ODE of the form $\frac{dx}{dt} = f(x)$, the stability of equilibria x^* is determined by the sign of the derivative $f'(x^*)$. (a) Stable: $f'(x^*) < 0$. (b) Unstable: $f'(x^*) > 0$.

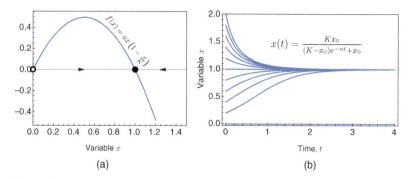

Figure 9.2 Logistic differential equation (9.4). (a) The function $f(x)$ with respect to x and (b) solutions of the ODE with different initial values. Parameters: $a = 2$ and $K = 1$.

hand, if $f'(x^*) > 0$, the linear approximation will go away from 0, so that x^* is an unstable equilibrium of the original model. Although the solution of the linear approximation approaches infinity, this does not necessarily mean that the original model would approach infinity. Once the solution of the linear approximation is far from 0, it is not anymore a good approximation of the original model.

Example 9.2 Investigate the logistic ODE (9.4) and the stability of its equilibria:

By solving $f(x) = ax(1 - \frac{x}{K}) = 0$, we obtain $x^* = 0$ or $x^* = K$ so that (9.4) has two equilibria, which correspond to the locations in which the curve in Figure 9.2a crosses the horizontal axis. By differentiation, we obtain $f'(x) = a - 2a\frac{x}{K}$. Therefore,

$$f'(0) = a > 0 \Rightarrow \text{the equilibrium } x^* = 0 \text{ is unstable}$$

$$f'(K) = a - 2a\frac{K}{K} = -a < 0 \Rightarrow \text{the equilibrium } x^* = K \text{ is stable}$$

(9.13)

From the local stability of equilibria and the sign of f, we see that the solution of the logistic ODE converges to K from all positive initial conditions (see Figure 9.2b for an illustration). We obtained this conclusion already in the end of Section 9.2.1 from the properties of the analytical solution (9.5). Note, however, that the stability analysis of equilibria can be performed even when no analytical solution to the ODE can be found.

For higher dimensional models, the eigenvalues of the Jacobi matrix evaluated at the equilibrium can be used to determine the stability of the equilibrium. If all eigenvalues have a negative real part, the equilibrium is stable, whereas an eigenvalue with a positive real part means that the equilibrium is unstable in the direction of the corresponding eigenvector.

9.2.3 Solving Differential Equations Numerically

There are many different methods for solving differential equations numerically. For generality, in this section, we consider the nonautonomous ODE

$$\frac{d}{dt}x = f(t,x), \text{ in which } f : \mathbb{R} \times \mathbb{R}^n \to \mathbb{R}^n \tag{9.14}$$

with the initial condition $x(t_0) = x_0$. The presented methods naturally apply also for autonomous ODEs of type (9.3), in which the function f does not depend explicitly on t.

Usually, numerical methods aim to provide approximations x_k of the solution at specific points $t_1, t_2, \ldots$, so that $x_k \approx x(t_k)$. We can integrate both sides of the ODE (9.14) to obtain

$$x(t_{k+1}) - x(t_k) = \int_{t_k}^{t_{k+1}} x'(t)dt = \int_{t_k}^{t_{k+1}} f(t, x(t))dt \tag{9.15}$$

We can use the rectangle rule to approximate the integral

$$\int_{t_k}^{t_{k+1}} f(t, x(t))dt \approx (t_{k+1} - t_k)f(t_k, x(t_k)) \tag{9.16}$$

so that we obtain $x(t_{k+1}) \approx x(t_k) + h_k f(t_k, x(t_k))$, in which $h_k = t_{k+1} - t_k$. We thus obtain the explicit Euler method

$$x_{k+1} = x_k + h_k f(t_k, x_k) \tag{9.17}$$

In other words, from x_k, we take a step of length h_k in the direction of the tangent $f(x_k)$. This is a first-order Taylor method. It is simple to use but very inefficient. In order to obtain sufficient accuracy, the step size h needs to be very small, resulting in long computation time.

In order to obtain better numerical solution methods, we can use the Taylor series

$$x(t_{k+1}) = x(t_k) + hx'(t_k) + \frac{h^2}{2!}x''(t_k) + \frac{h^3}{3!}x'''(t_k) + \cdots + \frac{h^n}{n!}x^{(n)}(t_k) + \cdots \tag{9.18}$$

Note that the Euler method is obtained by using only the first two terms on the right-hand side of (9.18). That is easy, as $x'(t_k) = f(t_k, x(t_k))$. The so-called Runge–Kutta methods use various differences to approximate the higher derivatives of (9.18). A widely used method is the fourth-order Runge–Kutta method (RK4)

$$x_{k+1} = x_k + \frac{h}{6}(K_1 + 2K_2 + 2K_3 + K_4) \tag{9.19}$$

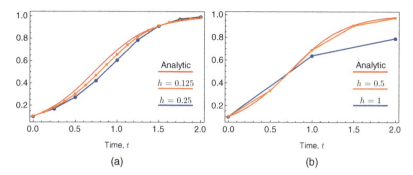

Figure 9.3 Analytical and numerical solutions of the logistic ODE (9.4). (a) Euler method and (b) fourth-order Runge–Kutta method (RK4). Parameters: $a = 3$, $K = 1$, and $x_0 = 0.1$. The step sizes are chosen large so that differences are visible, but they are too large for practical use.

in which

$$K_1 = f(t_k, y_k)$$
$$K_2 = f(t_k + \tfrac{1}{2}h, y_k + \tfrac{h}{2}K_1)$$
$$K_3 = f(t_k + \tfrac{1}{2}h, y_k + \tfrac{h}{2}K_2)$$
$$K_4 = f(t_{k+1}, y_k + hK_3) \tag{9.20}$$

It is efficient enough for practical use. Figure 9.3 illustrates the application of these two methods for the logistic ODE (9.4). The RK4 method calculates the value of the function f four times in each step. Therefore, when comparing the performance of the RK4 method with the Euler method, the step size in the Euler method should be one fourth of the one used in the RK4 method. For $h = 1$ in the RK4 method, the error at $t = 1$ is 0.052, which is of the same order of magnitude as the error 0.087 for the Euler method with $h = 1/4 = 0.25$. For $h = 0.5$ in the RK4 method, the error at $t = 1$ is 0.0034, which is considerably smaller than the error 0.039 for the Euler method with $h = 0.5/4 = 0.125$. For other illustrations of numerical solutions of ODEs, see Figures 9.6 and 9.7.

9.3 Mechanistic Derivation of Ordinary Differential Equations

In the phenomenological (top-down) modelling approach, one chooses a model that fits well with the available data and has good mathematical properties, such as, it has an analytical solution. Many classical models describing population growth have been chosen using such an approach. In phenomenological models, parameters do not usually have individual-level interpretations. Therefore, in such models, it is difficult to investigate how alterations in individual behavior affect population dynamics. On the other hand, because of their mathematical properties, model analysis is often relatively easy, and population-level predictions can be made.

In the mechanistic (bottom-up) modelling approach, the model is derived from individual-level processes (Geritz and Kisdi, 2004, Brännström and Sumpter, 2005, Eskola and Geritz, 2007). The mathematical model is obtained by straightforward book-keeping, i.e. by taking into account the effect of all relevant individual-level processes on the population-level variables. In this kind of models, all parameters have an interpretation on the individual level. Investigating the effects of alterations in individual behavior is in principle straightforward. This is especially important, when modelling evolution by natural selection in mathematical models (Geritz et al., 1998, Rueffler et al., 2006). On the other hand, the model is not necessarily mathematically simple.

9.3.1 Elementary Unimolecular Reaction (EUR)

Definition 9.3 In an elementary unimolecular reaction (EUR),

- there are plenty of molecules A, and they are stochastically identical and independent.
- there exists a constant probability k per time unit for a specific individual molecule to experience a specific transformation.

We use the notation

$$A \xrightarrow{k} \text{products} \tag{9.21}$$

in which the parameter k is the reaction constant (reaction speed).

Let $A(t)$ denote the concentration of A-molecules at time t. According to the Definition 9.3, we have

$$A(t + \Delta t) = A(t) - k\Delta t A(t) \quad \Rightarrow \quad \frac{A(t + \Delta t) - A(t)}{\Delta t} = -kA(t) \tag{9.22}$$

When we let $\Delta t \to 0$, we obtain the ODE

$$\frac{d}{dt}A(t) = -kA(t), \quad A(0) = A_0 \tag{9.23}$$

This is a linear ODE, and it has the solution

$$A(t) = A_0 e^{-kt} \tag{9.24}$$

Let **T** denote the time that is needed for a specific individual molecule A to react. The probability that a specific molecule has not reacted until time t is $P(\mathbf{T} > t) = \frac{A(t)}{A_0} = e^{-kt}$, and therefore, the cumulative distribution function of **T** is

$$F_\mathbf{T}(t) = P(\mathbf{T} \leqslant t) = 1 - P(\mathbf{T} > t) = 1 - e^{-kt}, \quad \text{for } t \geqslant 0 \qquad (9.25)$$

This means that the distribution of the waiting time **T** is exponential, $\mathbf{T} \sim Exp(k)$. Consequently, the expected value of **T** is $E[\mathbf{T}] = \frac{1}{k}$.

Example 9.3 In radioactive decay, an unstable atomic nucleus loses energy by radiation. Such an event can be considered as an EUR. When the collection of radioactive atoms is large enough, the reaction speed can be estimated relatively accurately. In radioactive decay, the reaction speed is typically measured using half-life time τ, which is the expected time for the amount of radioactive material to decay to half of its original amount.

$$A(\tau) = A_0 e^{-k\tau} = \frac{1}{2}A_0 \Rightarrow \tau = \frac{1}{k}\ln 2 \qquad (9.26)$$

Example 9.4 In ecological models, natural death is often modeled as an EUR. However, one should note that in reality, the age and physical condition of an individual considerably affect the death rate.

9.3.2 Elementary Bimolecular Reaction (EBR)

Definition 9.4 In an elementary bimolecular reaction (EBR),

- there are plenty of molecules A and B. Molecules A are stochastically identical and independent. Also molecules B are stochastically identical and independent.
- upon encounter, molecules A and B can react
- encounters of molecules A and B happen in relation to the product of the molecule concentrations (law of mass action).

We use the notation

$$A + B \xrightarrow{k} \text{products} \qquad (9.27)$$

in which the parameter k is the reaction constant (reaction speed).

In the EBR $A + B \xrightarrow{k} C$, the molecule concentrations satisfy the ODE

$$\begin{cases} \frac{dA}{dt} = -kAB; & A(0) = A_0 \\ \frac{dB}{dt} = -kAB; & B(0) = B_0 \\ \frac{dC}{dt} = kAB; & C(0) = C_0 \end{cases} \qquad (9.28)$$

The solution of (9.28) when $A_0 \neq B_0$ is

$$\begin{cases} A(t) = A_0 e^{-kB_0 t} \dfrac{A_0 - B_0}{A_0 e^{-kB_0 t} - B_0 e^{-kA_0 t}} \\ B(t) = B_0 e^{-kA_0 t} \dfrac{A_0 - B_0}{A_0 e^{-kB_0 t} - B_0 e^{-kA_0 t}} \end{cases} \quad (9.29)$$

Example 9.5 There is a huge amount of chemical reactions in which two different molecules react with each other, such as combustion. In ecological models, resource consumption can often be considered as an EBR, in which upon successful encounter (from the consumer point-of-view), a consumer and resource form a molecule consisting of the consumer individual handling the resource item.

9.3.3 Elementary Bimolecular Reaction of Two Identical Molecules

A special case of the EBR is when the two reacting molecules are similar. In such case, we use the notation

$$A + A \xrightarrow{k} B + R \quad (9.30)$$

The reaction (9.30) results in the ODE

$$\begin{aligned} \frac{dA}{dt} &= -kA^2 \\ \frac{dB}{dt} &= \tfrac{1}{2} kA^2 \\ \frac{dR}{dt} &= \tfrac{1}{2} kA^2 \end{aligned} \quad (9.31)$$

The solution for A is

$$A(t) = \frac{A_0}{A_0 k t + 1} \quad (9.32)$$

9.3.4 Reaction Networks

The reaction

$$A_1 + \cdots + A_k \longrightarrow \text{products} \quad (9.33)$$

is called a reaction of order n if the corresponding differential equations are of form

$$\left| \frac{dA_i}{dt} \right| = \cdots = \left| \frac{dA_k}{dt} \right| = c A_1^{m_1} \cdots A_k^{m_k} \quad (9.34)$$

in which $c > 0$ is some constant and $m_1 + \cdots + m_k = n$.

The EUR (9.21) is of order 1 and the EBRs (9.27) and (9.30) are of order 2. They are effectively the only existing elementary reactions. An elementary reaction of three molecules would namely require the simultaneous encounter of three molecules, whose probability is an order of magnitude smaller than the encounter of two molecules. In practice, reactions that involve several molecules actually consist of several consecutive elementary unimolecular and/or bimolecular reactions. A system involving several elementary reactions is called a reaction network or a reaction mechanism.

Example 9.6 Consider a population, in which individuals A can reproduce for example by division. In addition, aggressive encounters of two individuals can result in the death of one of the individuals. The corresponding reactions are

$$A \xrightarrow{\alpha} A + A \quad \text{(reproduction)}$$
$$A + A \xrightarrow{\delta} A \quad \text{(aggressive encounters)} \quad (9.35)$$

The ODE corresponding to these reactions is

$$\frac{dA}{dt} = -\alpha A + 2\alpha A - \delta A^2 + \frac{1}{2}\delta A^2 = \alpha A \left(1 - \frac{\delta}{2\alpha} A\right) = aA\left(1 - \frac{A}{K}\right) \quad (9.36)$$

which is the logistic ODE (9.4) with parameters $a = \alpha$ and $K = \frac{2\alpha}{\delta}$.

When the number of different molecules and the number of different reactions is large, the reaction network may become difficult to analyze. Two procedures may help in reducing the required number of equations and thus make model analysis easier. First, in many models, it is possible to recognize the presence of various conservation laws. For example, an ecological model may contain various kinds of individuals and living sites in different states. A living site may, for example, be free, occupied by one breeding individual or occupied by eggs, but the total number of sites remains constant. In such a case, some of the model variables may be expressed as a function of others, which reduces the number of necessary variables and equations in the ODE system. Second, depending on the model setup, one may be able to apply time-scale separation arguments. For example, some reactions may be fast compared with other reactions, or some model variables may be large compared with others. In such case, some model variables may turn out to be fast and others slow, so that they can be analyzed separately. On the fast time scale, slow variables remain constant, and one can concentrate on analyzing the dynamics of the fast variables, especially on finding their stable (quasi-)equilibria. On the slow time scale, the fast variables are all the time in their quasi-equilibria, which are functions of the slow variables. This way, one also obtains nonlinear terms into the ODE on the slow time scale. See Geritz and Gyllenberg (2012) for an example.

9.4 Classical Lotka–Volterra Differential Equation

The classical Lotka–Volterra predator–prey model is viewed by some as the starting point of mathematical biology. It was used to explain some unexpected effects of fishing to the fish population and catch. We present the model here as a classical example, providing a potential explanation to a puzzling question, which also serves as an example to illustrate various mathematical techniques, such as phase-plane analysis and determining the stability of an equilibrium using the eigenvalues of the Jacobi matrix.

9.4.1 Model Formation and History

In the 1920s, Humberto D'Ancona studied the fish catch in the Adriatic sea before and after WW1. During the war, the fishing effort was not as high as usually. One could expect that the smaller fishing activity would allow the fish population to grow larger so that the relative catch would be higher. However, he observed that during the war, the proportion of predatory fish (sharks and rays) increased. During those times, prey fish, such as sardines, were more preferred by the fishers. After the war, the proportion of the predatory fish decreased again. D'Ancona discussed with his father-in-law, the Italian mathematician Vito Volterra, and asked for an explanation. Volterra constructed the following predator–prey model (Volterra, 1926, 1927, Lotka, 1927).

Assume that in the absence of predator fish, the prey population R increases exponentially. The predator population, on the other hand, is assumed to decrease exponentially in the absence of prey fish. Predation is assumed to occur according to the law of mass action. A constant proportion of the biomass consumed by the predators is assumed to result in biomass increase of the predator population. These assumptions result in the ODE system

$$\begin{cases} \dfrac{dR}{dt} = \alpha R - \beta RX & \text{(prey fish)} \\ \dfrac{dX}{dt} = \gamma \beta RX - \delta X & \text{(predatory fish)} \end{cases} \quad (9.37)$$

which is called the Lotka–Volterra predator–prey model.

9.4 Classical Lotka–Volterra Differential Equation

Note that the heuristics behind (9.37) is given on the population level. Alternatively, the linear terms in (9.37) could be based on two separate unimolecular reactions,

$$R \xrightarrow{\alpha} R + R \text{ (prey reproduction)}$$
$$C \xrightarrow{\delta} \text{ (death)} \tag{9.38}$$

Note, however, that the EUR describing reproduction describes relatively well the reproduction of, e.g., bacteria, but not so well fish populations. In reality, juvenile fish are initially very small, and it takes quite some time for them to grow to such size that they could reproduce. The remaining terms in (9.37) corresponding to predation and reproduction of the predatory fish could be a result of the following EBR:

$$R + C \xrightarrow{\beta} C + \gamma C \text{ (predation+predator reproduction)} \tag{9.39}$$

In this format, however, the parameter γ should be a natural number, which is unrealistic considering the size difference of the prey and predatory fish. Various other individual-level mechanisms resulting in (9.37) could be constructed, but we do not consider them in further detail here.

9.4.2 Phase-Plane Analysis and Equilibria

Phase-plane analysis is a common method to analyze the dynamics of two-dimensional systems of ODE. Isoclines are curves connecting regions in which the direction of the vector field remains the same. When speaking of isoclines, people often mean isoclines corresponding to horizontal or vertical directions.

From the condition $\frac{dR}{dt} = 0$, we obtain $R = 0$ or $X = X^* = \frac{\alpha}{\beta}$. From the condition $\frac{dX}{dt} = 0$, we obtain $X = 0$ or $R = R^* = \frac{\delta}{\gamma\beta}$. Consequently, the isoclines of (9.37) are horizontal and vertical lines. Equilibrium points lie in the intersections of those lines, i.e. (0,0) and (R^*, X^*) (Figure 9.4a).

The arrows in Figure 9.4a describe the direction of the vector field. The Jacobi matrix is

$$J(R, X) = \begin{pmatrix} \frac{\partial}{\partial R}(\alpha R - \beta RX) & \frac{\partial}{\partial X}(\alpha R - \beta RX) \\ \frac{\partial}{\partial R}(\gamma \beta RX - \delta X) & \frac{\partial}{\partial X}(\gamma \beta RX - \delta X) \end{pmatrix} = \begin{pmatrix} \alpha - \beta X & -\beta R \\ \gamma \beta X & \gamma \beta R - \delta \end{pmatrix} \tag{9.40}$$

At the origin, the Jacobi matrix

$$J(0,0) = \begin{pmatrix} \alpha & 0 \\ 0 & -\delta \end{pmatrix} \tag{9.41}$$

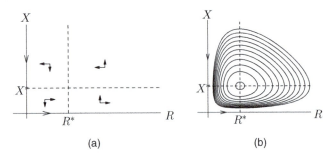

Figure 9.4 Lotka–Volterra predator–prey model. (a) Phase-plane plot illustrating the isoclines and direction of the vector field (9.37). (b) Isoclines together with orbits given by $H(R, X) = c$, in which the function H is given by (9.43).

is a diagonal matrix, and the diagonal elements are the eigenvalues. Since one of them is positive and the other is negative, the origin is a saddle. Let us next try to determine the stability of the positive equilibrium using the Jacobi matrix

$$J(R^*, X^*) = \begin{pmatrix} \alpha - \beta X^* & -\beta R^* \\ \gamma \beta X^* & \gamma \beta R^* - \delta \end{pmatrix} = \begin{pmatrix} 0 & -\beta R^* \\ \gamma \beta X^* & 0 \end{pmatrix} \quad (9.42)$$

The characteristic equation is $\lambda^2 + \gamma \beta^2 X^* R^* = \lambda^2 + \alpha\delta = 0$ so that the eigenvalues $\lambda_{1,2} = \pm i\sqrt{\alpha\delta}$ are purely imaginary (their real part is zero). In such a situation, the linearization does not answer the question whether the positive equilibrium (R^*, X^*) is stable or not. Therefore, we need to use other methods.

9.4.3 Constant of Motion

Let us now investigate how the value of the expression

$$H(R, X) = \delta \ln R - \gamma \beta R + \alpha \ln X - \beta X \quad (9.43)$$

changes along the orbits of the solution of the ODE system (9.37). By differentiation, we obtain

$$\frac{d}{dt} H(R(t), X(t)) = \left(\frac{\delta}{R} - \gamma\beta\right) \frac{d}{dt} R + \left(\frac{\alpha}{X} - \beta\right) \frac{d}{dt} X$$
$$= \left(\frac{\delta}{R} - \gamma\beta\right) R(\alpha - \beta X) + \left(\frac{\alpha}{X} - \beta\right) X(\gamma\beta R - \delta) = 0 \quad (9.44)$$

Since $\frac{d}{dt} H(R(t), X(t)) = 0$, the expression $H(R, X)$ remains constant along each orbit. In other words, the expression $H(R, X)$ is a constant of motion for the ODE system (9.37). Consequently, we can solve the orbits from the equation $H(R, X) = c$.

9.4 Classical Lotka–Volterra Differential Equation

We observe that $\frac{\partial}{\partial R}H(R,X) = \frac{\delta}{R} - \gamma\beta = 0$ if and only if $R = R^*$. Furthermore, $\frac{\partial}{\partial X}H(R,X) = \frac{\alpha}{X} - \beta = 0$ if and only if $X = X^*$. The second derivatives are $\frac{\partial^2}{\partial R^2}H(R,X) = -\frac{\delta}{R^2} < 0$, $\frac{\partial^2}{\partial X^2}H(R,X) = -\frac{\alpha}{X^2} < 0$, and $\frac{\partial^2}{\partial R\partial X}H(R,X) = 0$. Therefore, the function H is concave and obtains its unique maximum value at (R^*, X^*). Consequently, all curves satisfying $H(R,X) = c$ are closed orbits. We can thus conclude that the Lotka–Volterra model (9.37) has an infinite number of cyclic orbits located around the positive equilibrium (Figure 9.4b). These orbits are stable, but not asymptotically stable: two orbits with similar initial conditions will remain close to each other, but the orbits will not approach each other.

9.4.4 Average Population Densities

Consider now an arbitrary cyclic orbit of the Lotka–Volterra model (9.37). Let T denote the length of this orbit, i.e. $R(t + n \cdot T) = R(t)$ and $X(t + n \cdot T) = X(t)$ for all $t \in \mathbb{R}$ and $n \in \mathbb{N}$. Therefore,

$$0 = \ln R(T) - \ln R(0) = \int_0^T \frac{d}{dt}\ln R(t)dt = \int_0^T \frac{\frac{d}{dt}R(t)}{R(t)}dt$$

$$= \int_0^T \alpha - \beta X(t)dt = \alpha T - \beta \int_0^T X(t)dt \quad (9.45)$$

which means that the average population density of the predator population is

$$\bar{X} = \frac{1}{T}\int_0^T X(t)dt = \frac{\alpha}{\beta} = X^* \quad (9.46)$$

Analogously, we obtain the average population density of the prey population

$$\bar{R} = \frac{1}{T}\int_0^T R(t)dt = \frac{\delta}{\gamma\beta} = R^* \quad (9.47)$$

We can thus conclude that in the Lotka–Volterra model (9.37) the average population densities are the same as in the positive equilibrium (R^*, X^*).

9.4.5 Effect of Fishing on the Population Densities

Let us now add the effect of fishing into the Lotka–Volterra model (9.37). Assume that the catch is proportional to the fishing effort and the population densities, and the fishing effort for prey fish is $\mu > 0$ and $\varphi > 0$ for the predatory fish. Under these assumptions, the model (9.37) becomes

$$\begin{cases} \frac{dR}{dt} = \alpha R - \beta RX - \mu R \\ \frac{dX}{dt} = \gamma\beta RX - \delta X - \varphi X \end{cases} \quad (9.48)$$

which is the Lotka–Volterra model (9.37) with parameters $\alpha^* = \alpha - \mu$ and $\delta^* = \delta + \varphi$. Based on the analysis above, the average population densities in the model including fishing is

$$\overline{R} = \frac{\delta + \varphi}{\gamma \beta}$$
$$\overline{X} = \frac{\alpha - \mu}{\beta} \qquad (9.49)$$

These results lead into the surprising conclusion that fishing increases the average prey population density and decreases the average predator population density, giving some explanation to the observations by Humberto D'Ancona.

9.5 Model of Killer T-Cell and Cancer Cell Dynamics

Cancer often involves abnormal growth of cancer cells, potentially leading into malignant tumors. Since the molecular makeup of the disease can vary considerably between patients, better treatment results may be obtained by personalized treatment. ODE modelling may provide better understanding of cancer dynamics and thus provide aid in making treatment decisions. In this section, we investigate the model of killer T-cell and cancer cell dynamics by Halkola et al. (2020).

9.5.1 Model Definition

First, we list the mechanisms according to which resources, cancer cells, and killer T-cells affect each other in the model. These relations are illustrated in Figure 9.5.

9.5.1.1 Resource Dynamics

In this model, R denotes the resource concentration. Resources follow chemostat dynamics. There is thus a constant inflow of medium with resource concentration $\hat{R}$, and the resource concentration of the outflowing medium equals R. The parameter λ describes the speed of the flow. Resource consumption by cancer cells takes place according to the law of mass action. When the cancer cell density is low, the proliferation consumption rate of cancer cells with proliferation strategy $s_i \geq 0$ is $\alpha(s_i)$, assumed to be of the form

$$\alpha(s_i) = \alpha_0 + a \frac{s_i/b}{1 + s_i/b} \qquad (9.50)$$

The function $\alpha(s_i)$ is an increasing function of s_i, and its value approaches $\alpha_0 + a$ for large s_i. The parameter α_0 is the baseline consumption rate,

9.5 Model of Killer T-Cell and Cancer Cell Dynamics

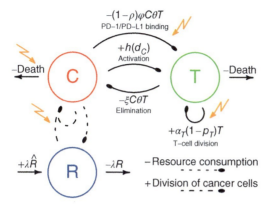

Figure 9.5 Presentation of the model as relations between different cell types (T for active killer T-cells and C for cancer cells) and resources (R). Positive terms increase the density and negative terms decrease the density with the ellipse pointing the affected entity. Cancer cell death includes both normal cell death and inhibition caused by targeted treatments and chemotherapies. T-cell death includes normal cell death, self-regulation, and death caused by chemotherapy. Cancer cells consume resources to maintain their proliferative capacity. Killer T-cells get activated when they encounter cancer cells' antigens, but cancer cells can also inhibit T-cell activation by binding their ligands (such as PD-L1) to inhibitory receptors on T-cell surface (such as PD-1). Lightning bolt arrows point the relations that are affected by different treatments. Source: Halkola et al. (2020).

and b is a half-saturation constant, i.e. the proliferation strategy value at which $\alpha(b) = \alpha_0 + \frac{1}{2}a$. Furthermore, cancer cells restrict each other access to resources so that the effective consumption rate is obtained by multiplying $\alpha(s_i)$ by the factor

$$1 - w\frac{C}{1+C} \tag{9.51}$$

which is a decreasing function of the total cancer cell density C and in which w is the maximum proportion of restriction. With these assumptions, we obtain the ODE for the resource concentration R given in (9.52).

$$\frac{dR}{dt} = \lambda(\hat{R} - R) - \left(1 - w\frac{C}{1+C}\right)\sum_j \alpha(s_j)RC_j \tag{9.52}$$

in which C_j is the density of the cancer cell subpopulation with proliferation strategy s_j and $C = \sum_j C_j$.

9.5.1.2 Cancer Cell Dynamics

We assume that cancer cells attempt division with a rate that is directly proportional to their resource usage, with the conversion coefficient γ. However, attempted divisions may be unsuccessful either due to treatment effects or limitations of space. A cytostatic drug may cause the cell to die during the attempted cell division with probability p_i. Finally, cells that have avoided this risk of death divide successfully with probability $1 - C/K$.

In addition to the natural death rate μ_i, targeted drugs and active killer T-cells cause mortality. Targeted drugs increase the death rate by β_i, and active killer T-cells that enter the microenvironment of cancer (θT) may kill cancer cells with the subpopulation-specific rates ξ_i. With these assumptions, we obtain the ODE for the cancer cell densities C_i given in (9.53).

$$\frac{dC_i}{dt} = \gamma \alpha(s_i)\left(1 - w\frac{C}{1+C}\right) RC_i \left((1-p_i)\left(1-\frac{C}{K}\right) - p_i\right)$$
$$- \mu_i C_i - \beta_i C_i - \xi_i C_i \theta T \quad (9.53)$$

Cancer cells deliver antigens to the blood stream when they die, with total rate

$$d_C = \sum_j \left(\epsilon \mu_j + \gamma \alpha(s_j)\left(1 - w\frac{C}{1+C}\right) Rp_j + \beta_j + \xi_j \theta T\right) C_j \quad (9.54)$$

Here, $\epsilon < 1$ specifies the proportion of delivered antigens by one normal cancer cell apoptosis compared with death caused by drugs or killer T-cells.

9.5.1.3 Killer T-Cell Dynamics

Antigens cause the activation of antigen-specific killer T-cells from naïve T-cells with the rate $h(d_C)$, which is assumed to be of sigmoid form

$$h(d_C) = m\left(1 - \frac{1}{1+(d_C/u)^\nu}\right) \quad (9.55)$$

where m is the maximum rate of killer T-cell activation, u is a half-saturation constant ($h(u) = m/2$), and ν affects the slope of $h(d_C)$ at u.

We assume that the amount of naïve T-cells is large compared with the rate $h(d_C)$, so that the amount of naïve T-cells can be assumed to be constant. Active killer T-cells proliferate with the rate α_T, and they go through normal cell death at the rate μ_T. In addition, immune system uses self-regulation according to bimolecular reaction causing T-cell death with reaction constant δ. Cytotoxic drugs also affect killer T-cell division, causing death with probability p_T. The proportion of active killer T-cells present in the microenvironment of cancer is assumed to be the constant θ for simplicity. Some cancer cells are able to make active killer T-cells ineffective, with the subpopulation-specific intensities φ_i. This effect can be prevented with probability ρ_i by immunotherapy.

As a result, we obtain the following ODE for the active killer T-cells (T):

$$\frac{dT}{dt} = h(d_C) + (\alpha_T(1-p_T) - \mu_T)T - \alpha_T p_T T$$
$$- \frac{\delta}{2} T^2 - \sum_j (1-\rho_j)\varphi_j C_j \theta T \quad (9.56)$$

9.5.2 Model Dynamics Without Treatment

Let us first investigate representative cases of model dynamics with two cancer subpopulations without treatment. Here, we only investigate such cases, in which cancer growth is possible. The upper panels of Figure 9.6 show cell densities with respect to time using the same initial conditions, $R(0) = \hat{R} = 1.5$, $C_1(0) = C_2(0) = 0.05$, and $T(0) = 0.02$, in each panel. The lower panels are specific phase-plane plots. A full phase-plane plot of an ODE system with four variables would be four dimensional, and plotting such figures on a two dimensional paper would be complicated. Therefore, we choose the sum of the cancer densities $C_1 + C_2$ and the T-cell density T as axes of our phase-plane plots and assume that the resource density is at the equilibrium satisfying $\frac{dR}{dt} = 0$ in (9.52) when determining the isoclines.

Figure 9.6a,d illustrates a case in which killer T-cells do not activate fast enough for their density to increase. Instead, the T-cell density decreases, allowing the fast increase of cancer cells toward a maximum level. Both Figure 9.6b,e and c,f illustrate a situation in which the T-cell density at least initially increases, and an equilibrium with substantial amount of T-cells (and cancer cells) exists. In the latter case (Figure 9.6c,f), however, that equilibrium is not stable, and population dynamics approaches a stable limit cycle. In each case, the subpopulation C_1 with $s_1 = 1$ divides

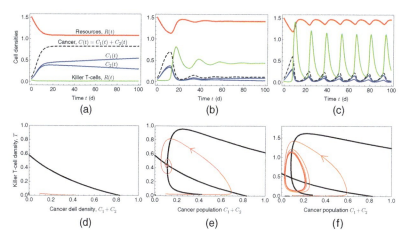

Figure 9.6 Three representative cases of dynamic competition between active killer T-cell (T) and cancer cell populations without treatment. Cell densities with respect to time (a–c) and phase-plane plots (d–f). (a, d) Active killer T-cells decrease, leading to fast increase of cancer cells toward a maximum level. (b, e) The amount of active killer T-cells increases without treatment, leading to decreased amount of cancer cells. The cancer cell amounts alternate, but they are approaching a fixed steady state. (c, f) Cell dynamics approaches a cyclic attractor.

faster and eventually dominates the less aggressive subpopulation C_2 with $s_2 = 0.95$. Figure 9.6 illustrates that patient-specific parameters may cause qualitatively different behavior, and in some cases, the immune system may be able to control the cancer even without treatment. Periodic behavior has been found both in cancer (Fortin and Mackey, 1999) and in the immune system (Stark et al., 2007), so likely it is also possible in the competition between immune system and cancer cells.

9.5.3 Treatment Effects

For patients without an immune response (low T-cell density, Figure 9.6a,d) cancer cells may naturally reach high densities. Even for patients with such cancer dynamics as illustrated in Figure 9.6b,e and c,f, without treatment, cancer cells may reach high densities, which can lead to severe symptoms and even death of the patient. Using the model investigated here, Halkola et al. (2020) investigated different treatment options, such as repeated immunotherapy, a combination of targeted therapy and immunotherapy, and a combination of chemotherapy and immunotherapy. These investigations suggest that patient-specific treatment may provide substantial benefits for the patient.

Here, we investigate a specific case with bistability in the model dynamics without treatment. In other words, depending on the initial conditions, model dynamics may reach either an equilibrium with a low cancer cell density or an equilibrium with a substantial amount of cancer cells (Figure 9.7b). Figure 9.7a illustrates the model dynamics

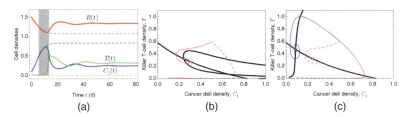

Figure 9.7 Threshold immunotherapy given when the cancer cell density exceeds 0.5. (a) Cell densities with respect to time. The gray rectangle marks the treatment period. Dashed curves illustrate model dynamics if no treatment would be given. (b, c) Phase-plane plots showing the isoclines (black lines) of the killer T-cell density T and cancer cell density C, (b) during no treatment and (c) during treatment. The solid trajectories shown with respect to time in (a) are also shown in the phase-plane plots (red curves). The trajectory is plotted with a solid curve when the actual treatment status is the same as in the phase-plane plot. In addition, the blue trajectory in (c) illustrates model dynamics if immunotherapy was given even after the cancer cell density drops below the threshold.

with respect to time starting from a situation with a low T-cell density, which is not enough to prevent the cancer cell density to increase to large values. Without treatment, the latter equilibrium would be reached. Now, we consider the results of immunotherapy, which is given when the cancer cell density exceeds 0.5. The timing of this treatment is illustrated with a gray rectangle in Figure 9.7a. During the treatment, the drug concentration is assumed constant, resulting in a constant treatment effect. In this case, the treatment almost completely prevents PD-1/PD-L1 binding and thus prevents cancer cells from making active killer T-cells ineffective ($\rho_i = 0.999$). If such treatment was given for a long time, model dynamics would reach an equilibrium with rather high T-cell density ($T \approx 0.47$, $C \approx 0.09$, blue curve in Figure 9.7c). However, even a relatively short treatment period (solid curve in Figure 9.7c) takes the model dynamics away from the basin of attraction of the equilibrium with high cancer cell density. Therefore, when the treatment is ended and model dynamics is described by Figure 9.7b, dynamics approach the equilibrium with a low cancer cell density ($T \approx 0.32$, $C \approx 0.24$) and substantial T-cell density. This example illustrates that for some patients, a short period of treatment may be enough, while others need a long sequence of treatment.

9.6 Conclusion

In this chapter, we have given an overview of modelling with ODEs. We discussed both model derivation and model analysis with examples. In the mechanistic modelling approach, the model is derived from individual-level processes, and the mathematical model is obtained by straightforward book-keeping. Here, we discussed the mechanistic derivation of ODEs using elementary reactions. From the model analysis side, we discussed both analytical and numerical solving of ODEs. In addition, we explained how to determine the local stability of equilibria. In our examples, we also used phase-plane analysis to illustrate model dynamics. The presented examples illustrate the wide applicability of ODE modelling in systems biology.

Acknowledgments

The author wishes to thank Stefan Geritz for his enthusiasm about mechanistic modelling. The author also wishes to thank the coauthors of the article on which the cancer model section is based for collaboration, especially Anni Halkola and Tero Aittokallio.

References

Tatsuya Akutsu, Satoru Miyano, and Satoru Kuhara. Inferring qualitative relations in genetic networks and metabolic pathways. *Bioinformatics*, 16(8):727–734, 2000.

D. Angeli, J. E. Ferrell, and E. D. Sontag. Detection of multistability, bifurcations, and hysteresis in a large class of biological positive-feed back systems. *Proceedings of the National Academy of Sciences of the United States of America*, 101(7):1822–1827, 2004.

Åke Brännström and D. J. T. Sumpter. The role of competition and clustering in population dynamics. *Proceedings of the Royal Society of London Series B*, 272:2065–2072, 2005. doi: 10.1098/rspb.2005.3185.

H. Eskola and S. A. H. Geritz. On the mechanistic derivation of various discrete-time population models. *Bulletin of Mathematical Biology*, 69:329–346, 2007.

Pascal Fortin and Michael C. Mackey. Periodic chronic myelogenous leukaemia: spectral analysis of blood cell counts and aetiological implications. *British Journal of Haematology*, 104(2):336–345, 1999. ISSN 00071048. doi: 10.1046/j.1365-2141.1999.01168.x.

Stefan Geritz and Mats Gyllenberg. A mechanistic derivation of the DeAngelis-Beddington functional response. *Journal of Theoretical Biology*, 314:106–108, 2012.

S. A. H. Geritz and É. Kisdi. On the mechanistic underpinning of discrete-time population models with complex dynamics. *Journal of Theoretical Biology*, 228:261–269, 2004.

S. A. H. Geritz, É. Kisdi, G. Meszéna, and J. A. J. Metz. Evolutionarily singular strategies and the adaptive growth and branching of the evolutionary tree. *Evolutionary Ecology*, 12:35–57, 1998.

M. Gyllenberg and K. Parvinen. Necessary and sufficient conditions for evolutionary suicide. *Bulletin of Mathematical Biology*, 63:981–993, 2001. doi: 10.1006/bulm.2001.0253.

M. Gyllenberg, K. Parvinen, and U. Dieckmann. Evolutionary suicide and evolution of dispersal in structured metapopulations. *Journal of Mathematical Biology*, 45:79–105, 2002. doi: 10.1007/s002850200151.

Anni S. Halkola, Kalle Parvinen, Henna Kasanen, Satu Mustjoki, and Tero Aittokallio. Modelling of killer T-cell and cancer cell subpopulation dynamics under immuno- and chemotherapies. *Journal of Theoretical Biology*, 488:110136, 2020. doi: 10.1016/j.jtbi.2019.110136.

I. A. Hanski. *Metapopulation Ecology*. Oxford University Press, Oxford, 1999.

A. Hastings. Can spatial variation alone lead to selection for dispersal. *Theoretical Population Biology*, 24:244–251, 1983.

A. Lotka. Fluctuations in the abundance of a species considered mathematically. *Nature*, 119:12, 1927.

J. A. J. Metz and M. Gyllenberg. How should we define fitness in structured metapopulation models? Including an application to the calculation of ES dispersal strategies. *Proceedings of the Royal Society of London Series B*, 268:499–508, 2001.

Bernt Øksendal. *Stochastic Differential Equations: An Introduction with Applications*. Springer-Verlag, Berlin, Heidelberg, 2003.

C. Rueffler, Martijn Egas, and J. A. J. Metz. Evolutionary predictions should be based on individual-level traits. *American Naturalist*, 168:E148–E162, 2006.

Jaroslav Stark, Cliburn Chan, and Andrew J. T. George. Oscillations in the immune system. *Immunological Reviews*, 216(1):213–231, 2007. ISSN 01052896. doi: 10.1111/j.1600-065X.2007.00501.x.

V. Volterra. Fluctuations in the abundance of a species considered mathematically. *Nature*, 118:558–560, 1926.

V. Volterra. Fluctuations in the abundance of a species considered mathematically. *Nature*, 119:12–13, 1927.

10

Network Modelling Methods for Precision Medicine

Elio Nushi[1], Victor-Bogdan Popescu[2], Jose-Angel Sanchez Martin[3], Sergiu Ivanov[4], Eugen Czeizler[2,6], and Ion Petre[5,6]

[1] Department of Computer Science, University of Helsinki, Helsinki, Finland
[2] Department of Information Technology, Åbo Akademi University, Turku, Finland
[3] Department of Computer Science, Technical University of Madrid, Madrid, Spain
[4] IBISC Laboratory, Université Paris-Saclay, Université Évry, Paris, France
[5] Department of Mathematics and Statistics, University of Turku, Turku, Finland
[6] National Institute of Research and Development in Biological Sciences, Bucharest, Romania

10.1 Introduction

Network medicine is a promising recent approach in which the goal is to analyze the dysregulation of a disease through its specific molecular interactions (Saqi et al. (2016), Tian et al. (2012)). The key analytic power of this approach is that knowledge about disease drivers and specific pathway deregulations can be combined with mechanistic knowledge of drug mechanisms to identify optimal drug combinations and do this dynamically throughout the evolution of the disease.

An exciting aspect of this approach is that it can, in principle, be applied in a personalized way, taking into account patient-specific aspects such as co-morbidities, previous treatments, and the patient's own molecular data (such as her mutations, gene expression anomalies, and corrupted signaling pathways). Therefore, a disease can be seen as part of a patient's own molecular and clinical context through the cumulative effect of various deregulations and anomalies. Also, drug therapies can be seen as external interventions aiming to compensate for the effects of these anomalies in the patient-specific disease network. The focus is on identifying tailored drug combinations uniquely suited to that patient's disease network in the current step of her disease progression.

Systems Biology Modelling and Analysis: Formal Bioinformatics Methods and Tools, First Edition. Edited by Elisabetta De Maria.
© 2023 John Wiley & Sons, Inc. Published 2023 by John Wiley & Sons, Inc.

This chapter introduces several network modelling methods and demonstrates their potential applicability in personalized medicine. We survey several network centrality measures, aiming to identify parts of the network that are unusual in the context of its topology; we discuss their definitions and the intuition of their significance. We also discuss two systems controllability methods: network controllability and maximum dominating sets. The aim of these methods is to identify efficient interventions to change the network's configuration and in principle to change from a setup associated with disease to the one associated with a healthy state.

We demonstrate how these network modelling methods can be used in precision medicine for identifying targeted, personalized drug combinations. Our case study is multiple myeloma, an incurable cancer of the blood. We analyze a dataset consisting of the genetic mutations of three different patients. For each of them, we construct its own personalized protein–protein directed interaction networks, and we analyze them with some of the methods in this chapter to extract personalized predictions of optimal drug combinations. We compare these results with the standard therapy lines in multiple myeloma.

We also include a brief discussion on the availability of software tools supporting this line of research.

The chapter is written in a tutorial style to facilitate the adoption of these methods.

10.2 Network Modelling Methods

We discuss in this section a number of network analysis methods: network centrality measures (including degree centralities, proximity centralities, path centralities, and spectral centralities) and two systems controllability methods (network controllability and minimum dominating sets). We apply several of these methods to a medical case study in Section 10.3.

10.2.1 Network Centrality Methods

Real-world networks often include a large number of nodes and connections, but the importance of the nodes is generally not the same. The simplest way to measure the importance of a given node is to compute its degree: the number of incident edges. However, in many cases, more sophisticated approaches are required to produce meaningful measures of importance. In the most general sense, a centrality measure can be defined in the following way.

Definition 10.1 *(Centrality)* Let $G = (V, E)$ be a network. A *centrality measure* is any function $f : V \to \mathbb{R}$.

This definition imposes no constraints on f, but most centrality measures take into account the structural properties of the network G – node connectivity, edge weights, etc. Based on which structural properties they take into account, centrality measures can be grouped into the following categories:

- *degree centralities*: measures based on the degree of a node;
- *proximity centralities*: measures based on how close a node is to the other nodes in the network;
- *path centralities*: measures based on the role the node plays in paths that traverse it;
- *spectral centralities*: measures related to the algebraic properties of the adjacency matrix of the network (in particular its eigenvectors and eigenvalues).

While the majority of centralities focus on individual nodes, one can define measures focused on other structures: edges, subsets of nodes, etc. Most of these measures are straightforward derivations from node-based centralities (e.g. edge and group betweenness in Brandes (2008)), and we will not discuss them here. Furthermore, we only consider unweighted networks, and we focus on structural measures, which do not take into account any possible dynamical states.

In the rest of this subsection, we discuss some well-known and often used centrality measures in detail. In particular, we give the intuitive motivations, the formal definitions, and the highlighted properties of the network. Furthermore, we give references to algorithms for computing the centrality measures and briefly discuss their time complexities. Finally, we describe *network centrality indices* – wide scores measuring the centralization of a network as a whole and allowing for comparisons between networks.

Although we aim at a comprehensive overview of the state of the art, we do not always provide a fully detailed discussion of all subjects. For in-depth treatment, we refer the reader to Koschützki (2007), Brandes and Erlebach (2005), and Newman (2010).

In the following sections, we will frequently refer to undirected and directed star-topology networks. The k-node undirected star-topology network is $U_k^\star = (V, E)$ (also known as the full bipartite graph $K_{1,k}$), where $V = \{1, \ldots, k\}$ and $E = \{\{i, 1\} \mid 2 \leq i \leq k\}$. The k-node directed star-topology network is $G_k^\star = (V, E')$, where V is the same as in $U_k^\star$ and $E' = \{(i, 1) \mid 2 \leq i \leq k\}$. Note that E consists of unordered pairs of vertices, while E' consists of ordered pairs. We will also write $U^\star$ and $G^\star$ to refer to the general notion of undirected and directed star-topology networks, respectively.

10.2.1.1 Running Example

We will use the directed network in Figure 10.1 to illustrate the centrality measures and the related concepts. This is a scale-free random network generated using the Python library *NetworkX* (Hagberg et al., 2008, 2021c) with the following line of code: networkx.scale_free_graph(10, alpha=.7, beta=.2, gamma=.1, seed=3).

10.2.1.2 Degree Centralities

Degree centrality was first introduced in Shaw (1954) to study the structure and behavior of groups of individuals in a society. It measures the importance of a node by directly counting the adjacent edges.

Definition 10.2 *(Degree Centrality)* Let $G = (V, E)$ be a network.

- If G is undirected, then the *degree centrality* is the function $C_D : V \to \mathbb{R}$ assigning to every node its degree: $C_D(v) = deg(v)$.
- If G is directed, then the *in-degree centrality* is the function $C_D^- : V \to \mathbb{R}$, assigning to every node its in-degree: $C_D^-(v) = deg^-(v)$. The *out-degree centrality* $C_D^+ : V \to \mathbb{R}$ assigns to every node its out-degree $C_D^+(v) = deg^+(v)$. The (full) degree centrality is the function $C_D(v) = C_D^+(v) + C_D^-(v)$.

It follows from these definitions that $C_D(v)$ (or $C_D^+(v)$ and $C_D^-(v)$) is large when the node v is adjacent to a high number of nodes. The extreme cases are $C_D(v) = k - 1$, in which v is connected to all other nodes in a k-node network and $C(v) = 0$, when v is isolated in the network.

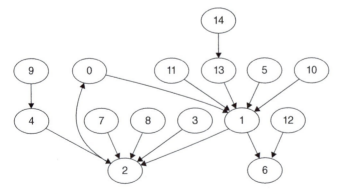

Figure 10.1 Network serving as a running example to illustrate centralities and the related concepts.

The degree centrality is strongly related to the number of nodes in a network. As a trivial example, consider the degree centrality of the central node in the star-topology network $U_k^\star$: $C_D^+(1) = k - 1$. The centrality of node 1 is larger in larger networks with this topology, even though the intuitive idea of the importance of this node is essentially the same: it is connected to all other nodes of the network in both cases. Normalized degree centrality may be used to better capture the independence of the notion of centrality on the size of the network.

Definition 10.3 *(Normalized Degree Centrality)* Let $G = (V, E)$ be a (either directed or undirected) network. The *normalized degree centrality* is the function $\tilde{C}_D : V \to \mathbb{R}$ defined as follows:

$$\tilde{C}_D(v) = \frac{C_D(v)}{|V| - 1}$$

Normalized degree centrality $\tilde{C}_D(v)$ therefore gives the ratio of the nodes adjacent to v, and it follows from the definition that $0 \leq \tilde{C}_D(v) \leq 1$. $\tilde{C}_D(v)$ can be thought of as the probability of v to be connected to another node w picked at random, an intuition that can be useful when generating random networks with a fixed degree distribution.

Example 10.1 The following table gives the degree, in-degree, and out-degree centralities for the example network in Figure 10.1.

v	0	1	2	3	4	5	6	7	8	9	10	11	12	13	14
$C_D(v)$	2	7	6	1	2	1	2	1	1	1	1	1	1	2	1
$C_D^-(v)$	1	5	5	0	1	0	2	0	0	0	0	0	0	1	0
$C_D^+(v)$	1	2	1	1	1	1	0	1	1	1	1	1	1	1	1

The complexity of computing the degree centrality of every individual node of a network depends linearly on the number of nodes and edges: $O(|V| + |E|)$ (e.g. (Das et al., 2018)). Computing the centrality of any given node may be of the complexity $O(|V|)$ or $O(|E|)$ depending on the data structure used to represent the connections.

The degree centrality is a useful tool not only for identifying "targets" in a given network but also for deciding which nodes may be discarded without impacting the quality of the model. This is common in network biology, see Hahn and Kern (2004) and Koschützki and Schreiber (2008). However, degree centrality is a strongly local measure, mostly focusing on individual nodes and their immediate neighborhoods. In practice, this means that many nodes may often have close degree centralities, requiring finer measures to discern relevant features (e.g. (Kang et al., 2011)).

10.2.1.3 Proximity Centralities

In this section, we discuss closeness, harmonic, and eccentricity centralities.

The degree centrality measure is a very straightforward approach to evaluate the importance of a node in a network. However, the nodes that are connected by a small number of edges to many others in a network are also important: their "influence" can reach many other nodes quickly. Counting the neighbors of a node clearly does not suffice to asses this kind of closeness to other nodes. The important aspect is rather having a small average distance to the other nodes in the network. This leads to the following definition of the closeness centrality measure as the reciprocal of farness (Bavelas, 1950, Leavitt, 1949, Sabidussi, 1966).

Definition 10.4 *(Closeness Centrality)* Let $G = (V, E)$ be an undirected network. The *closeness centrality* is the function $C_C : V \to \mathbb{R}$ defined as follows:

$$C_C(v) = \frac{1}{\sum_{u \in V} d(u, v)}$$

As in the case of degree centralities, the size of the network has an impact on the closeness centrality of its nodes: the larger the network, the more paths it contains, and the lower closeness centralities tend to become. To make closeness centralities more uniform, Beauchamp (1965) proposed the normalized version of this measure.

Definition 10.5 *(Normalized Closeness Centrality)* Let $G = (V, E)$ be a connected undirected network. The *normalized closeness centrality* is the function $\tilde{C}_C : V \to \mathbb{R}$ is defined as follows:

$$\tilde{C}_C(v) = \frac{|V| - 1}{\sum_{u \in V} d(u, v)}$$

Normalized closeness centrality can be thought of as the inverse of the mean of the distances to v from all other nodes. The bounds on the values of the normalized closeness centrality $\tilde{C}_C$ are the same as the bounds on the values of C_C: $0 < \tilde{C}_C(v) \leq 1$, but $\tilde{C}_C(v)$ reaches its maximal value 1 for any node v adjacent to all the other nodes of a given network.

A major drawback of closeness centrality is that it does not yield meaningful values on disconnected networks. Indeed, if no path connects nodes

u and v, by definition $d(u, v) = \infty$, and therefore, a single isolated node would make the sums of distances in the definition of closeness infinite, and the closeness centralities themselves will all become 0. There are several different ways in which this can be addressed:

- Restrict the notion of closeness centrality to strongly connected graphs (or strongly connected components of arbitrary graphs). This avoids the problem of having pairs of nodes (u, v) whose distance is infinite, on the grounds of v being unreachable from u.
- Restrict the sum of distances in the definition of closeness centrality to pairs of reachable nodes, addressing the same issue of infinite distances.
- Replace any infinite distances with a large enough constant, as proposed in Rochat (2009) and Csárdi and Nepusz (2006).

Closeness centrality can also be defined for directed networks. To do this, we can consider in the definition either the distances $d(u, v)$ from all ancestors of v to v or the distances $d(v, u)$ from v to all its descendants. This is important when using the closeness centrality measure as a proxy for the notion of either a node that is reachable (and modifiable) from many directions or a node that is influential in being able to reach many other nodes. One example of such a definition is the *Lin index* (Lin, 1976). The software package *NetworkX* computes the distances from the nodes that reach v (Hagberg et al., 2021a) and normalizes with respect to the number of these nodes. The difficulties with infinite distances for pairs of unreachable nodes also persist in the directed case, with possible solutions similar to those for the undirected case.

Example 10.2 The following table gives the closeness centralities and the normalized closeness centralities, rounded to two digits after the decimal point, for the nodes of the example network in Figure 10.1. The lines $C_C^u(v)$ and $\tilde{C}_C^u(v)$ take this network to be undirected, meaning that the directed edges appearing in the figure can be traversed both ways. The calculations for directed networks were done with *NetworkX* using the approach explained above.

v	0	1	2	3	4	5	6	7	8	9	10	11	12	13	14
$C_C(v)$	0.03	0.04	0.05	0.00	1.00	0.00	0.03	0.00	0.00	0.00	0.00	0.00	0.00	1.00	0.00
$\tilde{C}_C(v)$	0.40	0.48	0.60	0.00	1.00	0.00	0.36	0.00	0.00	0.00	0.00	0.00	0.00	1.00	0.00
$C_C^u(v)$	0.03	0.05	0.04	0.03	0.03	0.03	0.03	0.03	0.03	0.02	0.03	0.03	0.02	0.03	0.02
$\tilde{C}_C^u(v)$	0.48	0.64	0.58	0.38	0.40	0.40	0.42	0.38	0.38	0.29	0.40	0.40	0.30	0.42	0.30

Computing the closeness centrality of a node v requires finding the shortest paths to v from all other nodes of the network. As constructing all shortest paths to one particular node v is of complexity $O(|E|)$, using, e.g., a breadth-first search, computing the closeness centrality for all the nodes within a network is of complexity $O(|V| \cdot |E|)$. This means that computing the exact value of closeness centrality is impractical for many biological networks, which often contain thousands of nodes and tens of thousands of connections. It turns out that, in practice, one often only needs the first k nodes with the highest closeness centrality, without requiring the actual centrality values. Such rankings can be computed in reasonable time even for very large networks, see, e.g. Bergamini et al. (2019).

Due to its non-locality, closeness centrality is a finer tool for structural network analysis than degree centrality. For example, closeness fares better in identifying influential groups of nodes that may not individually have high-degree centrality. Closeness centrality has been shown to perform particularly well in biological network analysis. For example, Ma and Zeng (2003) shows that a slightly modified closeness measure allows for associating 8 of the top 10 metabolites of the metabolic network of *Escherichia coli* with the glycolysis and citric acid cycle pathways.

A modification of closeness centrality, addressing the difficulty of infinite distances, consists in swapping the summation out of the denominator, effectively transforming what is an inverse arithmetic mean in normalized closeness centrality into inverse *harmonic* mean. This new centrality measure was first discussed in Marchiori and Latora (2000), then independently introduced in Dekker (2005) under the name "valued centrality," and finally gained its current name of harmonic centrality in Rochat (2009).

Definition 10.6 *(Harmonic Centrality)* Let $G = (V, E)$ be a (either directed or undirected) network. The *harmonic centrality* is the function $C_H : V \to \mathbb{R}$ defined as follows:

$$C_H(v) = \sum_{u \in V \setminus \{v\}} \frac{1}{d(u, v)}$$

Note that all nodes u that are not connected to v do not contribute to $C_H(v)$ because $d(u, v) = \infty$, meaning that $1/d(u, v) = 0$.

As with closeness centrality, the same definition of harmonic centrality can be used for directed networks, in which case the order of nodes in the denominator $d(u, v)$ becomes important. If v has in-degree 0, $C_H(v) = 0$ by direct computation of the formula in the definition.

To avoid an increase in harmonic centrality only because of the increase in the size of the network, one defines the normalized version.

Definition 10.7 *(Normalized Harmonic Centrality)* Let $G = (V, E)$ be a (either directed or undirected) network. The *normalized harmonic centrality* is the function $\tilde{C}_H : V \to \mathbb{R}$ defined as follows:

$$\tilde{C}_H(v) = \frac{1}{|V|-1} \sum_{u \in V \setminus \{v\}} \frac{1}{d(u,v)}$$

It follows that $0 \leq \tilde{C}_H(v) \leq 1$ (and $0 \leq C_H(v) \leq |V|-1$) both in directed and undirected networks. $\tilde{C}_H(v) = 0$ for isolated vertices, while $\tilde{C}_H(v) = 1$ ($C_H(v) = |V|-1$) for the center of a star network, in both directed and undirected cases, because both in $U_k^\star$ and $G_k^\star$, every node is connected to the central node 1.

Note that central nodes in large connected components will have greater values of harmonic centrality than central nodes in small connected components. Furthermore, nodes in disconnected networks will tend to have lower harmonic centralities than nodes in connected networks.

Example 10.3 The following table gives the harmonic centralities and the normalized harmonic centralities, rounded to two digits after the decimal point, for the nodes of the example network in Figure 10.1. The lines $C_H^u(v)$ and $\tilde{C}_H^u(v)$ take this network to be undirected, meaning that the directed edges appearing in the figure can be traversed both ways.

v	0	1	2	3	4	5	6	7	8	9	10	11	12	13	14
$C_H(v)$	5.42	7.58	8.33	0.00	1.00	0.00	6.37	0.00	0.00	0.00	0.00	0.00	0.00	1.00	0.00
$\tilde{C}_H(v)$	0.39	0.54	0.60	0.00	0.07	0.00	0.45	0.00	0.00	0.00	0.00	0.00	0.00	0.07	0.00
$C_H^u(v)$	7.50	10.33	9.67	6.00	6.67	6.25	6.92	6.00	6.00	4.82	6.25	6.25	4.95	6.92	4.95
$\tilde{C}_H^u(v)$	0.54	0.74	0.69	0.43	0.48	0.45	0.49	0.43	0.43	0.34	0.45	0.45	0.35	0.49	0.35

The computational complexities related to the harmonic centrality are the same as those of the closeness centrality because of the similarities in the definitions of the two measures. As finding the shortest path is of complexity $O(|E|)$, computing the harmonic centrality in a given directed or undirected network $G = (V, E)$ is of complexity $O(|V| \cdot |E|)$. For very large networks, approximate calculation strategies can be used, or alternatively, the direct computation of centrality can be replaced by finding the top k nodes with the highest centrality value, similarly to Bergamini et al. (2019).

Like closeness centrality, harmonic centrality has great potential for analysis of biological networks because it captures the intuition of the influence of a node decaying with the distance while also naturally handling disconnected networks. Online resources for systems biology offer tools to compute harmonic centrality (e.g. (Zhang et al., 2016)), and this centrality measure is used in analyzing simulations of growth of biological networks (Paul and Kollmannsberger, 2020). We remark, however, that several papers use the term "harmonic centrality" to refer to a rather different centrality measure, e.g. Ren et al. (2015) and Mao et al. (2020).

Another modification to closeness centrality we will briefly consider in this subsection was introduced in Dangalchev (2006). This work goes beyond harmonic centrality and adds an exponential to the denominator:

$$D(v) = \sum_{u \in V \setminus \{v\}} \frac{1}{2^{d(u,v)}}$$

Like harmonic centrality, D treats disconnected networks naturally. In addition, D interacts conveniently with various operations on graphs, in particular with different kinds of graph unions (Dangalchev, 2006, Section 2). Finally, this centrality measure can be generalized to the following form (Dangalchev, 2011):

$$D'(v) = \sum_{u \in V \setminus \{v\}} \alpha^{d(u,v)}$$

where $\alpha \in (0,1)$. Clearly, for $\alpha = \frac{1}{2}$, $D'(v) = D_C(v)$, and as α increases between 0 and 1, D' moves from local (mostly immediate neighbors count) to global (even long-distance connections count).

Even though the centrality measures we surveyed so far in this subsection are all based on the notion of closeness, one should underline that they are not true extensions of the closeness centrality. Indeed, the paper (Yang and Zhuhadar, 2011) shows that, even on a seven-node tree, these centrality measures yield close, but different values.

The last centrality measure based on the length of paths to a given node that we discuss is the eccentricity centrality. It formalizes the intuition that important nodes are those from which any other node is quickly reachable. It was originally introduced in Hage and Harary (1995), but we give here the mathematical definition from Koschützki (2007).

Definition 10.8 (*Eccentricity Centrality*) Let $G = (V, E)$ be a strongly connected (either directed or undirected) network. The *eccentricity centrality* is the function $C_e : V \to \mathbb{R}$ defined as follows:

$$C_e(v) = \frac{1}{\max_{u \in V} d(v, u)}$$

The value in the denominator of C_e is the longest *shortest path* in G starting at v and is usually referred to as the *eccentricity* of v.

The same definition of eccentricity centrality can be used for undirected and directed networks. In the latter case, the order of v and u in d becomes important, and the directed network is often required to be strongly connected (Koschützki, 2007). In networks that are not strongly connected, the shortest path is computed only to nodes reachable from v. Moreover, if the out-degree of v is 0, then by definition $C_e(v) = 0$.

The bounds on eccentricity centrality are the same as for normalized closeness centrality: $0 \leq C_e(v) \leq 1$. C_e reaches its maximal value for every node that is directly connected to other nodes, as is the case of the center of a star-topology network or any node of a complete network.

Unlike closeness or harmonic centralities, eccentricity centrality does not directly depend on the size of the network, which is why normalization is not generally considered for this measure.

Example 10.4 The following table gives the eccentricity centralities for the nodes of the example network in Figure 10.1. As the example network is not strongly connected, the centralities $C_e(v)$ are computed as inverses of the lengths of the longest shortest paths to reachable nodes. The line labeled with $C_e^u(v)$ takes this network to be undirected, meaning that the directed edges appearing in the figure can be traversed both ways.

v	0	1	2	3	4	5	6	7	8	9	10	11	12	13	14
$C_e(v)$	$\frac{1}{2}$	$\frac{1}{2}$	$\frac{1}{3}$	$\frac{1}{4}$	$\frac{1}{4}$	$\frac{1}{3}$	0	$\frac{1}{4}$	$\frac{1}{4}$	$\frac{1}{5}$	$\frac{1}{3}$	$\frac{1}{3}$	1	$\frac{1}{3}$	$\frac{1}{4}$
$C_e^u(v)$	$\frac{1}{3}$	$\frac{1}{3}$	$\frac{1}{3}$	$\frac{1}{4}$	$\frac{1}{4}$	$\frac{1}{4}$	$\frac{1}{4}$	$\frac{1}{4}$	$\frac{1}{4}$	$\frac{1}{5}$	$\frac{1}{4}$	$\frac{1}{4}$	$\frac{1}{5}$	$\frac{1}{4}$	$\frac{1}{5}$

Similarly to closeness centrality, computing the eccentricity centrality of a node v requires finding the shortest paths to all other nodes in the network, meaning that computing $C_e(v)$ is of time complexity $O(|V| \cdot |E|)$.

Like closeness centrality, eccentricity centrality is able to capture well the notion of importance of a node as a function of its connections to the other nodes. For example, Wuchty and Stadler (2003) uses several centrality measures to analyze the networks of *E. coli* and *Saccharomyces cerevisiae* and shows that both closeness and eccentricity centralities produce very similar rankings of the top metabolites. On the other hand, eccentricity centrality was not able to distinguish essential from non-essential proteins in the PPI network of *S. cerevisiae*.

10.2.1.4 Path Centrality: Betweenness

Betweenness centrality measures the importance of a node by counting in how many connections between other nodes it is implicated. For example, the central node 1 of $U_k^\star$ is involved in all shortest paths between all other nodes. The idea of betweenness, i.e. being situated between other nodes, was introduced in the discussion of point centrality in Bavelas (1948), and the first formal definition was given in Freeman (1977).

Given a (either directed or undirected) network, we denote the set of all shortest paths (also referred to as *geodesics*) between nodes u and w by ρ_{uw} and by $\rho_{uw}(v)$ the set of those shortest paths from ρ_{uw} which pass through v. We further denote $g_{uw} = |\rho_{uw}|$ and $g_{uw}(v) = |\rho_{uw}(v)|$. Finally, we use the following notation:

$$p_{uw}(v) = \begin{cases} \dfrac{g_{uw}(v)}{g_{uw}}, & \rho_{uw} \neq \emptyset \\ 0, & \text{otherwise} \end{cases}$$

$p_{uw}(v)$ can be seen as the probability of finding v in a shortest path between u and w chosen at random from g_{uw}, in the case in which w is reachable from u Freeman (1977, 1978).

Definition 10.9 *(Betweenness Centrality)* Let $G = (V, E)$ be a (either directed or undirected) network. The *betweenness centrality* is the function $C_B : V \to \mathbb{R}$ defined as follows:

$$C_B(v) = \sum_{\substack{u,w \in V \setminus \{v\} \\ u \neq w}} p_{uw}(v)$$

Note that in this case too, there are subtle distinctions between directed and undirected networks. Thus, while in the case of undirected networks, ρ_{uw} (as well as $\rho_{uw}(v)$) is conceptually the same as ρ_{wu} ($\rho_{wu}(v)$, resp.), and consequently, the pair is not considered separately; this is not the case of directed networks. Hence, while the betweenness centrality function involves a $(k-1)(k-2)/2$ summation for undirected networks, i.e. the number of unordered pairs of nodes distinct from v, in the case of directed networks, it consists of a $(k-1)(k-2)$ summation.

Betweenness centrality reaches its minimal value 0 for isolated nodes and its maximal value for nodes v situated on *all* shortest paths between all other nodes of the network. For a k-node network with $k \geq 2$, this maximal value equals the number of pairs of nodes different from v: $C_B(v) = (k-1)(k-2)/2$ for undirected networks and $C_B(v) = (k-1)(k-2)$ for the directed ones. In the case of undirected networks, this value will be reached for the central node of a k-node star-topology network $U_k^\star$.

In the case of directed networks, this maximal value will be reached for the central node of the k-node star topology network, in which there are two symmetric edges between the central node and the non-central nodes: $\overrightarrow{G}_k^\star = (V, E)$, with $V = \{1, \ldots, k\}$ and $E = \{(1, i), (i, 1) \mid 2 \le i \le k\}$.

In fact, not only C_B reaches its maximal value for the central node of a star-topology network but the existence of a node for which C_B reaches its maximal value is sufficient to guarantee that the network is a star.

Lemma 10.1 Let $k \ge 2$ and let $U = (V, E)$ be a k-node undirected network ($|V| = k$) having a node $v \in V$ for which $C_B(v) = (k-1)(k-2)/2$. Then, U is isomorphic to $U_k^\star$.

Proof: The fact that $C_B(v) = (k-1)(k-2)/2$ means that v appears in *all* shortest paths between all other nodes. This implies on the one hand that v is connected to all other nodes and on the other hand that there are no other edges in U. Indeed, if v were not connected to some of the other nodes, then there would exist a pair of nodes $u, w \in V$ with the property $g_{uw}(v) = 0$, meaning that $C_B(v) < (k-1)(k-2)/2$. On the other hand, if there existed an edge between two vertices $u, w \in V$, then the only shortest path from u to w would not traverse v, which again would mean that $g_{uw}(v) = 0$ and $C_B(v) < (k-1)(k-2)/2$. □

A similar result for directed networks imposes that any directed network G having a node with maximal betweenness centrality should be isomorphic to $\overrightarrow{G}_k^\star$.

Lemma 10.2 Let $k \ge 2$ and let $G = (V, E)$ be a k-node directed network ($|V| = k$) having a node $v \in V$ for which $C_B(v) = (k-1)(k-2)$. Then, G is isomorphic to $\overrightarrow{G}_k^\star$.

Note that in the case of the star-topology directed network $G_k^\star$ in which there are arcs going from all non-central nodes to the central node 1 and no arcs going out of 1, the betweenness centrality of all nodes will be 0 because all paths in this network have length 1.

To avoid the increase in betweenness centrality due only to the increase in the size of the network, one typically defines its normalized version. Unlike normalized closeness and harmonic centralities, in the case of betweenness, one needs to normalize with respect to the number of pairs of nodes rather than the number of nodes.

Definition 10.10 *(Normalized Betweenness Centrality)* Let $G = (V, E)$ be a (either directed or undirected) network. The *normalized betweenness centrality* is the function $\tilde{C}_B : V \to \mathbb{R}$ defined as follows:

(a) $\tilde{C}_B(v) = \dfrac{2}{(|V| - 1)(|V| - 2)} \sum\limits_{\substack{u,w \in V \setminus \{v\} \\ u \neq w}} p_{uw}(v),$ for undirected networks,

(b) $\tilde{C}_B(v) = \dfrac{1}{(|V| - 1)(|V| - 2)} \sum\limits_{\substack{u,w \in V \setminus \{v\} \\ u \neq w}} p_{uw}(v),$ for directed networks.

Normalized betweenness centrality ranges between 0 for isolated vertices and 1 for central notes of star-topology networks.

Example 10.5 The following table gives the betweenness centralities and the normalized betweenness centralities, rounded to two digits after the decimal point, for the nodes of the example network in Figure 10.1. The lines $C_B^u(v)$ and $\tilde{C}_B^u(v)$ take this network to be undirected, meaning that the directed edges appearing in the figure can be traversed both ways.

v	0	1	2	3	4	5	6	7	8	9	10	11	12	13	14
$C_B(v)$	12.00	23.00	21.00	0.00	4.00	0.00	0.00	0.00	0.00	0.00	0.00	0.00	0.00	4.00	0.00
$\tilde{C}_B(v)$	0.07	0.13	0.12	0.00	0.02	0.00	0.00	0.00	0.00	0.00	0.00	0.00	0.00	0.02	0.00
$C_B^u(v)$	0.00	68.00	54.00	0.00	13.00	0.00	13.00	0.00	0.00	0.00	0.00	0.00	0.00	13.00	0.00
$\tilde{C}_B^u(v)$	0.00	0.75	0.59	0.00	0.14	0.00	0.14	0.00	0.00	0.00	0.00	0.00	0.00	0.14	0.00

Enumerating all shortest paths between all pairs of nodes different from a given node v is rather expensive. The seminal work (Freeman, 1977) suggests computing powers of the adjacency matrix of the network to count the shortest paths, according to the methods detailed in Harary et al. (1965). However, Brandes (2001) proposes a less expensive algorithm running in time $O(|V| \cdot |E|)$, and avoiding extra non-optimal paths that matrix multiplication would reveal. A more recent work (Nasre et al., 2014) proposes an even faster incremental algorithm, which updates the betweenness centralities of all nodes when a new edge is added. This algorithm runs in time $O(m|V|)$, where m is the maximal length of the shortest path in the network. Finally, the paper (Borassi and Natale, 2019) proposes a random approximation algorithm that computes betweenness centralities in time $|E|^{1/2+o(1)}$ with high probability, much faster than exact algorithms.

Betweenness centrality has been successfully applied in biological network analysis to identify significant nodes. For example, the work (Sahoo et al., 2016) used betweenness centrality to identify proteins significant in the context of cancer growth. In Yu et al. (2007), the authors rely on betweenness centrality to identify bottlenecks in PPI networks. Moreover, the authors show that these nodes have a high tendency of representing essential proteins. The work (Joy et al., 2005) shows that nodes with high betweenness centrality in yeast PPI networks are more likely to be essential and that the evolutionary age of proteins is positively correlated with their betweenness centrality.

We conclude this subsection by mentioning a well-known extension of the betweenness centrality: *percolation centrality*. This measure was introduced in Piraveenan et al. (2013) with the goal of taking into account the dynamical states of the nodes of the network. Intuitively, percolation centrality extends the definition of betweenness centrality by multiplying $p_{uw}(v)$ by *percolation*: a factor derived from the degree to which the nodes in the paths from u to w are involved in the spread of some value (information, infection, etc.) through the network. As percolation centrality considers network states, we do not discuss it in this section.

10.2.1.5 Spectral Centralities

In this section, we discuss eigenvector-based centralities: the eigenvector centrality, Katz centrality, and PageRank centrality.

The idea of using eigenvectors for assessing the importance of nodes in networks was first introduced in Bonacich (1987) and further discussed in several fundamental works, e.g. Ruhnau (2000) and Newman (2008) with the goal of taking into account the fact that not all connections in a network are equal – being connected to a highly central node has more impact than being connected to a more peripheral node. Before defining eigenvector centrality, we introduce some additional notations.

Given a (either directed or undirected) network $G = (V, E)$, we denote its adjacency matrix by $\mathbf{A}$, in which $A_{uv} = 1$ if and only if G contains an edge from node u to node v. For a centrality measure $C : V \to \mathbb{R}$, we will denote by $\mathbf{c}$ the vector obtained by computing C for every node of G, i.e. $\mathbf{c}_v = C(v)$.

For a fixed node v, the idea that the contributions of its neighbors are proportional to their own centralities can be expressed as follows:

$$C(v) = \frac{1}{\lambda} \sum_{u \in V} A_{uv} C(u)$$

where $\lambda > 0$ is a real constant. This relation can be rewritten in the matrix form:

$$\lambda \mathbf{c} = \mathbf{A} \mathbf{c} \qquad (10.1)$$

which effectively defines **c** as an eigenvector and λ as an eigenvalue of the adjacency matrix **A** of the network G. Since **A** is a real square matrix (its elements are 0 and 1 in the case of unweighted networks), according to the Perron–Frobenius theorem of linear algebra, **A** has a unique largest real eigenvalue λ and one can choose the corresponding eigenvector **c** to be strictly positive, which allows defining a meaningful centrality measure (Newman, 2008).

Definition 10.11 *(Eigenvector Centrality)* Let $G = (V, E)$ be a (either directed or undirected) network and **A** its adjacency matrix. Let λ be the largest real eigenvalue of **A** and **c** a corresponding eigenvector with strictly positive components. An *eigenvector centrality* is a function $C_E : V \to \mathbb{R}$ defined as follows: $C_E(v) = \mathbf{c}_v$.

Eigenvalue centrality is sometimes referred to as eigencentrality or eigenvector prestige. The same definition can be used for directed networks, in which case the adjacency matrix **A** may not be diagonally symmetric.

We will refer to eigenvectors **c** satisfying the constraints of the previous definition as *principal* eigenvectors of **A** and of the network G. In fact, it follows from (10.1) that any other vector $\alpha\mathbf{c}$, $\alpha > 0$, is also a principal eigenvector of **A**. This means that the components of **c** (and therefore the values of C_E) indicate relative centralities of the nodes of G with respect to one another rather than some absolute centrality score.

A standard way of fixing a preferred principal eigenvector **c** to avoid ambiguity is by normalizing **c**. Several ways to normalize eigenvectors for centrality exist; we focus here on three norms studied in Ruhnau (2000) and we discuss implications of these normalizations for the eigenvector centrality index – a network-wide measure of centralization. The results from Ruhnau (2000) also rely on the work of (Papendieck and Recht, 2000), which studies maximal entries in the principal eigenvector of a graph.

Normalizing an n-vector **a** typically consists in dividing its components by a p-norm, which is commonly defined as follows:

$$\|\mathbf{a}\|_p = \begin{cases} \left(\sum_{i=1}^{n} a_i^p \right)^{\frac{1}{p}}, & 1 \leq p < \infty \\ \max_{1 \leq i \leq n} |a_i|, & p = \infty \end{cases}$$

One of the ways of normalizing the vector **c** from Definition 10.11 is by dividing all of its components by the ∞-norm, also known as the maximum norm.

Definition 10.12 (∞-*Norm Eigenvector Centrality*) Let $G = (V, E)$ be a (either directed or undirected) network and $\mathbf{c}$ be the principal eigenvector of G. The *∞-norm eigenvector centrality* is the function $C_E^{(\infty)} : V \to \mathbb{R}$ defined as follows:

$$C_E^{(\infty)}(v) = \frac{\mathbf{c}_v}{\|\mathbf{c}\|_\infty} = \frac{\mathbf{c}_v}{\max_{u \in |V|} \mathbf{c}_u}$$

It follows directly from this definition that $0 \leq C_E^{(\infty)}(v) \leq 1$ and that every connected network G has a node v^* for which $C_E^{(\infty)}(v^*) = 1$.

Another way of normalizing the eigenvector centrality is by using the 1-norm, also known as the sum norm.

Definition 10.13 (*1-Norm Eigenvector Centrality*) Let $G = (V, E)$ be a (either directed or undirected) network and $\mathbf{c}$ be the principal eigenvector of G. The *1-norm eigenvector centrality* is the function $C_E^{(1)} : V \to \mathbb{R}$ defined as follows:

$$C_E^{(1)}(v) = \frac{\mathbf{c}_v}{\|\mathbf{c}\|_1} = \frac{\mathbf{c}_v}{\sum_{u \in V} \mathbf{c}_u}$$

$C_E^{(1)}(v)$ can be seen as the proportion of centrality that v reaches in G. The bounds on the values of the 1-norm eigenvector centrality are as follows (Ruhnau, 2000):

$$0 \leq C_E^{(1)}(v) \leq \frac{1}{1 + (2\cos\frac{\pi}{n+1})^{-1}}$$

This centrality measure cannot ever reach 1 for any network with 2 or more connected nodes. The upper bound is only reached for the nodes of a network that only contains one edge (Papendieck and Recht, 2000). In the case of networks with more than two connected nodes, it is not known what is the actual maximal value $C_E^{(1)}$ can achieve, nor in which kind of network topologies this value can be achieved. For example, the central node of a star-topology network *does not* reach the maximal value.

One last way of normalizing the eigenvector centrality which we consider in this section is by using the 2-norm, also known as the Euclidean norm.

Definition 10.14 (*2-Norm Eigenvector Centrality*) Let $G = (V, E)$ be a (either directed or undirected) network and $\mathbf{c}$ be the principal eigenvector of G. The *2-norm eigenvector centrality* is the function $C_E^{(2)} : V \to \mathbb{R}$ defined as follows:

$$C_E^{(2)}(v) = \frac{\mathbf{c}_v}{\|\mathbf{c}\|_2} = \frac{\mathbf{c}_v}{\sqrt{\sum_{u \in V} \mathbf{c}_u^2}}$$

The bounds on the 2-norm eigenvector centrality are $0 \leq C_E^{(2)}(v) \leq \frac{1}{\sqrt{2}}$. The maximal value $\frac{1}{\sqrt{2}}$ is only reached by the center of the star topology network (Ruhnau, 2000). One may define a derived measure ranging from 0 to 1 by multiplying $C_E^{(2)}$ by $\sqrt{2}$.

Finding eigenvectors and eigenvalues of a given square matrix is closely related to finding the roots of polynomials. It follows from the Abel–Ruffini theorem that there exists no algorithm computing exactly the eigenvectors and eigenvalues for square matrices of size greater than 4 (e.g. (Golub and van der Vorst, 2000)). Therefore, iterative algorithms are usually used, one of the most popular being the power iteration method (used, e.g. in *NetworkX* (Hagberg et al., 2008)). The complexity of such iterative methods is generally between $O(|V|^2)$ and $O(|V|^3)$, while the convergence rate ranges from linear to cubic (Demmel, 1997).

Eigenvalue centrality measures are a fine instrument for measuring the importance of nodes in a network because they acknowledge the difference in impact between a connection to a high-centrality neighbor and a connection to a low-centrality one. With eigenvalue centrality, a node with a smaller number of "high-quality" connections may outrank a node with a larger number of "low-quality" connections (Newman, 2008). The work (Estrada, 2006) applies closeness, betweenness, and eigenvector centralities to identify essential proteins in PPI networks and concludes that spectral centralities show the best performance. The paper (Negre et al., 2018) uses eigenvector centrality to pinpoint key amino acids in terms of their relevance in the allosteric regulation. The study (Melak and Gakkhar, 2015) uses different centrality measures to identify potential drug targets of Mycobacterium tuberculosis, the etiological agent of tuberculosis (TB), and show that eigenvalue centrality fares best.

Another centrality measure based on the algebraic properties of the adjacency matrix is *Katz centrality* (also referred to as *Katz prestige* or *Katz status index*). This method was proposed in Katz (1953), and it gives a centrality score by taking into consideration all the nodes of a given network. According to a reasoning similar to the one made for closeness centrality and its variants (Section 10.2.1.3), a node is of high importance if it is connected to many other nodes, but nodes situated farther away count less toward the total centrality score.

Definition 10.15 *(Katz Centrality)* Let $G = (V, E)$ be a (either directed or undirected) loop-free network. Let **A** be its adjacency matrix and λ be its largest positive eigenvalue. The *Katz centrality* is the function $C_K : V \to \mathbb{R}$ defined as follows:

$$C_K(v) = \sum_{i=1}^{\infty} \sum_{u \in V} \alpha^i (\mathbf{A}^i)_{uv},$$

where α is a constant, $0 \leq \alpha \leq 1/\lambda$, and $(\mathbf{A}^i)_{uv}$ is the element in row u and column v of the ith power of $\mathbf{A}$.

Since $(\mathbf{A}^i)_{uv}$ is non-zero if and only if there exists a path of length exactly i between u and v, Katz centrality can be interpreted as a generalization of degree centrality (Koschützki, 2007). Indeed, the role of the factor α^i is to scale down the contributions of longer paths, and without it, $C_K(v)$ essentially becomes the *reachability index*: the number of nodes from which v can be reached. If α is close to 0, the contributions of longer paths are essentially discarded, and C_K approaches a form of degree centrality. On the other hand, as α approaches $1/\lambda$, C_K approaches eigenvector centrality (Newman, 2010).

Let $\mathbf{c}_K$ be the vector collecting Katz centralities of a node in a given network: $(\mathbf{c}_K)_v = C_K(v)$. Then, $\mathbf{c}_K$ can be expressed in a more compact matrix multiplication form in the following way (Koschützki, 2007):

$$\mathbf{c}_K = ((\mathbf{I} - \alpha A^T)^{-1} - \mathbf{I})\mathbf{1}$$

where $\mathbf{I}$ is the identity matrix of the same size as $\mathbf{A}$, $\mathbf{A}^T$ is the transpose of $\mathbf{A}$, $(\cdot)^{-1}$ denotes matrix inversion, and $\mathbf{1}$ is a $|V|$-vector whose components are all 1.

Because of the similarities with eigenvector centrality, exact Katz centrality can be computed by similar algorithms with similar running time complexities. In particular, the power iteration method can be applied (e.g. (Hagberg et al., 2021b)). Faster approximate algorithms exist, in particular (Foster et al., 2001) presents an iterative algorithm with time complexity $O(|V| + |E|)$, given a constant desired precision.

Typical applications of Katz centrality concern directed networks, in particular directed acyclic networks, in which eigenvector centrality appears less useful (Newman, 2010). Katz centrality has found promising applications in neuroscience. The work (Fletcher and Wennekers, 2018) shows that Katz centrality is the best predictor of firing rate given the network structure, with almost perfect correlation in all cases studied. The paper (Mantzaris et al., 2013) uses Katz centrality to analyze the fMRI data of human brain activity during learning and shows how key brain regions contributing to the process can be discovered.

A third eigenvalue- and eigenvector-related centrality measure we briefly discuss in this section is PageRank, one of the algorithms used by Google to measure the importance of web pages (Google, 2011). Multiple variants of PageRank have been proposed. We start with the definition from Langville and Meyer (2005).

Definition 10.16 *(PageRank)* Let $G = (V, E)$ be a directed network and $\mathbf{A}$ its adjacency matrix. The *PageRank* centrality is the function $C_P : V \to \mathbb{R}$ satisfying the following equation:

$$C_P(v) = \sum_{u \in V} A_{uv} \frac{C_P(u)}{deg^+(u)}$$

where $deg^+(u)$ is the out-degree of the node u.

According to this definition, the centrality $C_P(v)$ depends on the centralities of the nodes u from which there is an edge going to v ($A_{uv} \neq 0$). The contribution of each of these nodes u to $C_P(v)$ is inversely proportional to the number of edges going out of u. Thus, nodes connected to many other nodes contribute less to each of their neighbors than nodes that are connected to fewer nodes.

The complete definition of PageRank centrality (which was used in the first versions of the Google search engine) includes two additional parameters:

$$C_P^{(0)}(v) = \alpha \sum_{u \in V} A_{uv} \frac{C_P(u)}{deg^+(u)} + \beta$$

where α can be seen as a decay factor and β as a vector of initial centralities (source ranks), scaled by α. We take this definition from Franceschet (2021), which gives a slightly simplified and generalized version of the original definition from Page et al. (1999).

The PageRank centrality is related to eigenvector centrality in the way the adjacency matrix of the network is exploited to define the relative centrality scores. Since the formulae for computing PageRank are more complex, using iterative algorithms is often preferred, especially for large networks (Langville and Meyer, 2005, Page et al., 1999). Several specific and faster algorithms have been proposed for computing PageRank. In particular, the paper (Del Corso et al., 2005) proposes an algorithm consisting in reducing PageRank computation to computing solutions of linear systems, while the article (Bahmani et al., 2010) proposes Monte Carlo methods for incremental computation of PageRank.

PageRank is a relatively recent centrality measure, but it has already shown some promising performance in the analysis of biological networks. In the work (Iván and Grolmusz, 2010), the authors computed PageRank centralities for the metabolic network of the Mycobacterium tuberculosis and the PPI networks of melanoma patients, and in both cases, important proteins received a high centrality score.

Example 10.6 The following table gives eigenvector (C_E^X), Katz (C_K^X), and PageRank (C_P^X) centralities for the nodes of the example network in Figure 10.1, as computed with *NetworkX* (Hagberg et al., 2021c) and rounded to two digits after the decimal point.

v	0	1	2	3	4	5	6	7	8	9	10	11	12	13	14
C_E^X	0.50	0.50	0.50	0.00	0.00	0.00	0.50	0.00	0.00	0.00	0.00	0.00	0.00	0.00	0.00
C_K^X	0.26	0.35	0.36	0.23	0.25	0.23	0.29	0.23	0.23	0.23	0.23	0.23	0.23	0.25	0.23
C_P^X	0.19	0.25	0.20	0.02	0.03	0.02	0.14	0.02	0.02	0.02	0.02	0.02	0.02	0.03	0.02

10.2.2 System Controllability Methods

One of the important aspects when dealing with a biomedical network is to be able to influence part of or even the totality of its nodes. This objective is not restricted to biomedical networks but is one of the earliest and most studied network theory problems, generally known as the (target) network controllability problem. This is intrinsically an optimization problem, as any network can be controlled from a sufficiently large controlling set, e.g. the entire set of nodes. However, one is interested in finding the minimal set of nodes needed in order to achieve such control. Early works in the field come from the 1960s; however, it was in early 2010s that one of the key results from the field has been provided in Liu et al. (2011), namely, the possibility of efficiently determining the minimum set of nodes needed to control an entire network. This result sparked a new interest in the field, with a strong emphasis on possible applications in the biomedical field.

The prospect of enforcing control over a biomedical network has been investigated within two main methodological approaches: that of (structural) network controllability, see, e.g. Liu et al. (2011), Czeizler et al. (2018), Guo et al. (2018), and Kanhaiya et al. (2017) and that of dominating sets (Wuchty, 2014, Nacher and Akutsu, 2016, Zhang et al., 2015). In the current section, we review both approaches, detailing the theoretical, algorithmic, and biomedical applicability aspects of these methods.

10.2.2.1 Network Controllability

In general terms, we say that a directed network, or generally any dynamical system, is controllable from a set of input nodes if, with a suitable selection of input values for these nodes, the entire network can be driven from any initial state to any desired final state within a finite time. The state of a node is given by its numerical value; in the context of biomedical networks, this could be, for example, the expression level of a gene/protein. From one time point to another, this value/state evolves depending (linearly) on the values/states of its neighbors; this is why such systems are also known as linear time invariant dynamical systems (LTIS), as their evolution can be described by the system of linear differential equations:

$$\frac{dx(t)}{dt} = Ax(t) \qquad (10.2)$$

where $x(t) = (x_1(t), ..., x_n(t))^T$ is the n-dimensional vector describing the system's state at time t and $A \in \mathbb{R}^{n \times n}$ is the time-invariant *state transition matrix*, describing how each of these states are influencing the dynamics of the system. Namely, for any nodes i, j, $1 \leq i, j \leq n$ within the network, the entry $a_{i,j}$ of matrix A either documents the weight of the influence of node j over the node i, if there exists a (directed) edge (i, j), or is equal to 0 otherwise. By convention, from now on during this section, all vectors are considered to be column vectors so that the matrix–vector multiplications are well defined.

Assume that we allow the system to be influenced through an m-dimensional input controller, i.e. an input vector u of real functions, $u : \mathbb{R} \to \mathbb{R}^m$ acting upon some m nodes of the network. Consider the subset of input nodes $I \subseteq \{1, 2, ..., n\}$, $I = \{i_1, ..., i_m\}$, $1 \leq m \leq n$, as the nodes of the network on which the external input is applied to; such nodes are also known as *driver/driven* nodes. The system (10.2) becomes

$$\frac{dx(t)}{dt} = Ax(t) + B_I u(t) \qquad (10.3)$$

where $B_I \in \mathbb{R}^{n \times m}$ is the characteristic matrix associated with the subset I, $B_I(r, s) = 1$ if $r = i_s$ and $B_I(r, s) = 0$ otherwise, for all $1 \leq r \leq n$ and $1 \leq s \leq m$.

It is often the case, particularly in the biomedical field that it is enough to enforce control only over a subset of the network nodes to get the desired change in the network dynamics. The associated optimization problem is known as the *target controllability problem* or more generally as the *output controllability problem*. Thus, consider a subset of target nodes $T \subseteq \{1, 2, ..., n\}$, $T = \{t_1, ..., t_l\}$, $m \leq l \leq n$, thought as a subset of the nodes of the linear dynamical system, whose dynamics we aim to control (as defined below) through a suitable choice of input nodes and of an input vector. The subset of target nodes can also be defined through its characteristic matrix $C_T \in \mathbb{R}^{l \times n}$, defined as $C_T(r,s) = 1$ if $t_r = s$ and $C_T(r,s) = 0$ otherwise, for all $1 \leq r \leq l$ and $1 \leq s \leq n$. We will denote by the triplet (A, I, T) the targeted linear (time-invariant) dynamical system defined by matrix A, with input set I, and target set T. In case T is the entire network, we can omit it from the notation.

Definition 10.17 *(Target Controllability of Linear Networks)* Given a target linear dynamical system (A, I, T) (or similarly a linear network), we say that this system is *target controllable* (or simply controllable if the target is the entire network) if for any $x(0) \in \mathbb{R}^n$ and any $\alpha \in \mathbb{R}^l$, there is an input vector $u : \mathbb{R} \to \mathbb{R}^m$ such that the solution $\tilde{x}$ of (10.3) eventually coincides with α on its T-components, i.e. $C_T\tilde{x}(\tau) = \alpha$, for some $\tau \geq 0$.

Note that the general case of *output controllability* is defined similarly, with the only difference that instead of a 0–1 characteristic matrix C_T, we have an arbitrary matrix $C \in \mathbb{R}^{n \times m}$, with $m \leq n$, defining the output $y \in \mathbb{R}^m$ of the LTIS as a linear combination of the solution $x(t)$ of Eq. (10.3).

Intuitively, the system being target controllable means that for any input state x_0 and any desired final state α of the target nodes, there is a suitable input vector u driving the target nodes to α. Obviously, the input vector u depends on x_0 and α.

It is known from Kalman (1963) that the system (A, I) is controllable from some input controller $u(t)$ if and only if the rank of the matrix $(B_I \mid AB_I \mid A^2B_I \mid \ldots \mid A^{n-1}B_I)$, known as the *controllability matrix*, is equal to n, the number of nodes in the network. The operator $\mid$ denotes here the simple matrix concatenation operation. This criterion is known as the *Kalman's criterion for full controllability* (Kalman, 1963), and it is easily extendable to the case of target (or generally, output) controllability:

Theorem 10.1 *A targeted linear dynamical system with inputs (A, I, T) is controllable if and only if the rank of its controllability matrix $(C_T B_I \mid C_T AB_I \mid C_T A^2 B_I \mid \ldots \mid C_T A^{n-1} B'_I)$ is equal with $|T|$ (Kalman, 1963).*

Intuitively, the controllability matrix describes all weighted paths from the input nodes to the target nodes in the directed graph associated with the linear dynamical system. This leads to the notion of *control path* from an input node to a target node and an input node *controlling* a target node. This line of thought can be further developed into a structural formulation of the targeted controllability. Thus, although the Kalman controllability criterion seems to suggest that the (target) controllability problem is strictly related to a particular valuation of the state transition matrix A, it turns out that this is actually a network property. To explain this, we define two matrices C, D of the same size to be *equivalent* if they have the same set of non-zero entries. The key concept here is that of *structural controllability*:

Definition 10.18 *(Structural Target Controllability of Linear Networks)* We say that a given targeted linear dynamical system (A, I, T) (or similarly a linear network) is *structural target controllable* if there exists a matrix A' equivalent with A such that the system (A', I, T) is target controllable. That is, (A, I, T) becomes controllable by replacing A with any suitable equivalent matrix A', or similarly, by replacing the weights of the network's edges with any other non-zero values.

The key result, proved in Lin (1974) and Shields and Pearson (1976), is that if a system (A, I) is structurally controllable, then it is controllable for all, except a *thin* set, of equivalent matrices A'. We recall that a thin subset of the n-dimensional complex space is nowhere dense and has Lebesgue measure 0. This indicates that the controllability problem can indeed be reduced to its structural version, known as the *structural (target) controllability problem*, which is ultimately a property of the network's connectivity rather than of the effective edge weights. This problem can be defined both as a decision problem (is the linear time-invariant dynamical system (A, I, T) structurally target controllable?) and as an optimization problem (given a matrix A of size $n \times n$ and a subset of target nodes $T \subseteq \{1, 2, \ldots, n\}$, find the minimal m, $m \leq n$, and a suitable choice for the size-m set of input nodes I such that (A, I, T) is structurally target controllable). We are most interested in the latter formulation of the problem.

From an algorithmic standpoint, the Kalman criterion for controllability from Theorem 10.1 can lead to an efficient/polynomial time algorithm for verifying whether a given network can be (target) controlled from a given input controller, acting upon some of the network nodes. However, it does not lead to an efficient way of computing the composition of such a minimal controller, or even its (minimal) size. As it turns out, the (target) control optimization problem has been open for more than 20 years, before it was shown in Liu et al. (2011) that there exists a low polynomial time algorithm (i.e. cubic in the size of the network) that can provide both the size and the composition of the minimum input controller needed to control an entire network. It was thus surprising that the corresponding result for the target control optimization problem was proved by Czeizler et al. (2018) to be NP-hard, meaning that any exact optimization algorithm would have to run in exponential time. Several approximation algorithms have been developed for this latter problem, with good results when applied to both real-word networks, e.g. biomedical, social, electrical, etc., and artificial networks (Kanhaiya et al., 2017, Czeizler et al., 2018, Gao et al., 2014).

Another generalization coming from the practical application of network controllability in pharmacology and biomedicine considers the case when the input controller should, or it is desired to, be selected mostly from a subset $P \subseteq \{1,2,\ldots,n\}$ of the network nodes. For example, in network pharmacology studies, it is more advantageous to consider controllers consisting of those genes/proteins for which there already exist approved drugs known to target that particular element. This leads to the so-called *input-constrained targeted structural controllability* problem:

Definition 10.19 *(Input-Constrained, Structural Target Controllability)* Given a linear dynamical system defined by a matrix $A \in \mathbb{R}^{n \times n}$, a set of target nodes T and a set of preferred nodes P, the *input-constrained structural targeted controllability* problem asks to find a smallest sized input set I whose intersection with P is maximal, such that the targeted linear dynamical system with inputs (A, I, T) is controllable, i.e. such that the rank of the matrix $(C_T B_I \mid C_T A B_I \mid C_T A^2 B_I \mid \ldots \mid C_T A^{n-1} B_I)$ is equal with $|T|$.

Although this optimization problem is also NP-hard (as a generalization of the previous case), several efficient approximation algorithms have been introduced in Kanhaiya et al. (2017) and Popescu et al. () and analyzed particularly with respect to biomedical networks.

10.2.2.2 Minimum Dominating Sets

The second frequently used controllability technique, applied especially in the case of undirected networks, relies on the notion of domination.

Definition 10.20 *(Dominating Set)* Given a size n network, we say that a subset $D \subseteq \{1,2,\ldots,n\}$ of its nodes is *dominating* the network if any node within the network is either in D or it is adjacent to a node in D.

Definition 10.21 *(Minimum dominating set problem)* Given a (undirected) network, the *minimum dominating set problem (MDS)* asks to find a dominating set of minimum cardinality. Such MDS can thus be considered as an efficient first-hand controlling set for the respective network.

From an algorithmic point of view, MDS is a classical NP-hard problem (Garey and Johnson, 1990). However, as before, there are many efficient approximation algorithms providing efficient solutions even in the case of large networks (Hedar and Ismail, 2012, Alon, 1990, Nacher and Akutsu, 2014).

The MDS approach has been applied in connection to various fields such as the controllability of biological networks (Wuchty, 2014, Nacher and Akutsu, 2016, Zhang et al., 2015), design and analysis of wireless computer networks (Yu et al., 2013, Wu et al., 2006), the study of social networks (Daliri Khomami et al., 2018, Wang et al., 2011), etc. Several generalizations of MDS have also been considered:

Definition 10.22 *(Minimum k-dominating set problem)* Given a (undirected) network, a set D of its nodes is *k-path dominating* if any node from the network is either in D or it is connected to a node in D through a path of length at most k. The *minimum k-dominating set problem* (MkDS), also known as the *d-hop dominating* problem, asks to find a *k*-path dominating set of minimal cardinality.

Definition 10.23 *(Red–Blue (k-)Dominating Problem)* Given a network and two subsets of its nodes, Red and Blue, the *Red–Blue (k-)domination problem* asks to find a minimum subset of the Blue nodes (*k*-path) dominating all the Red nodes.

Such generalizations were considered in Penso and Barbosa (2004), Nguyen et al. (2020), Coelho et al. (2017), and Abu-Khzam et al. (2011), respectively. All these generalizations, while preserving the algorithmic complexity of the original MDS, provide sometimes a closer connection to practical applications.

Compared to the structural network controllability formalism, the dominating set methodology enforces a reachability type of control. Indeed, in the latter case, the dominating nodes are controlling the network either by direct interactions, in the case of MDS, or by paths of length at most k, for MkDS. This is different from within the structural network controllability formalism, where each controller node is asked to enforce an independent control over its dominating nodes. This requirement imposes that each controlled node, i.e. *target* node, is positioned at the end of a different lengthed path starting from the controller, i.e. the *driver* node.

10.2.3 Software

In this section, we briefly present several applications and libraries that can be used for generating, visualizing, or analyzing graphs and networks. All

10.2.3.1 NetworkX

NetworkX (Hagberg et al., 2008) is a Python package for the creation, manipulation, and study of the structure, dynamics, and functions of complex networks (Hagberg et al., 2021c). As all of the networks described in this chapter (except the simple demonstrative network from Figure 10.1) represent personalized protein–protein interaction networks and have been generated externally, we have focused on the usage of the *NetworkX* package solely on analyzing them. The package provides out-of-the-box implemented algorithms and functions for identifying all of the centrality measures presented in this chapter. It is worth noting that *NetworkX* also provides several algorithms for different layout routines, and basic interconnections with dedicated graph visualization packages, such as *Matplotlib* (Matplotlib, 2021) and *Graphviz* (Ellson et al., 2021). However, graph visualization does not represent the main goal of *NetworkX* and its creators recommend using a dedicated and fully featured tool instead.

An alternative to *NetworkX* is the *igraph* collection (The igraph core team, 2021), which provides network analysis libraries and packages for R, Mathematica and C/C++ in addition to Python.

We used the *NetworkX* package to compute the centrality measures corresponding to the nodes in all of the networks presented in this chapter. Specifically, the functions used for ranking the nodes according to the corresponding centrality method described in Section 10.2.1 are presented in Table 10.1.

Table 10.1 *NetworkX* functions used for computing the centrality methods.

Method	Function	Parameters
Degree centrality	*in_degree_centrality*	—
Closeness centrality	*closeness_centrality*	Default
Harmonic centrality	*harmonic_centrality*	Default
Eccentricity centrality	*eccentricity*	Default
Betweenness centrality	*betweenness_centrality*	Default
Eigenvector-based prestige	*eigenvector_centrality*	Default

Additionally, we used the *dominating_set* function with the default parameters to run the MDS analysis described in Section 10.2.2 for each network. Although the package does not provide a direct equivalent for the MkDS algorithm, it allows for the initial network manipulation (i.e. parsing the network and adding specific edges) required to transform the default MDS analysis into an MkDS one. The methodology and results are presented in detail in Section 10.3.3.

10.2.3.2 Cytoscape

Cytoscape (Shannon et al., 2003) is a standalone software platform implemented in Java for visualizing complex networks, together with attribute data integration (Shannon et al., 2021). *Cytoscape* was initially developed for biological research and biological network visualization and analysis. In addition to the basic functionality, *Cytoscape* provides extended functionality through the use of the so-called apps, which can be community developed and add support for additional analysis methods, layouts, or database connections, among others.

Another well-established desktop network visualization software is *Gephi* (Bastian et al., 2021), a Java application with the same capabilities, including the usage of plugins for extended functionality.

We have used the *Cytoscape* application to render all of the network visualizations presented throughout this chapter. The layout of the nodes was automatically calculated and applied using the default preferred layout algorithm, while the general design of the networks (e.g. labels, colors, or sizes) is based on the default style. A screenshot with the visualization of a network can be seen in Figure 10.2.

10.2.3.3 NetControl4BioMed

NetControl4BioMed ((Kanhaiya et al., 2018), updated version in Popescu et al. (2021)), is a C# (.NET Core) web application for the generation and structural target controllability analysis of protein–protein interaction networks, freely available at https://netcontrol.combio.org/. To this end, the application integrates and combines multiple protein and protein–protein interaction databases which, based on the user input, are used in the process.

We have used the *NetControl4BioMed* platform to generate the personalized protein–protein interaction networks presented in this chapter. The details of the network generation method (e.g. used interaction databases or algorithm parameters) are presented in Section 10.3.1. A screenshot with the home page of the application can be seen in Figure 10.3.

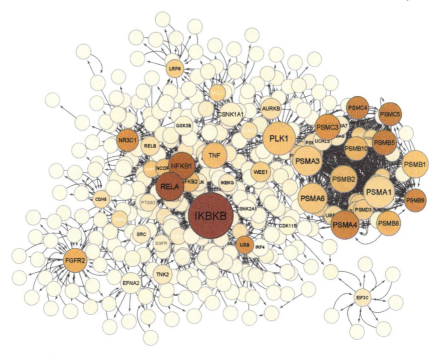

Figure 10.2 Example of network visualization with *Cytoscape*. The color of a node is proportional to its in-degree (darker nodes have higher in-degree), while the size of a node is proportional to its out-degree (larger nodes have higher out-degree).

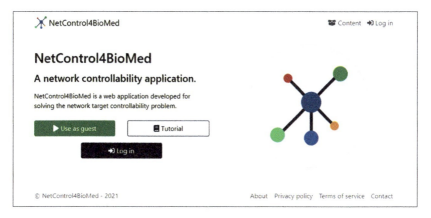

Figure 10.3 Screenshot with the home page of *NetControl4BioMed*.

Additionally, we used the network analysis section of the application, with the default parameters, to run the structural target controllability analysis described in Section 10.2.2 for each network. The methodology and results are presented in detail in Section 10.3.3.

10.3 Applications of Network Modelling in Personalized Medicine

We demonstrate our approach to personalized medicine on three multiple myeloma patients through the analysis of customized networks built around the mutated genes of each patient, the disease-specific survivability-essential genes, and the genes targeted by drugs in the standard therapy for multiple myeloma.

10.3.1 Constructing Personalized Disease Networks

We used the patient data documented in Lohr et al. (2014), which include information about the evolution of the mutation, the mutated genes, and the stage of treatment, as well as details about the patient characteristics, such as age, race, and gender, among others. In this study, we focus on the mutated genetic information for tumor samples 28, 38, and 191.

We used a list of 70 multiple myeloma-specific essential genes presented in Matthews et al. (2017), Tiedemann et al. (2012), and Krönke et al. (2014) and shown in Table 10.2.

We used the multiple myeloma standard treatment drugs described in Multiple Myeloma Research Foundation (2019), Engelhardt et al. (2019), and Mateos et al. (2018) and their corresponding drug targets from DrugBank (Wishart et al., 2017). The list thus obtained is presented in Table 10.3, while the associated drugs and the standard drug therapies are documented in Tables 10.4 and 10.5, respectively.

We used the *NetControl4BioMed* application, briefly presented in Section 10.2.3.3, to build a personalized protein–protein interaction network for each multiple myeloma patient around the seed genes defined by the patient-specific mutated genes, the disease-specific survivability-essential genes, and the drug target genes corresponding to the standard treatment drugs. We used the interaction data from the *KEGG*, *OmniPath*, *InnateDB*, and *SIGNOR* databases. The networks include all paths of length at most three between the seed proteins that could be formed with these interactions. We added to the network all intermediary nodes that were not part of the set of seed nodes. An overview of the generated networks is presented in Table 10.6.

Table 10.2 Disease-specific survivability-essential genes for multiple myeloma (Matthews et al., 2017, Tiedemann et al., 2012, Krönke et al., 2014).

AGTRAP	EIF3C	KIFC2	PLK1	RGAG1	TRIM68
AURKB	EIF4A3	LEPROT	PRPF8	RPL27	TUBGCP6
CARS	GNRH2	MAF	PSMA1	RPL38	UBB
CCND2	GPR77	MCL1	PSMA3	RRM1	UBQLNL
CDK11	HIP1	MED14	PSMA4	RSF1	ULK3
CDK11A	IK	MED15	PSMA6	SF3A1	USP36
CDK11B	IKBKB	NDC80	PSMC3	SLC25A23	USP8
CKAP5	IKZF1	NFKB1	PSMC4	SNRPA1	WBSCR22
COPB2	IKZF3	NFKB2	PSMC5	SNW1	WEE1
CSNK1A1	IRF4	NUF2	RAB11A	TNK2	XPO1
CUL9	KIF11	PCDH18	RELA	TPMT	
EFNA2	KIF18A	PIM2	RELB	TRIM21	

Table 10.3 Targets of the drugs used in standard therapy lines for multiple myeloma.

ANXA1	GSR	NR0B1	PSMB2	SLAMF7	TUBB
CD38	HSD11B1	NR1I2	PSMB5	TNF	XPO1
CDH5	NFKB1	NR3C1	PSMB8	TNFSF11	
CRBN	NOLC1	PSMB1	PSMB9	TOP2A	
FGFR2	NOS2	PSMB10	PTGS2	TUBA4A	

We used the *NetworkX* package, briefly presented in Section 10.2.3.1, to rank the genes in each of the generated networks based on the centrality measures described in Section 10.2.1. Tables 10.7–10.12 present, for each network, the essential genes that have the corresponding centrality measure higher than the median among all other genes in the network.

10.3.2 Analysis Methods

In this section, we describe the methodology applied for the analysis of the data presented in Section 10.3.1 and using the centrality measures presented in Section 10.2.1 and the controllability methods presented in Section 10.2.2.

Table 10.4 Main drugs for treating multiple myeloma.

Bortezomib	PSMB1;PSMB5
Carfilzomib	PSMB1;PSMB10;PSMB2;PSMB5;PSMB8;PSMB9
Carmustine	GSR
Cisplatin	A2M;ATOX1;MPG;TF
Cyclophosphamide	NR1I2
Dacetuzumab	CD40
Daratumumab	CD38
Dexamethasone	ANXA1;NOS2;NR0B1;NR1I2;NR3C1
Doxorubicin	NOLC1;TOP2A
Elotuzumab	SLAMF7
Etoposide	TOP2A;TOP2B
Ixazomib	PSMB5
Lenalidomide	CDH5;CRBN;PTGS2;TNFSF11
Liposomal doxorubicin	TOP2A;TOP2B
Oprozomib	LMP7;PSMB5
Panobinostat	HDAC1;HDAC2;HDAC3;HDAC6;HDAC7;HIF1A;VEGF
Plerixafor	CXCR4
Pomalidomide	CRBN;PTGS2;TNF
Prednisone	HSD11B1;NR3C1
Selinexor	XPO1
Thalidomide	CRBN;FGFR2;NFKB1;PTGS2;TNF
Vincristine	TUBA4A;TUBB

All analyses follow a similar flow, aiming to identify, through the different controllability methods, a subset of "important" drug target genes in the network, ranked according to the number of essential genes that they control. Once a set has been identified, the corresponding drugs are ranked based on a very similar criterion, taking into account the number of essential genes that their drug target control. The three top ranked drugs are then reported as a personalized drug combination therapy customized to the patient and compared with the standard lines of therapy.

Firstly, we ran each type of analysis on the complete networks and datasets, considering as targets the essential genes presented in Table 10.2 and as preferred inputs the drug target genes in Table 10.3. Then, to reduce

Table 10.5 Lines of therapy for treating multiple myeloma.

First line of therapy

Lenalidomide; bortezomib; dexamethasone (RVD)

Bortezomib; cyclophosphamide; dexamethasone (VKD)

Bortezomib; thalidomide; dexamethasone (VTD)

Bortezomib; melphalan; prednisone

Vincristine; doxorubicin; dexamethasone (VAD)

Melphalan; dexamethasone

Daratumumab; bortezomib; thalidomide; dexamethasone

Carfilzomib; thalidomide; dexamethasone (KTD)

Second line of therapy

Carfilzomib; lenalidomide; dexamethasone (KRD)

Ixazomib; lenalidomide; dexamethasone

Elotuzumab; lenalidomide; dexamethasone

Bendamustine; lenalidomide; dexamethasone

Third line of therapy

Pomalidomide; dexamethasone

Panobinostat; bortezomib; dexamethasone

Daratumumab

Table 10.6 Summary of the generated personalized protein–protein interaction networks.

Network	G	N	E	CC	D	AD
Tumor sample 28	36	360	1486	4	12	6.02
Tumor sample 38	117	446	1732	5	12	5.77
Tumor sample 191	218	515	1955	3	12	5.74

G, number of mutated genes in the sample; N, number of nodes in the network; E, number of edges; CC, number of connected components; D, the network diameter; AD, the network average degree.

the noise in the data, for each network and each centrality measure, we focused on the subgraphs formed by the top ranked essential genes and the drug targets that can reach them through a path of length of 3 or less and all the interactions between them. For each such subgraph, all analyses consider as targets the corresponding essential genes and as preferred inputs the corresponding drug target genes.

Table 10.7 Top ranked genes based on their in-degree centrality, for each network.

Tumor sample 28		Tumor sample 38		Tumor sample 191	
IKBKB	AURKB	IKBKB	CCND2	RELA	CCND2
RELA	XPO1	RELA	XPO1	IKBKB	XPO1
NFKB1	EIF3C	NFKB1	EIF3C	NFKB1	EIF3C
PSMC4	TNK2	PSMC4	TNK2	PSMC4	TNK2
PSMC5	CSNK1A1	PSMC5	CSNK1A1	PSMC5	RPL38
PSMA4	KIF11	PSMA4	KIF11	UBB	CSNK1A1
UBB	EFNA2	UBB	CDK11B	PSMA4	SNW1
PSMC3	CDK11B	PSMC3	SNW1	PSMC3	CDK11B
NFKB2	SNW1	NFKB2	EFNA2	NFKB2	KIF11
PLK1	USP8	PLK1	IKZF3	PLK1	IKZF3
PSMA6	IKZF3	MCL1	MED14	MCL1	IKZF1
PSMA3	RSF1	PSMA6	USP8	PSMA6	EFNA2
MCL1	IKZF1	WEE1	RSF1	WEE1	RRM1
WEE1	RPL38	PSMA3	IKZF1	PSMA3	USP8
PSMA1	TRIM21	RELB	RPL38	RELB	TRIM21
CCND2	MAF	PSMA1	TRIM21	PSMA1	RSF1
RELB		AURKB		AURKB	RPL27

We used the *NetControl4BioMed* application, briefly presented in Section 10.2.3.3, to run the structural target controllability analyses with the previously described setup. We used the default parameters, with a set maximum path length of 3. The analysis outputs a set of genes that can control the entire target set, from which only the controlling drug targets are selected and further considered.

Similarly, we used the *NetworkX* package, presented in Section 10.2.3.1, to run the minimum dominating set analyses with the previously described setup. An additional step was required in order to enable the default function to perform the required minimum k-dominating set analysis. To this end, we transformed the analyzed networks by adding a direct edge between each pair of nodes indirectly connected by a path of length of k or less. We set $k = 3$ and used the default parameters. The analysis outputs a set of genes that can dominate the entire target set, from which only the dominating drug targets are selected and further considered.

Table 10.8 Top ranked genes based on their closeness centrality, for each network.

Tumor sample 28		Tumor sample 38		Tumor sample 191	
UBB	CSNK1A1	IKBKB	PSMA4	RELA	PSMA4
IKBKB	PSMA4	UBB	XPO1	UBB	CSNK1A1
RELA	XPO1	RELA	CSNK1A1	IKBKB	PSMC3
NFKB1	PLK1	NFKB1	PSMC3	NFKB1	XPO1
NFKB2	SNW1	NFKB2	SNW1	NFKB2	SNW1
RELB	PSMC3	RELB	TRIM21	RELB	PSMC4
WEE1	PSMC4	PSMA3	PIM2	WEE1	AURKB
PSMA3	PSMC5	MCL1	PSMC4	PSMA3	PSMC5
MCL1	TRIM21	WEE1	PSMC5	MCL1	PIM2
RSF1	KIF11	RSF1	PSMA1	RSF1	PSMA1
CCND2	PSMA1	CCND2	KIF11	CCND2	TRIM21
PSMA6	PIM2	PLK1	TNK2	PSMA6	KIF11
		PSMA6	AURKB	PLK1	TNK2

Table 10.9 Top ranked genes based on their harmonic centrality, for each network.

Tumor sample 28		Tumor sample 38		Tumor sample 191	
IKBKB	PSMC3	IKBKB	PSMC5	RELA	PSMC3
RELA	PSMA6	RELA	PLK1	IKBKB	PLK1
NFKB1	MCL1	NFKB1	CCND2	NFKB1	PSMA6
UBB	CCND2	UBB	RSF1	UBB	RSF1
NFKB2	RSF1	NFKB2	XPO1	NFKB2	CCND2
PSMA4	PLK1	RELB	PSMA1	RELB	PSMA1
RELB	XPO1	PSMA4	CSNK1A1	PSMA4	AURKB
PSMA3	PSMA1	PSMA3	SNW1	WEE1	CSNK1A1
WEE1	CSNK1A1	WEE1	TRIM21	PSMA3	XPO1
PSMC4	SNW1	MCL1	PIM2	PSMC4	SNW1
PSMC5		PSMC3	AURKB	MCL1	PIM2
		PSMA6	TNK2	PSMC5	TRIM21
		PSMC4	KIF11		

Table 10.10 Top ranked genes based on their eccentricity, for each network.

Tumor sample 28		Tumor sample 38		Tumor sample 191	
IKBKB	RELB	IKBKB	EFNA2	IKBKB	EFNA2
RELA	MCL1	RELA	RELB	RELA	MCL1
NFKB1	EFNA2	NFKB1	MCL1	NFKB1	TNK2
PSMA4	TNK2	PLK1	TNK2	PLK1	CCND2
PLK1	CCND2	PSMA4	CCND2	PSMA4	EIF3C
PSMC3	EIF3C	PSMC3	EIF3C	PSMC5	CDK11B
PSMC5	CDK11B	PSMC5	CDK11B	PSMC4	XPO1
PSMC4	XPO1	PSMC4	XPO1	PSMC3	SNW1
PSMA6	IRF4	PSMA6	IRF4	PSMA6	RPL38
PSMA1	SNW1	PSMA1	SNW1	PSMA1	IRF4
PSMA3	USP8	PSMA3	PIM2	PSMA3	PIM2
UBB	PIM2	UBB	MED14	UBB	USP8
NFKB2	HIP1	NFKB2	USP8	NFKB2	IKZF1
WEE1	KIF11	WEE1	RSF1	CSNK1A1	HIP1
CSNK1A1	RSF1	AURKB	HIP1	WEE1	KIF11
AURKB		CSNK1A1	KIF11	AURKB	IKZF3
				RELB	RAB11A

10.3.3 Results

In this section, we present and discuss the results of the analyses described in Section 10.3.2, with the aim of suggesting personalized treatments.

10.3.3.1 Structural Controllability Analysis

We applied the input-constrained structural target controllability method on the complete networks and sets of essential and drug target genes. The drug target genes in the obtained controlling sets are presented in Table 10.13.

Next, we applied the method once more on the subgraphs described in Section 10.3.2 and corresponding to each centrality measure. The drug target genes in the controlling sets obtained for each measure are presented in Tables 10.14–10.19.

Next, we studied the results in the context of drug therapy. For each network and centrality measure, the drug target genes were matched with all the multiple myeloma standard treatment drug targeting them, while the

Table 10.11 Top ranked genes based on their betweenness centrality, for each network.

Tumor sample 28		Tumor sample 38		Tumor sample 191	
IKBKB	PSMA6	IKBKB	RELB	IKBKB	RELB
NFKB1	EIF3C	RELA	CCND2	RELA	UBB
RELA	RELB	NFKB1	EIF3C	NFKB1	PSMA6
PLK1	PSMA1	PLK1	PSMA1	PLK1	PSMA1
CSNK1A1	PSMA4	PSMC3	PSMA4	NFKB2	PSMC5
NFKB2	PSMC5	NFKB2	PSMC5	WEE1	PSMA4
PSMA3	EFNA2	WEE1	EFNA2	CSNK1A1	EFNA2
WEE1	USP8	PSMA3	MED14	CDK11B	PSMC4
PSMC3	PIM2	CDK11B	USP8	PSMA3	RAB11A
CDK11B	PSMC4	CSNK1A1	PIM2	MCL1	USP8
MCL1	MED14	TNK2	PSMC4	AURKB	MED14
TNK2	RPL38	AURKB	RSF1	EIF3C	MAF
AURKB	MAF	MCL1	SNW1	PSMC3	PIM2
XPO1	RSF1	XPO1	RPL38	TNK2	SNW1
CCND2	SNW1	PSMA6	MAF	XPO1	RPL38
UBB		UBB		CCND2	RSF1

Table 10.12 Top ranked genes based on their eigenvector-based prestige, for each network.

Tumor sample 28		Tumor sample 38		Tumor sample 191	
PSMC4	NFKB1	PSMC4	NFKB2	PSMC4	CCND2
PSMC5	CCND2	PSMC5	CCND2	PSMC5	WEE1
PSMA4	NFKB2	PSMA4	WEE1	PSMA4	MCL1
PSMC3	WEE1	PSMC3	MCL1	PSMC3	RELB
PSMA6	MCL1	PSMA6	RELB	PSMA6	RSF1
PSMA3	RELB	PSMA3	PLK1	PSMA3	AURKB
PSMA1	RSF1	PSMA1	RSF1	PSMA1	PLK1
UBB	SNW1	UBB	SNW1	UBB	SNW1
IKBKB	XPO1	IKBKB	XPO1	IKBKB	XPO1
RELA	PLK1	RELA	AURKB	RELA	KIF11
		NFKB1		NFKB1	RPL27
				NFKB2	PIM2

Table 10.13 Drug target genes in the controlling set obtained by the structural target controllability analysis.

Tumor sample 28		Tumor sample 38		Tumor sample 191	
ANXA1	NOLC1	ANXA1	PSMB1	ANXA1	PSMB2
NFKB1		NOLC1	TNF		

Table 10.14 Drug target genes in the controlling set obtained by the structural target controllability analysis corresponding to the top ranked essential genes based on their in-degree centrality.

Tumor sample 28		Tumor sample 38		Tumor sample 191	
ANXA1	NOLC1	ANXA1	TNF	ANXA1	NOLC1
NFKB1	TNF	NFKB1		NFKB1	TNF

Table 10.15 Drug target genes in the controlling set obtained by the structural target controllability analysis corresponding to the top ranked essential genes based on their closeness centrality.

Tumor sample 28		Tumor sample 38		Tumor sample 191	
ANXA1	PSMB8	FGFR2	PSMB5	ANXA1	PSMB1
NFKB1	TNF	NFKB1	TNF	NFKB1	TNF
		NR3C1		NR3C1	XPO1

Table 10.16 Drug target genes in the controlling set obtained by the structural target controllability analysis corresponding to the top ranked essential genes based on their harmonic centrality.

Tumor sample 28		Tumor sample 38		Tumor sample 191	
ANXA1	PTGS2	ANXA1	TNF	ANXA1	PSMB1
NFKB1	TNF	NFKB1	XPO1	NFKB1	TNF
PSMB1	XPO1	PSMB10			

Table 10.17 Drug target genes in the controlling set obtained by the structural target controllability analysis corresponding to the top ranked essential genes based on their eccentricity.

Tumor sample 28		Tumor sample 38		Tumor sample 191	
ANXA1	PSMB1	ANXA1	NR3C1	ANXA1	NFKB1
NFKB1	TNF	NFKB1			

Table 10.18 Drug target genes in the controlling set obtained by the structural target controllability analysis corresponding to the top ranked essential genes based on their betweenness centrality.

Tumor sample 28		Tumor sample 38		Tumor sample 191	
ANXA1	TNF	ANXA1	PTGS2	ANXA1	PTGS2
NFKB1		NFKB1	TNF	NFKB1	TNF
		PSMB10			

Table 10.19 Drug target genes in the controlling set obtained by the structural target controllability analysis corresponding to the top ranked essential genes based on their eigenvector-based prestige.

Tumor sample 28		Tumor sample 38		Tumor sample 191	
NFKB1	XPO1	ANXA1	PSMB5	ANXA1	TNF
PSMB1		NFKB1	TNF	NFKB1	
		PSMB2			

latter was sorted according to the number of top ranked essential genes controlled through one or more of their drug target genes. Then, we selected the three top drugs as our proposed drug combination for that patient's treatment. Informally, this approach aims for a reinforced influential effect over as many essential genes as possible through a minimal combination of at most three drugs. The chosen drugs are documented in Tables 10.20, 10.22, 10.24, 10.26, 10.28, and 10.30. Following an identical procedure, we also suggest a personalized drug combination in Table 10.32 based on the conclusions reached by the controllability analysis on the whole tumor network. The associated most effective standard drug therapy for each tumor sample and centrality measure are given in Tables 10.21, 10.23, 10.25, 10.27, 10.29, 10.31, and 10.33.

Table 10.20 Proposed drug therapy and the number of essential genes it controls, for the structural target controllability analysis corresponding to the top ranked essential genes based on their in-degree.

Tumor sample	Proposed therapy	Controlled EG
MM-0028-Tumor	Dexamethasone; thalidomide; doxorubicin	29
MM-0038-Tumor	Dexamethasone; thalidomide; pomalidomide	24
MM-0191-Tumor	Dexamethasone; thalidomide; pomalidomide	28

Table 10.21 Most effective standard drug therapy and the number of essential genes it controls, for the structural target controllability analysis corresponding to the top ranked essential genes based on their in-degree.

Tumor sample	Standard therapy	Controlled EG
MM-0028-Tumor	Bortezomib; thalidomide; dexamethasone	27
MM-0038-Tumor	Bortezomib; thalidomide; dexamethasone	24
MM-0191-Tumor	Bortezomib; thalidomide; dexamethasone	28

Table 10.22 Proposed drug therapy and the number of essential genes it controls, for structural target controllability analysis corresponding to the top ranked essential genes based on their closeness.

Tumor sample	Proposed therapy	Controlled EG
MM-0028-Tumor	Thalidomide; dexamethasone; pomalidomide	21
MM-0038-Tumor	Thalidomide; pomalidomide; dexamethasone	23
MM-0191-Tumor	Thalidomide; dexamethasone; selinexor	24

Table 10.23 Most effective standard drug therapy and the number of essential genes it controls, for the structural target controllability analysis corresponding to the top ranked essential genes based on their closeness.

Tumor sample	Standard therapy	Controlled EG
MM-0028-Tumor	Bortezomib; thalidomide; dexamethasone	21
MM-0038-Tumor	Bortezomib; thalidomide; dexamethasone	24
MM-0191-Tumor	Bortezomib; thalidomide; dexamethasone	24

Table 10.24 Proposed drug therapy and the number of essential genes it controls, for the structural target controllability analysis corresponding to the top ranked essential genes based on their harmonic centrality.

Tumor sample	Proposed therapy	Controlled EG
MM-0028-Tumor	Thalidomide; dexamethasone; selinexor	21
MM-0038-Tumor	Dexamethasone; thalidomide; pomalidomide	22
MM-0191-Tumor	Dexamethasone; thalidomide; bortezomib	22

Table 10.25 Most effective standard drug therapy and the number of essential genes it controls, for the structural target controllability analysis corresponding to the top ranked essential genes based on their harmonic centrality.

Tumor sample	Standard therapy	Controlled EG
MM-0028-Tumor	Bortezomib; thalidomide; dexamethasone	20
MM-0038-Tumor	Bortezomib; thalidomide; dexamethasone	22
MM-0191-Tumor	Bortezomib; thalidomide; dexamethasone	22

Table 10.26 Proposed drug therapy and the number of essential genes it controls, for the structural target controllability analysis corresponding to the top ranked essential genes based on their eccentricity.

Tumor sample	Proposed therapy	Controlled EG
MM-0028-Tumor	Thalidomide; pomalidomide; bortezomib	25
MM-0038-Tumor	Dexamethasone; prednisone; thalidomide	3
MM-0191-Tumor	Thalidomide; dexamethasone	26

Table 10.27 Most effective standard drug therapy and the number of essential genes it controls, for the structural target controllability analysis corresponding to the top ranked essential genes based on their eccentricity.

Tumor sample	Standard therapy	Controlled EG
MM-0028-Tumor	Bortezomib; thalidomide; dexamethasone	26
MM-0038-Tumor	Bortezomib; thalidomide; dexamethasone	3
MM-0191-Tumor	Bortezomib; thalidomide; dexamethasone	26

Table 10.28 Proposed drug therapy and the number of essential genes it controls, for the structural target controllability analysis corresponding to the top ranked essential genes based on their betweenness.

Tumor sample	Proposed therapy	Controlled EG
MM-0028-Tumor	Thalidomide; pomalidomide; dexamethasone	26
MM-0038-Tumor	Thalidomide; pomalidomide; dexamethasone	26
MM-0191-Tumor	Dexamethasone; thalidomide; pomalidomide	27

Table 10.29 Most effective standard drug therapy and the number of essential genes it controls, for the structural target controllability analysis corresponding to the top ranked essential genes based on their betweenness.

Tumor sample	Standard therapy	Controlled EG
MM-0028-Tumor	Bortezomib; thalidomide; dexamethasone	26
MM-0038-Tumor	Bortezomib; thalidomide; dexamethasone	26
MM-0191-Tumor	Bortezomib; thalidomide; dexamethasone	27

Table 10.30 Proposed drug therapy and the number of essential genes it controls, for the structural target controllability analysis corresponding to the top ranked essential genes based on their eigenvector centrality.

Tumor sample	Proposed therapy	Controlled EG
MM-0028-Tumor	Thalidomide; selinexor; bortezomib	3
MM-0038-Tumor	Dexamethasone; thalidomide; pomalidomide	19
MM-0191-Tumor	Thalidomide; dexamethasone; pomalidomide	21

Table 10.31 Most effective standard drug therapy and the number of essential genes it controls, for the structural target controllability analysis corresponding to the top ranked essential genes based on their eigenvector centrality.

Tumor sample	Standard therapy	Controlled EG
MM-0028-Tumor	Bortezomib; thalidomide; dexamethasone	2
MM-0038-Tumor	Bortezomib; thalidomide; dexamethasone	19
MM-0191-Tumor	Bortezomib; thalidomide; dexamethasone	21

Table 10.32 Proposed drug therapy and the number of essential genes it controls, for the structural target controllability analysis corresponding to the essential genes in the patient network.

Tumor sample	Proposed therapy	Controlled EG
MM-0028-Tumor	Thalidomide; doxorubicin; dexamethasone	31
MM-0038-Tumor	Thalidomide; pomalidomide; doxorubicin	28
MM-0191-Tumor	Dexamethasone; carfilzomib	25

Table 10.33 Most effective standard drug therapy and the number of essential genes it controls, for the structural target controllability analysis corresponding to the essential genes in the patient network.

Tumor sample	Standard therapy	Controlled EG
MM-0028-Tumor	Bortezomib; thalidomide; dexamethasone	29
MM-0038-Tumor	Bortezomib; thalidomide; dexamethasone	29
MM-0191-Tumor	Lenalidomide; bortezomib; dexamethasone	24

It is immediate to see that thalidomide and dexamethasone are predicted to be extremely suitable for treating the unique disease circumstances afflicting all the patients. These two drugs are commonly preferred as a first choice when approaching multiple myeloma cases and are part of two therapies in combination with either bortezomib or carfilzomib frequently used in latest medical practice, namely, VTD and KTD, respectively. These two combinations are supported by successful outcomes in different studies, such as (Roussel et al., 2020, Wester R, 2019).

The third drug combination spot is usually taken by pomalidomide. Although this drug is mostly considered in later stages of the treatment in the context of traditional care, the analysis predicts this drug to have a strong reinforcement impact on the activity of thalidomide for these specific cases, as it shares the same controlled essential genes with the latter. Our approach also identifies several outliers to the predominant three-drug combination, namely, selinexor, doxorubicin, prednisone, bortezomib, and carfilzomib. The first two drugs are not considered until the last stages of the treatment in standard therapy lines. Although the conclusions collected by this study for these three patients align with this traditional approach, they still provide grounds to consider them over other late stage drugs if the treatment progresses to an evolved phase. On the other hand, the remaining three are relevant in starting medical diagnosis, with bortezomib being especially remarkable. Following a more traditional approach, these three drugs may be given preference over pomalidomide for the first prescriptions.

The suggested drug therapies are predicted to control most of the considered essential genes in the whole network and each of the centrality subnetworks. Consequently, it is expected for them to have a significant favorable impact on the condition of our targeted patients. Furthermore, the prescribed sequences of drugs are very close to the known therapies frequently used in the current state of the art treatment. All of these events provide a strong foundation for the feasibility of this method in personalized medicine.

10.3.3.2 Minimum Dominating Set Analysis

We applied the input-constrained minimum three-dominating set method on the complete networks and sets of essential and drug target genes. The drug target genes in the obtained dominating sets are shown in Table 10.34.

Next, we applied the method once more on the subgraphs described in Section 10.3.2 and corresponding to each centrality measure. The drug target genes in the dominating set obtained for each measure are presented in Tables 10.35–10.40.

The examination of these results follows an analogous procedure to the one introduced for target controllability. For each network and centrality measure, as well as the network associated with the whole minimum dominating set, we ascertain a drug therapy including the top ranked drugs in terms of reached essential genes. The chosen drugs are documented in

Table 10.34 Drug target genes in the dominating set obtained by the minimum dominating set analysis.

Tumor sample 28		Tumor sample 38		Tumor sample 191	
ANXA1	PSMB5	ANXA1	PSMB5	ANXA1	PSMB5
NFKB1	PSMB9	NFKB1	PSMB8	NFKB1	PTGS2
NOLC1	PTGS2	NOLC1	PSMB9	NOLC1	TNF
NR3C1	TNF	NR3C1	PTGS2	NR3C1	TNFSF11
PSMB1	TNFSF11	PSMB1	TNF		
PSMB2	TOP2A	PSMB10	TNFSF11		
		PSMB2	TOP2A		

10.3 Applications of Network Modelling in Personalized Medicine

Table 10.35 Drug target genes in the dominating set obtained by the minimum dominating set analysis corresponding to the top ranked essential genes based on their in-degree centrality.

Tumor sample 28		Tumor sample 38		Tumor sample 191	
ANXA1	PSMB5	ANXA1	NR3C1	ANXA1	NR3C1
NFKB1	PSMB9	NFKB1	TNF	NFKB1	TNF
NOLC1	PTGS2	NOLC1	TNFSF11	NOLC1	TNFSF11
NR3C1	TNF				
PSMB1	TNFSF11				
PSMB2	TOP2A				

Table 10.36 Drug target genes in the dominating set obtained by the minimum dominating set analysis corresponding to the top ranked essential genes based on their closeness centrality.

Tumor sample 28		Tumor sample 38		Tumor sample 191	
ANXA1	TNF	ANXA1	TNF	ANXA1	TNF
NFKB1	TNFSF11	NFKB1	TNFSF11	NFKB1	TNFSF11
NR3C1		NR3C1		NR3C1	

Table 10.37 Drug target genes in the dominating set obtained by the minimum dominating set analysis corresponding to the top ranked essential genes based on their harmonic centrality.

Tumor sample 28		Tumor sample 38		Tumor sample 191	
ANXA1	TNF	ANXA1	TNF	ANXA1	TNF
NFKB1		NFKB1	TNFSF11	NFKB1	TNFSF11
		NR3C1		NR3C1	

Table 10.38 Drug target genes in the dominating set obtained by the minimum dominating set analysis corresponding to the top ranked essential genes based on their eccentricity.

Tumor sample 28		Tumor sample 38		Tumor sample 191	
ANXA1	TNF	ANXA1	TNF	ANXA1	NR3C1
NFKB1	TNFSF11	NFKB1	TNFSF11	NFKB1	TNF
NR3C1		NR3C1		NOLC1	TNFSF11

Table 10.39 Drug target genes in the dominating set obtained by the minimum dominating set analysis corresponding to the top ranked essential genes based on their betweenness centrality.

Tumor sample 28		Tumor sample 38		Tumor sample 191	
ANXA1	PSMB5	ANXA1	PTGS2	ANXA1	PTGS2
NFKB1	PSMB9	NFKB1	TNF	NFKB1	TNF
NOLC1	PTGS2	NR3C1	TNFSF11	NR3C1	TNFSF11
NR3C1	TNF				
PSMB1	TNFSF11				
PSMB2	TOP2A				

Table 10.40 Drug target genes in the dominating set obtained by the minimum dominating set analysis corresponding to the top ranked essential genes based on their eigenvector-based prestige.

Tumor sample 28		Tumor sample 38		Tumor sample 191	
ANXA1	TNF	ANXA1	TNF	ANXA1	TNF
NFKB1		NFKB1		NFKB1	

Table 10.41 Proposed drug therapy and the number of essential genes it dominates, for the minimum dominating set corresponding to the top ranked essential genes based on their in-degree.

Tumor sample	Proposed therapy	Dominated EG
MM-0028-Tumor	Thalidomide; pomalidomide; dexamethasone	28
MM-0038-Tumor	Thalidomide; pomalidomide; dexamethasone	27
MM-0191-Tumor	Thalidomide; pomalidomide; dexamethasone	29

Tables 10.41, 10.43, 10.45, 10.47, 10.49, and 10.51. Furthermore, the drugs corresponding to the dominating set associated with all the essential genes in the original patient network are shown in Table 10.53. The associated most effective standard drug therapy for each tumor sample and centrality measure are given in Tables 10.42, 10.44, 10.46, 10.48, 10.50, 10.52, and 10.54.

Table 10.42 Most effective standard drug therapy and the number of essential genes it dominates, for the minimum dominating set corresponding to the top ranked essential genes based on their in-degree.

Tumor sample	Standard therapy	Dominated EG
MM-0028-Tumor	Bortezomib; thalidomide; dexamethasone	28
MM-0038-Tumor	Bortezomib; thalidomide; dexamethasone	27
MM-0191-Tumor	Bortezomib; thalidomide; dexamethasone	29

Table 10.43 Proposed drug therapy and the number of essential genes it dominates, for the minimum dominating set corresponding to the top ranked essential genes based on their closeness.

Tumor sample	Proposed therapy	Dominated EG
MM-0028-Tumor	Thalidomide; pomalidomide; dexamethasone	24
MM-0038-Tumor	Thalidomide; pomalidomide; dexamethasone	26
MM-0191-Tumor	Thalidomide; pomalidomide; dexamethasone	26

Table 10.44 Most effective standard drug therapy and the number of essential genes it dominates, for the minimum dominating set corresponding to the top ranked essential genes based on their closeness.

Tumor sample	Standard therapy	Dominated EG
MM-0028-Tumor	Bortezomib; thalidomide; dexamethasone	24
MM-0038-Tumor	Bortezomib; thalidomide; dexamethasone	26
MM-0191-Tumor	Bortezomib; thalidomide; dexamethasone	26

Table 10.45 Proposed drug therapy and the number of essential genes it dominates, for the minimum dominating set corresponding to the top ranked essential genes based on their harmonic centrality.

Tumor sample	Proposed therapy	Dominated EG
MM-0028-Tumor	Thalidomide; pomalidomide; dexamethasone	21
MM-0038-Tumor	Thalidomide; pomalidomide; dexamethasone	26
MM-0191-Tumor	Thalidomide; pomalidomide; dexamethasone	24

Table 10.46 Most effective standard drug therapy and the number of essential genes it dominates, for the minimum dominating set corresponding to the top ranked essential genes based on their harmonic centrality.

Tumor sample	Standard therapy	Dominated EG
MM-0028-Tumor	Bortezomib; thalidomide; dexamethasone	21
MM-0038-Tumor	Bortezomib; thalidomide; dexamethasone	26
MM-0191-Tumor	Bortezomib; thalidomide; dexamethasone	24

Table 10.47 Proposed drug therapy and the number of essential genes it dominates, for the minimum dominating set corresponding to the top ranked essential genes based on their eccentricity.

Tumor sample	Proposed therapy	Dominated EG
MM-0028-Tumor	Thalidomide; pomalidomide; dexamethasone	26
MM-0038-Tumor	Thalidomide; pomalidomide; dexamethasone	26
MM-0191-Tumor	Thalidomide; pomalidomide; dexamethasone	26

Table 10.48 Most effective standard drug therapy and the number of essential genes it dominates, for the minimum dominating set corresponding to the top ranked essential genes based on their eccentricity.

Tumor sample	Standard therapy	Dominated EG
MM-0028-Tumor	Bortezomib; thalidomide; dexamethasone	26
MM-0038-Tumor	Bortezomib; thalidomide; dexamethasone	26
MM-0191-Tumor	Bortezomib; thalidomide; dexamethasone	26

Table 10.49 Proposed drug therapy and the number of essential genes it dominates, for the minimum dominating set corresponding to the top ranked essential genes based on their betweenness.

Tumor sample	Proposed therapy	Dominated EG
MM-0028-Tumor	Thalidomide; pomalidomide; dexamethasone	27
MM-0038-Tumor	Thalidomide; pomalidomide; dexamethasone	27
MM-0191-Tumor	Thalidomide; pomalidomide; dexamethasone	27

Table 10.50 Most effective standard drug therapy and the number of essential genes it dominates, for the minimum dominating set corresponding to the top ranked essential genes based on their betweenness.

Tumor sample	Standard therapy	Dominated EG
MM-0028-Tumor	Bortezomib; thalidomide; dexamethasone	27
MM-0038-Tumor	Bortezomib; thalidomide; dexamethasone	27
MM-0191-Tumor	Bortezomib; thalidomide; dexamethasone	27

Table 10.51 Proposed drug therapy and the number of essential genes it dominates, for the minimum dominating set corresponding to the top ranked essential genes based on their eigenvector centrality.

Tumor sample	Proposed therapy	Dominated EG
MM-0028-Tumor	Thalidomide; pomalidomide; dexamethasone	20
MM-0038-Tumor	Thalidomide; pomalidomide; dexamethasone	21
MM-0191-Tumor	Thalidomide; pomalidomide; dexamethasone	24

Table 10.52 Most effective standard drug therapy and the number of essential genes it dominates, for the minimum dominating set corresponding to the top ranked essential genes based on their eigenvector centrality.

Tumor sample	Standard therapy	Dominated EG
MM-0028-Tumor	Bortezomib; thalidomide; dexamethasone	20
MM-0038-Tumor	Bortezomib; thalidomide; dexamethasone	21
MM-0191-Tumor	Bortezomib; thalidomide; dexamethasone	24

Table 10.53 Proposed drug therapy and the number of essential genes it dominates, for the minimum dominating set corresponding to the essential genes in the patient network.

Tumor sample	Proposed therapy	Dominated EG
MM-0028-Tumor	Thalidomide; pomalidomide; dexamethasone	31
MM-0038-Tumor	Thalidomide; pomalidomide; dexamethasone	31
MM-0191-Tumor	Thalidomide; pomalidomide; dexamethasone	31

Table 10.54 Most effective standard drug therapy and the number of essential genes it dominates, for the minimum dominating set corresponding to the essential genes in the patient network.

Tumor sample	Standard therapy	Dominated EG
MM-0028-Tumor	Bortezomib; thalidomide; dexamethasone	31
MM-0038-Tumor	Bortezomib; thalidomide; dexamethasone	31
MM-0191-Tumor	Bortezomib; thalidomide; dexamethasone	31

From the point of view of centrality analysis, the results are consistent over all the networks and centrality measures with the prescription of the three-drug combination of thalidomide, pomalidomide, and dexamethasone. The parallel outcomes yielded by this approach and the one focused on target controllability support our hypothesis of an interrelationship between the network topological properties and the genes influencing the disease. On these grounds, we propose hub genes ascertained by the centrality analysis of a disease network to be considered as a basis for the discovery of new essential genes and drug targets.

10.4 Conclusion

Network medicine is an exciting and promising field of research, with a high potential for personalized approaches. It brings together approaches in graph theory, network science, systems biology, bioinformatics, and medicine, opening the door to detailed patient- and disease-specific insights. We discussed in this chapter several basic methods of network modelling and their potential applicability in personalized medicine. We also demonstrated their potential on three multiple myeloma patient datasets. We showed how various methods (topological analysis and systems controllability) can be combined to predict optimal and personalized drug combination therapies. In some cases, they differ from the initial standard therapy lines routinely offered to multiple myeloma patients and resemble in part the options becoming available later in the disease progression. More studies (especially longitudinal studies) are needed to explore the full potential of these methods and convincingly demonstrate their applicability in the clinical practice.

The results we obtained on the three datasets differ slightly from method to method. This is not surprising, as each method identifies different nodes and paths in the graphs that are of interest from various computational points of view. Which ones work best in practice should be explored with other research instruments, and it is likely that the results may differ from case to case.

The results in network medicine are critically dependent on the quality of the patient networks being applied to. The patient data that can be included in such networks can be quite diverse, including genetic mutations, copy number variations, differential gene expression, co-morbidities, and concurrent treatments. All these data sources contribute to the set of nodes in the network. The interactions included in the network typically come from various interaction databases (some of which we discussed in this chapter). The data going into these databases are very diverse: some of them are experimental (but not always on human samples), some are inferred from other experiments, while others are deduced through various machine learning methods. The importance of curating these datasets or at least choosing carefully which to include in the analyses cannot be underestimated.

References

Faisal N. Abu-Khzam, Amer E. Mouawad, and Mathieu Liedloff. An exact algorithm for connected red–blue dominating set. *Journal of Discrete Algorithms*, 9 (3): 252–262, 2011. ISSN 1570-8667. doi: https://doi.org/10.1016/j.jda.2011.03.006. URL https://www.sciencedirect.com/science/article/pii/S1570866711000360. Selected papers from the 7th International Conference on Algorithms and Complexity (CIAC 2010).

Noga Alon. Transversal numbers of uniform hypergraphs. *Graphs and Combinatorics*, 6 (1): 1–4, March 1990. ISSN 1435-5914. doi: https://doi.org/10.1007/BF01787474.

Bahman Bahmani, Abdur Chowdhury, and Ashish Goel. Fast incremental and personalized PageRank. *Proceedings of the VLDB Endowment*, 4 (3): 173–184, December 2010. ISSN 2150-8097. doi: https://doi.org/10.14778/1929861.1929864.

Mathieu Bastian, Sebastien Heymann, and Mathieu Jacomy. Gephi – the open graph viz platform. https://gephi.org/, 2021. [Online; accessed 4-February-2021].

Alex Bavelas. A mathematical model for group structures. *Human Organization*, 7: 16–30, 1948.

Alex Bavelas. Communication patterns in task-oriented groups. *The Journal of the Acoustical Society of America*, 22 (6): 725–730, 1950.

Murray A. Beauchamp. An improved index of centrality. *Behavioral Science*, 10 (2): 161–163, 1965. doi: https://doi.org/10.1002/bs.3830100205. URL https://onlinelibrary.wiley.com/doi/abs/10.1002/bs.3830100205.

Elisabetta Bergamini, Michele Borassi, Pierluigi Crescenzi, Andrea Marino, and Henning Meyerhenke. Computing top-k closeness centrality faster in unweighted graphs. *ACM Transactions on Knowledge Discovery from Data*, 13 (5), September 2019. ISSN 1556-4681. doi: https://doi.org/10.1145/3344719.

Phillip Bonacich. Power and centrality: a family of measures. *American Journal of Sociology*, 92 (5): 1170–1182, 1987. ISSN 00029602, 15375390. URL http://www.jstor.org/stable/2780000.

Michele Borassi and Emanuele Natale. KADABRA is an adaptive algorithm for betweenness via random approximation. *ACM Journal of Experimental Algorithmics (JEA)*, 24, February 2019. ISSN 1084-6654. doi: https://doi.org/10.1145/3284359.

Ulrik Brandes. A faster algorithm for betweenness centrality. *The Journal of Mathematical Sociology*, 25 (2): 163–177, 2001. doi: https://doi.org/10.1080/0022250X.2001.9990249.

Ulrik Brandes. On variants of shortest-path betweenness centrality and their generic computation. *Social Networks*, 30 (2): 136–145, 2008. ISSN 0378-8733. doi: https://doi.org/10.1016/j.socnet.2007.11.001.

Ulrik Brandes and Thomas Erlebach, editors. *Network Analysis—Methodological Foundations*. Springer-Verlag, Berlin Heidelberg, 2005. doi: https://doi.org/10.1007/b106453.

Rafael Santos Coelho, Phablo Moura, and Yoshiko Wakabayashi. The k-hop connected dominating set problem: approximation and hardness. *Journal of Combinatorial Optimization*, 34, 2017. doi: https://doi.org/10.1007/s10878-017-0128-y.

Gábor Csárdi and Tamás Nepusz. The igraph software package for complex network research. *InterJournal Complex Systems:* 1965, 2006.

E. Czeizler, K. C. Wu, C. Gratie, K. Kanhaiya, and I. Petre. Structural target controllability of linear networks. *IEEE/ACM Transactions on Computational Biology and Bioinformatics*, 15 (4): 1217–1228, 2018. doi: https://doi.org/10.1109/TCBB.2018.2797271.

Mohammad Mehdi Daliri Khomami, Alireza Rezvanian, Negin Bagherpour, and Mohammad Reza Meybodi. Minimum positive influence dominating set and its application in influence maximization: a learning automata approach. *Applied Intelligence*, 48 (3): 570–593, March 2018. ISSN 1573-7497. doi: https://doi.org/10.1007/s10489-017-0987-z.

Chavdar Dangalchev. Residual closeness in networks. *Physica A: Statistical Mechanics and its Applications*, 365 (2): 556–564, 2006. ISSN 0378-4371. doi: https://doi.org/10.1016/j.physa.2005.12.020.

Chavdar Dangalchev. Residual closeness and generalized closeness. *International Journal of Foundations of Computer Science*, 22 (8): 1939–1948, 2011. doi: https://doi.org/10.1142/S0129054111009136.

Kousik Das, Sovan Samanta, and Madhumangal Pal. Study on centrality measures in social networks: a survey. *Social Network Analysis and Mining*, 8 (1): 13, 2018.

Anthony Dekker. Conceptual distance in social network analysis. *Journal of Social Structure*, 6: 1–34, 2005.

Gianna M. Del Corso, Antonio Gullí, and Francesco Romani. Fast PageRank computation via a sparse linear system. *Internet Mathematics*, 2 (3): 251–273, 2005. doi: 10.1080/15427951.2005.10129108.

James W. Demmel. *Chapter 7. Iterative Methods for Eigenvalue Problems*, pages 361–387. Society for Industrial and Applied Mathematics, 1997. doi: https://doi.org/10.1137/1.9781611971446.ch7.

John Ellson, Emden Gansner, Yifan Hu, and Stephen North. Graphviz – graph visualization software. http://www.graphviz.org/, 2021. [Online; accessed 4-February-2021].

Monika Engelhardt, Kwee Yong, Sara Bringhen, and Ralph Wäsch. Carfilzomib combination treatment as first-line therapy in multiple myeloma: where do we go from the carthadex (KTD)-trial update? *Haematologica*, 104 (11): 2128–2131, November 2019. ISSN 1592-8721. doi: https://doi.org/10.3324/haematol.2019.228684. 31666342[PMID], PMC6821633[PMCID], haematol.2019.228684[PII].

Ernesto Estrada. Virtual identification of essential proteins within the protein interaction network of yeast. *Proteomics*, 6 (1): 35–40, 2006.

Jack McKay Fletcher and Thomas Wennekers. From structure to activity: using centrality measures to predict neuronal activity. *International Journal of Neural Systems*, 28 (02): 1750013, 2018. doi: https://doi.org/10.1142/S0129065717500137.

Kurt Foster, Stephen Muth, John Potterat, and Richard Rothenberg. A faster Katz status score algorithm. *Computational & Mathematical Organization Theory*, 7: 275–285, 2001. doi: https://doi.org/10.1023/A:1013470632383.

Massimo Franceschet. Pagerank centrality. https://www.sci.unich.it/francesc/teaching/network/pagerank, 2021. [Online teaching material; accessed 23-February-2021].

Linton C. Freeman. A set of measures of centrality based on betweenness. *Sociometry*, 40 (1): 35–41, 1977. ISSN 00380431. URL http://www.jstor.org/stable/3033543.

Linton C. Freeman. Centrality in social networks conceptual clarification. *Social Networks*, 1 (3): 215–239, 1978. ISSN 0378-8733. doi: https://doi.org/10.1016/0378-8733(78)90021-7. URL http://www.sciencedirect.com/science/article/pii/0378873378900217.

Jianxi Gao, Yang-Yu Liu, Raissa M. D'Souza, and Albert-L A ¡szl A³ Barab A ¡si. Target control of complex networks. *Nature Communications*, 5 (1): 5415, November 2014. ISSN 2041-1723. doi: https://doi.org/10.1038/ncomms6415.

Michael R. Garey and David S. Johnson. *Computers and Intractability; A Guide to the Theory of NP-Completeness*. W. H. Freeman and Co., USA, 1990. ISBN 0716710455.

Gene H. Golub and Henk A. van der Vorst. Eigenvalue computation in the 20th century. *Journal of Computational and Applied Mathematics*, 123 (1): 35–65, 2000. ISSN 0377-0427. doi: https://doi.org/10.1016/S0377-0427(00)00413-1. URL https://www.sciencedirect.com/science/article/pii/S0377042700004131. Numerical Analysis 2000. Vol. III: Linear Algebra.

Google. Facts about Google and competition. https://web.archive.org/web/20111104131332/https://www.google.com/competition/howgooglesearchworks.html, November 2011. [Online; accessed 23-February-2021].

Wei-Feng Guo, Shao-Wu Zhang, Qian-Qian Shi, Cheng-Ming Zhang, Tao Zeng, and Luonan Chen. A novel algorithm for finding optimal driver nodes to target control complex networks and its applications for drug targets identification. *BMC Genomics*, 19 (1): 924, January 2018. ISSN 1471-2164. doi: https://doi.org/10.1186/s12864-017-4332-z.

Aric A. Hagberg, Daniel A. Schult, and Pieter J. Swart. Exploring network structure, dynamics, and function using NetworkX. In Gaël Varoquaux, Travis Vaught, and Jarrod Millman, editors, *Proceedings of the 7th Python in Science Conference*, pages 11–15, Pasadena, CA USA, 2008.

Aric A. Hagberg, Daniel A. Schult, and Pieter J. Swart. Documentation for networkx.algorithms.centrality.closeness_centrality. NetworkX – network analysis in Python. https://networkx.org/, 2021a. [Online; accessed 12-April-2021].

Aric A. Hagberg, Daniel A. Schult, and Pieter J. Swart. Documentation for networkx.algorithms.centrality.katz_centrality. NetworkX – network analysis in Python. https://networkx.org/, 2021b. [Online; accessed 23-February-2021].

Aric A. Hagberg, Daniel A. Schult, and Pieter J. Swart. NetworkX – network analysis in Python. https://networkx.org/, 2021c. [Online; accessed 4-February-2021].

Per Hage and Frank Harary. Eccentricity and centrality in networks. *Social Networks*, 17 (1): 57–63, 1995. ISSN 0378-8733. doi: https://doi.org/10.1016/0378-8733(94)00248-9. URL https://www.sciencedirect.com/science/article/pii/0378873394002489.

Matthew W. Hahn and Andrew D. Kern. Comparative genomics of centrality and essentiality in three eukaryotic protein-interaction networks. *Molecular Biology and Evolution*, 22 (4): 803–806, 2004. ISSN 0737-4038. doi: https://doi.org/10.1093/molbev/msi072.

Frank Harary, Robert Z. Norman, and Dorwin Cartwrigh. *Structural Models: An Introduction to the Theory of Directed Graphs*. John Wiley and Sons, Inc., New York, 1965.

Abdel-Rahman Hedar and Rashad Ismail. Simulated annealing with stochastic local search for minimum dominating set problem. *International Journal of Machine Learning and Cybernetics*, 3 (2): 97–109, June 2012. ISSN 1868-808X. doi: https://doi.org/10.1007/s13042-011-0043-y.

Gábor Iván and Vince Grolmusz. When the Web meets the cell: using personalized PageRank for analyzing protein interaction networks. *Bioinformatics*, 27 (3): 405–407, 2010. ISSN 1367-4803. doi: https://doi.org/10.1093/bioinformatics/btq680.

Maliackal Poulo Joy, Amy Brock, Donald E. Ingber, and Sui Huang. High-betweenness proteins in the yeast protein interaction network. *Journal of Biomedicine and Biotechnology*, 2005 (2): 96–103, June 2005. ISSN 1110-7243. doi: https://doi.org/10.1155/jbb.2005.96. URL https://europepmc.org/articles/PMC1184047.

R. E. Kalman. Mathematical description of linear dynamical systems. *Journal of the Society for Industrial and Applied Mathematics Series A Control*, 1 (2): 152–192, 1963. doi: https://doi.org/10.1137/0301010.

U. Kang, Spiros Papadimitriou, Jimeng Sun, and Hanghang Tong. Centralities in large networks: algorithms and observations. In *Proceedings of the 2011 SIAM International Conference on Data Mining*, pages 119–130, 2011. doi: https://doi.org/10.1137/1.9781611972818.11. URL https://epubs.siam.org/doi/abs/10.1137/1.9781611972818.11.

Krishna Kanhaiya, Eugen Czeizler, Cristian Gratie, and Ion Petre. Controlling directed protein interaction networks in cancer. *Scientific Reports*, 7 (1): 10327, September 2017. ISSN 2045-2322. doi: https://doi.org/10.1038/s41598-017-10491-y.

Krishna Kanhaiya, Vladimir Rogojin, Keivan Kazemi, Eugen Czeizler, and Ion Petre. NetControl4BioMed: a pipeline for biomedical data acquisition and analysis of network controllability. *BMC Bioinformatics*, 19, 2018. doi: https://doi.org/10.1186/s12859-018-2177-3.

Leo Katz. A new status index derived from sociometric analysis. *Psychometrika*, 18 (1): 39–43, 1953.

Dirk Koschützki. *Network Centralities*, chapter 4, pages 65–84. Wiley Online Library, 2007. ISBN 9780470253489. doi: https://doi.org/10.1002/9780470253489.ch4.

Dirk Koschützki and Falk Schreiber. Centrality analysis methods for biological networks and their application to gene regulatory networks. *Gene Regulation and Systems Biology*, 2: GRSB.S702, 2008. doi: https://doi.org/10.4137/GRSB.S702.

Jan Krönke, Slater N. Hurst, and Benjamin L. Ebert. Lenalidomide induces degradation of IKZF1 and IKZF3. *Oncoimmunology*, 3 (7), 2014. doi: https://doi.org/10.4161/21624011.2014.941742.

Amy N. Langville and Carl D. Meyer. A survey of eigenvector methods for web information retrieval. *SIAM Review*, 47 (1): 135–161, 2005.

Harold J. Leavitt. *Some Effects of Certain Communication Patterns Upon Group Performance*. PhD thesis, Massachusetts Institute of Technology. Dept. of Economics, 1949, 1949.

C. T. Lin. Structural controllability. *IEEE Transactions on Automatic Control*, AC-19 (3): 201–208, 1974.

Nan Lin. *Foundations of Social Research*. McGraw-Hill, 1976.

Yang-Yu Liu, Jean-Jacques Slotine, and Albert-L A¡szl A³ Barab A¡si. Controllability of complex networks. *Nature*, 473 (7346): 167–173, May 2011. ISSN 1476-4687. doi: https://doi.org/10.1038/nature10011.

Jens G. Lohr, Petar Stojanov, Scott L. Carter, Peter Cruz-Gordillo, Michael S. Lawrence, Daniel Auclair, Carrie Sougnez, Birgit Knoechel, Joshua Gould, Gordon Saksena, Kristian Cibulskis, Aaron McKenna, Michael A. Chapman, Ravid Straussman, Joan Levy, Louise M. Perkins, Jonathan J. Keats, Steven E. Schumacher, Mara Rosenberg, Gad Getz, and Todd R. Golub. Widespread genetic heterogeneity in multiple myeloma: implications for targeted therapy. *Cancer Cell*, 25 (1): 91–101, January 2014. ISSN 1878-3686 (Electronic); 1535-6108 (Linking). doi: https://doi.org/10.1016/j.ccr.2013.12.015.

Hong-Wu Ma and An-Ping Zeng. The connectivity structure, giant strong component and centrality of metabolic networks. *Bioinformatics*, 19 (11): 1423–1430, 2003. ISSN 1367-4803. doi: https://doi.org/10.1093/bioinformatics/btg177.

Alexander V. Mantzaris, Danielle S. Bassett, Nicholas F. Wymbs, Ernesto Estrada, Mason A. Porter, Peter J. Mucha, Scott T. Grafton, and Desmond J. Higham. Dynamic network centrality summarizes learning in the human brain. *Journal of Complex Networks*, 1 (1): 83–92, 2013. ISSN 2051-1310. doi: https://doi.org/10.1093/comnet/cnt001.

Yimin Mao, Le Chen, Jianghua Li, Anna Junjie Shangguan, Stacy Kujawa, and Hong Zhao. A network analysis revealed the essential and common downstream proteins related to inguinal Hernia. *PLoS ONE*, 15 (1): 1–21, 2020. doi: https://doi.org/10.1371/journal.pone.0226885.

Massimo Marchiori and Vito Latora. Harmony in the small-world. *Physica A: Statistical Mechanics and its Applications*, 285 (3): 539–546, 2000. ISSN 0378-4371. doi: https://doi.org/10.1016/S0378-4371(00)00311-3. URL https://www.sciencedirect.com/science/article/pii/S0378437100003113.

María-Victoria Mateos, Meletios A. Dimopoulos, Michele Cavo, Kenshi Suzuki, Andrzej Jakubowiak, Stefan Knop, Chantal Doyen, Paulo Lucio, Zsolt Nagy, Polina Kaplan, Ludek Pour, Mark Cook, Sebastian Grosicki, Andre Crepaldi, Anna M. Liberati, Philip Campbell, Tatiana Shelekhova, Sung-Soo Yoon, Genadi Iosava, Tomoaki Fujisaki, Mamta Garg, Christopher Chiu, Jianping Wang, Robin Carson, Wendy Crist, William Deraedt, Huong Nguyen, Ming Qi, and Jesus San-Miguel. Daratumumab plus Bortezomib, Melphalan, and Prednisone for untreated myeloma. *New England Journal of Medicine*, 378 (6): 518–528, 2018. doi: https://doi.org/10.1056/NEJMoa1714678. PMID: 29231133.

Matplotlib. Matplotlib: python plotting. https://matplotlib.org/, 2021. [Online; accessed 4-February-2021].

Geoffrey M. Matthews, Ricardo de Matos Simoes, Yiguo Hu, Michal Sheffer, et al. Characterization of lineage vs. context-dependent essential genes in multiple myeloma using CRISPR-Cas9 genome editing. *Cancer Research*, 77 (13), 2017. doi: https://doi.org/10.1158/1538-7445.AM2017-LB-118.

Tilahun Melak and Sunita Gakkhar. Comparative genome and network centrality analysis to identify drug targets of Mycobacterium tuberculosis H37Rv. *BioMed Research International*, 2015, 2015. doi: https://doi.org/10.1155/2015/212061.

Multiple Myeloma Research Foundation. Multiple myeloma treatment overview. https://themmrf.org/wp-content/uploads/2019/09/MMRF-Treatment-Overview.pdf, 2019. [Online; accessed 23-February-2021].

J. C. Nacher and T. Akutsu. Analysis of critical and redundant nodes in controlling directed and undirected complex networks using dominating sets. *Journal of Complex Networks*, 2 (4): 394–412, 2014. doi: https://doi.org/10.1093/comnet/cnu029.

Jose C. Nacher and Tatsuya Akutsu. Minimum dominating set-based methods for analyzing biological networks. *Methods*, 102: 57–63, 2016.
ISSN 1046-2023. doi: https://doi.org/10.1016/j.ymeth.2015.12.017. URL https://www.sciencedirect.com/science/article/pii/S1046202315300967. Pan-omics analysis of biological data.

Meghana Nasre, Matteo Pontecorvi, and Vijaya Ramachandran. Betweenness centrality – incremental and faster. In Erzsébet Csuhaj-Varjú, Martin Dietzfelbinger, and Zoltán Ésik, editors, *Mathematical Foundations of Computer Science 2014*, pages 577–588, Berlin, Heidelberg, Springer-Verlag, 2014. ISBN 978-3-662-44465-8.

Christian F. A. Negre, Uriel N. Morzan, Heidi P. Hendrickson, Rhitankar Pal, George P. Lisi, J. Patrick Loria, Ivan Rivalta, Junming Ho, and Victor S. Batista. Eigenvector centrality for characterization of protein allosteric pathways. *Proceedings of the National Academy of Sciences of the United States of America*, 115 (52), 2018. doi: https://doi.org/10.1073/pnas.1810452115.

Mark E. J. Newman. *Mathematics of Networks*, pages 1–8. Palgrave Macmillan, 2008. doi: https://doi.org/10.1057/978-1-349-95121-5_2565-1.

Mark E. J. Newman. *Networks: An Introduction*. Oxford University Press, USA, 2010.

M. Nguyen, Minh Hoàng Hà, and D. Nguyen. Solving the k-dominating set problem on very large-scale networks. *Computational Social Networks*, 7: 1–15, 2020.

Lawrence Page, Sergey Brin, Rajeev Motwani, and Terry Winograd. The PageRank citation ranking: bringing order to the web. Technical Report 1999-66, Stanford InfoLab, November 1999. URL http://ilpubs.stanford.edu:8090/422/. Previous number = SIDL-WP-1999-0120.

Britta Papendieck and Peter Recht. On maximal entries in the principal eigenvector of graphs. *Linear Algebra and its Applications*, 310 (1): 129–138, 2000. ISSN 0024-3795. doi: https://doi.org/10.1016/S0024-3795(00)00063-X. URL https://www.sciencedirect.com/science/article/pii/S002437950000063X.

Torsten Johann Paul and Philip Kollmannsberger. Biological network growth in complex environments: a computational framework. *PLoS Computational Biology*, 16 (11), e1008003, November 2020. doi: https://doi.org/10.1371/journal.pcbi.1008003.

Lucia D. Penso and Valmir C. Barbosa. A distributed algorithm to find k-dominating sets. *Discrete Applied Mathematics*, 141 (1): 243–253, 2004. ISSN 0166-218X. doi: https://doi.org/10.1016/S0166-218X(03)00368-8. URL https://www.sciencedirect.com/science/article/pii/S0166218X03003688. Brazilian Symposium on Graphs, Algorithms and Combinatorics.

Mahendra Piraveenan, Mikhail Prokopenko, and Liaquat Hossain. Percolation centrality: quantifying graph-theoretic impact of nodes during percolation in networks. *PLoS ONE*, 8(1) e53095, 2013. doi: https://dx.doi.org/10.1371%2Fjournal.pone.0053095.

Victor-Bogdan Popescu, Krishna Kanhaiya, Iulian Năstac, Eugen Czeizler, and Ion Petre. Identifying efficient controls of complex interaction networks using genetic algorithms, Scientific Reports 12, 1437, 2022. https://doi.org/10.1038/s41598-022-05335-3

Victor-Bogdan Popescu, Jose Angel Sanchez-Martin, Daniela Schacherer, Sadra Safadoust, Negin Majidi, Andrei Andronescu, Alexandru Nedea, Diana Ion, Eduard Mititelu, Eugen Czeizler, and Ion Petre. NetControl4BioMed: a platform for biomedical network synthesis and analysis. *Bioinformatics*, 37 (21): 3976–3978, 2021. http://10.1093/bioinformatics/btab570.

Jun Ren, Jianxin Wang, Min Li, and Fangxiang Wu. Discovering essential proteins based on PPI network and protein complex. *International Journal of Data Mining and Bioinformatics*, 12 (1): 24–43, 2015. doi: https://doi.org/10.1504/ijdmb.2015.068951.

Yannick Rochat. Closeness centrality extended to unconnected graphs: The harmonic centrality index. In *Applications of Social Network Analysis, ASNA2009*, 2009.

Murielle Roussel, Philippe Moreau, Benjamin Hebraud, Kamel Laribi, Arnaud Jaccard, Mamoun Dib, Borhane Slama, Véronique Dorvaux, Bruno Royer, Laurent Frenzel, Sonja Zweegman, Saskia K. Klein, Annemiek Broijl, Kon-Siong Jie, Jianping Wang, Veronique Vanquickelberghe, Carla de Boer, Tobias Kampfenkel, Katharine S. Gries, John Fastenau, and Pieter Sonneveld. Bortezomib, thalidomide, and dexamethasone with or without daratumumab for transplantation-eligible patients with newly diagnosed multiple myeloma (CASSIOPEIA): health-related quality of life outcomes of a randomised, open-label, phase 3 trial. *The Lancet Haematology*, 7 (12): e874–e883, 2020. ISSN 2352-3026. doi: https://doi.org/10.1016/S2352-3026(20)30356-2. URL https://www.sciencedirect.com/science/article/pii/S2352302620303562.

Britta Ruhnau. Eigenvector-centrality—a node-centrality? *Social Networks*, 22 (4): 357–365, 2000. ISSN 0378-8733. doi: https://doi.org/10.1016/S0378-8733(00)00031-9. URL https://www.sciencedirect.com/science/article/pii/S0378873300000319.

Gert Sabidussi. The centrality index of a graph. *Psychometrika*, 31 (4): 581–603, 1966.

R. Sahoo, T.S. Rani, and S.D. Bhavani. Chapter 17 – Differentiating cancer from normal protein–protein interactions through network analysis. In Quoc Nam Tran and Hamid R. Arabnia, editors, *Emerging Trends in Applications and Infrastructures for Computational Biology, Bioinformatics, and Systems Biology*, Emerging Trends in Computer Science and Applied Computing, pages 253–269. Morgan Kaufmann, Boston, MA, 2016. ISBN 978-0-12-804203-8. doi: https://doi.org/10.1016/B978-0-12-804203-8.00017-1. URL http://www.sciencedirect.com/science/article/pii/B9780128042038000171.

Mansoor Saqi, Johann Pellet, Irina Roznovat, Alexander Mazein, Stéphane Ballereau, Bertrand De Meulder, and Charles Auffray. Systems medicine: the future of medical genomics, healthcare, and wellness. *Molecular Biology Methods*, 1386: 43–60, 2016. ISSN 1940-6029 (Electronic); 1064-3745 (Linking).

P. Shannon, A. Markiel, O. Ozier, N. S. Baliga, J. T. Wang, D. Ramage, N. Amin, B. Schwikowski, and T. Ideker. Cytoscape: a software environment for integrated models of biomolecular interaction networks. *Genome Research*, 13: 2498–2504, 2003. doi: https://doi.org/10.1101/gr.1239303.

P. Shannon, A. Markiel, O. Ozier, N. S. Baliga, J. T. Wang, D. Ramage, N. Amin, B. Schwikowski, and T. Ideker. Cytoscape: an open source platform for complex network analysis and visualization. https://cytoscape.org/, 2021. [Online; accessed 4-February-2021].

Marvin E. Shaw. Group structure and the behavior of individuals in small groups. *The Journal of Psychology*, 38 (1): 139–149, 1954. doi: https://doi.org/10.1080/00223980.1954.9712925.

R. W. Shields and J. B. Pearson. Structural controllability of multi-input linear systems. *IEEE Transactions on Automatic Control*, 21 (2): 203–212, 1976.

The igraph core team. igraph – network analysis software. https://igraph.org/, 2021. [Online; accessed 4-February-2021].

Q. Tian, N. D. Price, and L. Hood. Systems cancer medicine: towards realization of predictive, preventive, personalized and participatory (P4) medicine. *Journal of Internal Medicine*, 271 (2): 111–121, February 2012. ISSN 1365-2796 (Electronic); 0954-6820 (Print); 0954-6820 (Linking).

Rodger E. Tiedemann, Yuan Xao Zhu, Jessica Schmidt, Chang Xin Shi, et al. Identification of molecular vulnerabilities in human multiple myeloma cells by RNA interference lethality screening of the druggable genome. *Cancer Research*, 72, 2012. doi: https://doi.org/10.1158/0008-5472.CAN-11-2781.

Feng Wang, Hongwei Du, Erika Camacho, Kuai Xu, Wonjun Lee, Yan Shi, and Shan Shan. On positive influence dominating sets in social networks. *Theoretical Computer Science*, 412 (3): 265–269, 2011. ISSN 0304-3975. doi: https://doi.org/10.1016/j.tcs.2009.10.001. URL https://www.sciencedirect.com/science/article/pii/S0304397509007221. Combinatorial Optimization and Applications.

R. Wester, B. van der Holt, and E. Asselbergs. Phase II study of carfilzomib, thalidomide, and low-dose dexamethasone as induction and consolidation in newly diagnosed, transplant eligible patients with multiple myeloma; the carthadex trial. *Haematologica*, 104 (11): 2265–2273, 2019. doi: https://doi.org/10.3324/haematol.2018.205476.

David S. Wishart, Yannick D. Feunang, An C. Guo, Ana Marcu, et al. DrugBank 5.0: a major update to the drugbank database for 2018. *Nucleic Acids Research*, 46: 1074–1082, 2017. doi: https://doi.org/10.1093/nar/gkx1037.

Jie Wu, Mihaela Cardei, Fei Dai, and Shuhui Yang. Extended dominating set and its applications in ad hoc networks using cooperative communication. *IEEE Transactions on Parallel and Distributed Systems*, 17 (8): 851–864, August 2006. ISSN 1045-9219. doi: https://doi.org/10.1109/TPDS.2006.103.

Stefan Wuchty. Controllability in protein interaction networks. *Proceedings of the National Academy of Sciences of the United States of America*, 111 (19): 7156–7160, 2014. ISSN 0027-8424. doi: https://doi.org/10.1073/pnas.1311231111. URL https://www.pnas.org/content/111/19/7156.

Stefan Wuchty and Peter F. Stadler. Centers of complex networks. *Journal of Theoretical Biology*, 223 (1): 45–53, 2003. ISSN 0022-5193. doi: https://doi.org/10.1016/S0022-5193(03)00071-7. URL https://www.sciencedirect.com/science/article/pii/S0022519303000717.

Rong Yang and Leyla Zhuhadar. Extensions of closeness centrality? In *Proceedings of the 49th Annual Southeast Regional Conference*, ACM-SE '11, pages 304–305, New York, NY, USA, 2011. Association for Computing Machinery. ISBN 9781450306867. doi: https://doi.org/10.1145/2016039.2016119.

Haiyuan Yu, Philip M. Kim, Emmett Sprecher, Valery Trifonov, and Mark Gerstein. The importance of bottlenecks in protein networks: correlation with gene essentiality and expression dynamics. *PLoS Computational Biology*, 3 (4): e59, April 2007. ISSN 1553-734X. doi: https://doi.org/10.1371/journal.pcbi.0030059. URL https://europepmc.org/articles/PMC1853125.

Jiguo Yu, Nannan Wang, Guanghui Wang, and Dongxiao Yu. Connected dominating sets in wireless ad hoc and sensor networks – a comprehensive survey. *Computer Communications*, 36 (2): 121–134, 2013. ISSN 0140-3664. doi: https://doi.org/10.1016/j.comcom.2012.10.005. URL https://www.sciencedirect.com/science/article/pii/S014036641200374X.

Xiao-Fei Zhang, Le Ou-Yang, Yuan Zhu, Meng-Yun Wu, and Dao-Qing Dai. Determining minimum set of driver nodes in protein-protein interaction networks. *BMC Bioinformatics*, 16 (1): 146, May 2015. ISSN 1471-2105. doi: https://doi.org/10.1186/s12859-015-0591-3.

Peng Zhang, Lin Tao, Xian Zeng, Chu Qin, Shangying Chen, Feng Zhu, Zerong Li, Yuyang Jiang, Weiping Chen, and Yu-Zong Chen. A protein network descriptor server and its use in studying protein, disease, metabolic and drug targeted networks. *Briefings in Bioinformatics*, 18 (6): 1057–1070, 08 2016. ISSN 1467-5463. doi: https://doi.org/10.1093/bib/bbw071.

11

Conclusion

Elisabetta De Maria

Université Côte d'Azur, CNRS, I3S, Sophia Antipolis, France

This book explores in detail the main formal techniques currently used to model biological systems, which are at the heart of systems biology. A large panel of formalisms is introduced, including both qualitative and quantitative approaches. Concrete and suited case studies are provided for each formalism, and this allows us to put in evidence the advantages/drawbacks of each of them. Furthermore, each formalism is presented along with formal tools to automate reasoning on the biological systems to be modeled. After browsing through all these powerful techniques, the reader has certainly come to the conclusion that there is no technique absolutely better than the other ones. The modeler has to choose from time to time the formalism that better fits the needs of biologists, according not only to the system to be modeled but also to the kind of questions to be addressed. Great attention must be paid to the granularity of representation of all the involved entities (concentration levels, time evolution, etc.). The palette of techniques selected in this book constitutes an effective tool kit greatly helping the modeler in choosing the most suited approach advisedly.

Having reached the end of this book, the reader is aware that formal methods play a crucial role in systems biology, being especially suitable to model and validate biological systems. This book substantiates the claim that formal methods are essential to perform deep analysis of biological systems: in particular, it presents several application cases and examples of involved questions that could not be answered by simple simulation techniques. Several semi-formal techniques exist, often based on graphical methods such as Unified Modelling Language (UML), but they are out of the scope of this book because they are not rigorous enough for the kind of in-depth analysis targeted here.

Systems Biology Modelling and Analysis: Formal Bioinformatics Methods and Tools,
First Edition. Edited by Elisabetta De Maria.
© 2023 John Wiley & Sons, Inc. Published 2023 by John Wiley & Sons, Inc.

Another important point to underline is the emerging role of artificial intelligence in systems biology. This book presents several examples in which formal methods are combined with techniques coming from artificial intelligence, and we believe that the integration of these two fields has to be explored more to understand biological networks in detail, keeping into account the huge data flow coming from next-generation molecular technologies. While a decade ago one of the main challenges in systems biology was to infer suitable parameter values to fit data in biological networks, the challenge has now moved to directly infer the topology (structure) of networks. A clever combination of formal methods and artificial intelligence algorithms can greatly help in finding (pieces of) networks whose expected behavior is described as a set of formal properties. Our feeling is that artificial intelligence is very powerful but cannot always be employed alone: as a matter of fact, it can be exploited to classify data and efficiently look for many unknown values, but it is often employed as a black box, that is, we get the answers but we do not know the reason for them. Formal methods complement artificial intelligence approaches, giving the keys to understand why some specific phenomena take place, why other ones are not possible, etc.

To conclude, formal methods are extraordinary tools to conceive and understand a variety of biological systems, and we believe that their use will be more and more unavoidable in the next years.

Index

a

AAN (Asynchronous Automata Network)   254, 258–259, 264–294
  attractor   271–275
  semantics   267–271
  stable state   271–275
  translation in ASP   276–290
ASP (Answer Set Programming)   6, 257–265, 275–299. *See also* logic programming language
  predicates   261–263
  rules   259–261
  scripting   263–264
  syntax   259–261
A. thaliana   209–211
attractor   280–290
  cycle enumeration   281–285
  enumeration   258–259, 285–288
  identification   255–256

b

bio-ambients   5
Bio-PEPA   60
bisimulation   52
BNS   232
Boolean network   5, 174–238, 252
  asynchronous updating modes   199–201

attractor   216–224
block-parallel updating modes   189–193
block-sequential updating modes   187–189
complexity   224–232
context-dependent updating modes   203–204
deterministic updating modes   187–199
dynamics   183–208
feedback cycle   217–224
fixed point   216–217
limit configuration   229–232
local-clocks updating modes   193–197
most permissive updating mode   205–208
non-deterministic updates   201–203
non-deterministic updating modes   199–208
reachability   227–229
specification   181–183
structural properties   216–224
trajectory   185–186
update   183–208
updating mode hierarchy   235–236

Systems Biology Modelling and Analysis: Formal Bioinformatics Methods and Tools, First Edition. Edited by Elisabetta De Maria.
© 2023 John Wiley & Sons, Inc. Published 2023 by John Wiley & Sons, Inc.

Boolean network (*contd.*)
  updating modes induces by Boolean networks with memory 197–199
BOOLSIM 232
BRN (Biological Regulatory Networks) 254–256, 258

c

CABEAN 232
causality 234
cell cycle 211–212
cell death 159–160
Charlie 19
chemical kinetics 54
Clingo 263–264, 276, 279–280, 287–291, 293–294
CoLoMoTo 235
concurrent system 36–38
contact map 79–90
COSMOS 330
CSL (Continuous Stochastic Logic) 22–23
CTL (Computation Tree Logic) 8, 317–318
CTL* (Computation Tree Logic*) 7, 316–317
CTMC (Continuous Time Markov Chain) 21, 55–56, 328
Cytoscape 390

d

datum curation 150–153
death map 159–163
duration of activity 233
dynamic modelling 264–275
dynamical systems 1–2

e

Escherichia coli 306, 326, 330
event 313–331
experiment 2–3

f

first-order language 307–308, 312–313
first-order theory 307–308, 312–313
formal methods 1–9, 425–426
formal model 2–3
formal verification 8–9. *See also* model checking

g

gene regulatory network 174–177
Gillespie's algorithm 21, 55–56
GINSim 227, 229, 292–295
G-protein in *Saccharomyces cerevisiae* 16, 21–22, 25–26, 28–30
GRN (Gene Regulatory Network) 319–322

h

HASL (Hybrid Automaton Stochastic Logic) 330
HSC (Hepatic Stellate Cell) 76–78, 91–113
  activation 97–99
  apoptosis and senescence of MFB 101–102
  behavior of inactivated HSC 102–104
  behavior of receptors 106–107
  behavior of TGFB1 proteins 95–96
  degradation od reactivated MFB 106
  differentiation 97–99
  inactivation of MFB 102
  model parameters 108
  model simulation 111–113
  model static analysis 109–110
  proliferation of activated HSC 99–100
  proliferation of MFB 100

proliferation of reactivated cell 105
renewal of quiescent HSC 96–97
hybrid automaton 7, 319–331
   FOCoRe (First-order Constant Reset hybrid automaton) 325–326
   IDA (Independent Dynamics hybrid automaton) 325–326
   o-minimal automaton 325–326. *See also* o-minimal theory
   semantics 324–326
   STORMED hybrid system 325–326
   syntax 323–324

i

Ifnb1 135–139, 142, 147
inferring rules 153–155
influence graph 181–183, 217–223
interaction map 1. *See also* interaction network
interaction network 1. *See also* interaction map

k

Kappa 71–76, 78–91
   complex 81–82
   ecosystem 73–75
   embedding between patterns 84–85
   interaction rule 86–88
   pattern 82–84
   semantics 72–73
   signature 79–81
   site graph 71–75, 78–91
   underlying reaction network 88–91
killer T-cell and cancer cell models 354–359
   cancer cell dynamics 355–356

   dynamics without treatment 357–358
   killer T-cell dynamics 356
   resource dynamics 354–355
   treatment effect 358–359
Kripke's structure 314–315. *See also* transition system

l

lambda phage (bacteriophage lambda) 292–293
law of mass action 54
logic 6
logic programming language 6
Lotka–Volterra predator-prey model 350–354
   constant of motion 352–353
   equilibrium 351–352
   ODE system 350–351
   phase-plane analysis 351–352
   population density 353–354
Lps (lipopolysaccharide) 135–149, 152–163
LTL (Linear Time Logic) 8, 317–318

m

Maude 131–135, 140, 142
MFB (myofibroblast) 80, 96–115
model checking 8–9, 22–23, 318. *See also* formal verification
   simulative model checking 25
modelling 4–7
MPBN 229

n

NetControl4BioMed 390–392
network centrality method 364–392
   centrality 365–366
   degree centrality 366–368
   path centrality 373–377
   proximity centrality 368–373

network centrality method (*contd.*)
  spectral centrality   377–383
network medicine   363–412
NetworkX   389–390
neural oscillator   322–323, 330
Numv   9

o

observability   3
ODE (Ordinary Differential Equation)   6, 55–56, 339–359
  EBR (Elementary Bimolecular Reaction)   347–350
  equilibrium   341–344
  EUR (Elementary Unimolecular Reaction)   346–347
  logistic differential equation   340–345
  mechanistic derivation   345–359
  solving analytically   340–341
  solving numerically   344–345
  stability   341–344
o-minimal theory   312–313
operability   3

p

parameter search   3
path quantifier   8
pathways   44. *See also* interaction network
PCTL (Probabilistic Computation Tree Logic)   8, 330
PEPA   60
personalized medicine   392–412
  analysis method   393–398
  personalized disease network   392–393
Petri net   5, 15–30
  CPN (Continuous Petri Net)   24
  deadlock   18
  extended Petri net   20
  FCPN (Fuzzy Continuous Petri Net)   29
  FSPN (Fuzzy Stochastic Petri Net)   27
  generation and analysis with PL   142–144
  Hybrid Petri net   6
  P-invariant   18
  place   17
  representation in PL   140–142
  SPN (Stochastic Petri Net)   20
  T-invariant   18
  token   17
  transition   17
  trap   18
pi-calculus (π-calculus)   5, 38–42
PINT   227, 229
PL (Pathway Logic)   127–168
  assistant   144–150
  language   134–140
  SMT8   155–163
PLTLc   26
  MC2   26
precedence   234
prediction   2–3
PRISM   9, 330
probabilistic automaton   327–328
process algebra   5, 35–61
  stochastic process algebra   59
process-hitting   5

q

qualitative pathway dynamics   46–50

r

reaction rule   5, 88–90. *See also* rule-based language
real-time probabilistic process   329
repressilator   331
rewriting logic   133–134

rewriting rule   70–72, 86–91
rule-based approach   69–113.
       See also reaction rule
rule-based language   7. *See also*
       reaction rule
Runge–Kutta method   24–25,
       344–345

s

SBML (Systems Biology Markup
       Language)   5
SCSG (Strongly Connected
       Subgraph)   273–274,
       284–285, 289
specification   7
SPIN   9
stable-state   278–280
state quantifier   8
stochastic automaton   329
stochastic differential equation
       6
strong bisimulation   52
system controllability method
       383–392
   minimum dominating set
       387–388
   network controllability
       384–387

systems biology   15, 35–36, 251–253,
       339–340, 425–426
   characteristics   1–3
   definition   1

t

TCTL (Timed Computation Tree
       Logic)   329
temporal logics   7–8, 316–318. *See
       also* CTL*, CTL, LTL, PLTLc,
       TCTL
TGFB1   77–78, 80–115
theorem prover   9
Thomas' network   5, 174–177,
       200–201, 220–223
timed automaton   7, 329
transition system   9, 37
   labeled probabilistic transition
       system   328
   labeled transition system   37,
       313–314
trap domain   253, 274

u

UPPAAL   330

z

Zeitgeber   212–213

Printed and bound by CPI Group (UK) Ltd, Croydon, CR0 4YY
27/02/2023

03195015-0001